Regenerative Agriculture

NIPA® GENX ELECTRONIC RESOURCES & SOLUTIONS P. LTD.
New Delhi-110 034

About the Editors

Dr. Haobijam James Watt is working as an Assistant Professor at the Department of Agricultural Economics and Extension (AEE), School of Agriculture, Lovely Professional University. He has completed his Ph.D. (Agricultural Extension) from Visva-Bharati, Santiniketan, West Bengal, post-graduate (M.Sc. in Agricultural Extension) from Bidhan Chandra Krishi Viswavidyalaya, West Bengal, and graduated from Tamil Nadu Agricultural University, Coimbatore. He has qualified ICAR-AIEEA-UG, ICAR-AIEEA-PG, ICAR-SRF-(PGS) exams, ICAR-ASRB-NET (2016, 2021), and UGC-NET 2016 (Adult. Edu./Con.Edu/Andragogy/N.F.Edu). He has been involved in the teaching of U.G. and P.G. courses in Agricultural Extension. He has published several research articles including book chapters in reputed national as well as international journals. His area of research lies in the field of livelihood, sustainability, entrepreneurship, and other areas.

Dr. Devina Seram is currently working as an Assistant Professor at the Department of Entomology, School of Agriculture, LPU. She has completed her Ph.D. (Agricultural Entomology) from Tamil Nadu Agricultural University, Coimbatore with a DST-INSPIRE fellowship, post-graduation (M.Sc. in Agricultural Entomology) from College of Post Graduate Studies (Umiam), Central Agricultural University (CAU), Imphal, Manipur and graduated from Assam Agricultural University with distinction. She was a recipient of the CAU Merit Scholarship and Gold Medallist in her master's degree. She has qualified ICAR-AIEEA-PG, ICAR-SRF-(PGS), and ICAR-ASRB-NET. She has also received several awards in her field of expertise including the National Dedication Day Award, Excellence in Research Award, and a few others. Her research specialization and interests include host plant resistance, insect resistance bioassay, storage entomology, and molecular biology. She has several research, review papers, and book chapters to her publication credit, has guided four masters' students, and currently guiding two students including one Ph.D. scholar.

Dr. Yudhishther Singh Bagal did his Ph.D. in Agricultural Extension and Communication from Sher-e-Kashmir University of Agricultural Sciences and Technology, Jammu, J&K, India. He has published several research articles including book chapters in reputed national as well as international journals and books respectively. Presently he is working as an Assistant Professor in the School of Agriculture, Lovely Professional University, Phagwara, Punjab, India.

Regenerative Agriculture

Haobijam James Watt
Assistant Professor
Department of Agricultural Economics and Extension
School of Agriculture Lovely Professional University
Phagwara, Punjab, India

Devina Seram
Assistant Professor,
Department of Entomology
School of Agriculture, Lovely Professional University
Phagwara, Punjab, India

Yudhishther Singh Bagal
Assistant Professor,
Department of Agricultural Economics and Extension
School of Agriculture, Lovely Professional University
Phagwara, Punjab, India

NIPA® GENX ELECTRONIC RESOURCES & SOLUTIONS P. LTD.
New Delhi-110 034

NIPA® GENX ELECTRONIC RESOURCES & SOLUTIONS P. LTD.

101,103, Vikas Surya Plaza, CU Block
L.S.C.Market, Pitam Pura, New Delhi-110 034
Ph : +91 11 27341616, 27341717, 27341718
E-mail:newindiapublishingagency@gmail.com
www: www.nipabooks.com

For customer assistance, please contact
Phone: + 91-11-27 34 17 17
Fax: + 91-11- 27 34 16 16

Print ISBN: 978-93-58877-27-4

ebook ISBN: 978-93-58874-62-4

Composed and Designed by NIPA

LOVELY PROFESSIONAL UNIVERSITY
Phagwara, Punjab - 144411, India

Dr. Ramesh Kumar Sadawarti
Dean - Agriculture
School of Agriculture, Lovely Professional University
Phagwara, Punjab, India

Foreword

The current agricultural model is facing an unsustainable state because of important issues such as climate change, resource depletion, and declining soil health. These challenges not only threaten global food security but also disrupt the delicate ecological balance that sustains life on Earth. In this context, the concept of "Regenerative Agriculture" emerges as a beacon of hope, offering a holistic and integrated set of practices that promote a more symbiotic relationship between humanity and the natural world. This meticulous book entitled "**Regenerative Agriculture**" is a collaborative effort of a dedicated team of researchers, scientists, and practitioners, which delves deep into the core philosophies of agricultural regeneration. Regenerative agriculture is a farming and land management method that seeks to revive and increase the well-being of ecosystems, while simultaneously enhancing soil health and biological diversity. It mainly emphasizes the use of natural processes to enhance soil health, capture carbon, and enhance resistance to environmental pressures, thus diverging from the traditional farming methods that mainly depend on agricultural chemicals and monoculture. To name a few, topics such as Integrated Farming Systems for the Future, Improving Water Retention and Nutrient Uptake, Mitigating Climate Change (CC) by Arresting Soil Organic Carbon, Recycling Wastes to Ameliorate Soil Biology, Economics of Regenerative Agriculture are highlighted. It provides a rich tapestry of the diverse techniques employed, from harnessing cutting-edge technological advancements in areas like on-farm cooling for fresh produce to reviving the invaluable insights gleaned from indigenous knowledge systems (ITK) for pest management.

This book transcends mere theoretical discussion since it explores the application of transformative practices, exploring critical topics often overlooked in conventional agriculture. It also investigates the economic viability of regenerative approaches, analyzes disease management strategies aligned with these principles, and discovers innovative waste recycling methods that can significantly enhance soil biology, which is the cornerstone of agricultural sustainability. This volume serves as more than just a book; it is a potent call to action. It extends an invitation to a broad audience, including fellow scientists, policymakers, students, and anyone with a stake in securing a more sustainable future for agriculture. By fostering a collaborative partnership with nature, we can unlock the immense potential of Regenerative Agriculture to ensure the long-term health of our planet and agricultural systems for generations to come. This book will equip readers with the knowledge and inspiration to become a vital contributor to this critical movement, and also provide practical guidance on its implementation. Whether you are a seasoned scientist seeking to expand your knowledge base, a policymaker crafting regulation that support sustainable practices, or a student embarking on a career dedicated to environmental well-being, this book offers valuable insights and inspiration. I urge you to delve into the following chapters with an open mind and a commitment to exploring a future where agriculture works in harmony with nature.

(Ramesh K Sadawarti)

Preface

Amidst the worldwide environmental challenges and apprehensions regarding the long-term viability of our agricultural food systems, a subtle revolution known as "Regenerative Agriculture" is emerging across the globe. It can be regarded as a revolutionary approach, which presents a hopeful route to restore our planet's health and provide sustenance to individuals and communities. The concept of Regenerative Agriculture is both straightforward and profound since its objective is not only to achieve sustainable food production but also to restore and enhance the well-being of the earth's environment. It signifies a fundamental change from traditional agricultural methods that have frequently caused harm to the environment and exhausted natural resources. Regenerative Agriculture, in contrast to perceiving agriculture as a purely extractive sector, acknowledges the interdependence of all living organisms and aims to operate in symbiosis with the natural ecosystem. This book entitled "**Regenerative Agriculture**" takes us on a journey deep into the core of agricultural land re-establishment. We explore the fundamental concepts, methods, and compelling case studies of farmers and educators who are at the forefront of advancing a more environment-friendly and adaptable future.

This book unveils the immense capacity of Regenerative Agriculture to bring about transformation through approaches toward tackling the complex challenges our planet is currently facing, including the restoration of degraded landscapes, the improvement of soil health, the promotion of biodiversity, and the mitigation of climate change. Utilizing current research, practical insights, and real-world illustrations, we delve into the potential of regenerative practices to not only provide sustainable nourishment for the expanding global population but also to restore the fundamental elements of life on our planet. At this crucial moment, where we are confronted with pressing demands to tackle climate change, biodiversity decline, and food insecurity, Regenerative Agriculture emerges as a promising solution. This statement serves as a reminder of our ability to restore the health of the land, foster a diverse range of species, and generate plentiful resources for all individuals. However, achieving this vision necessitates a combined endeavor – a mutual dedication to stewardship, innovation, and collaboration. Our intention for this book is to motivate readers from all spheres of disciplines to actively participate in this movement, regardless of whether they are farmers, consumers, policymakers, advocates, or students. Collectively, we have the ability to foster a future in which agriculture is not only sustainable but truly rejuvenating. Our future entails the flourishing of land, the prosperity of communities, and the harmonious coexistence of all living beings with nature.

Editors

Contents

Abbreviations

ADLI	: Agriculture Development Led Industrialization
AEC	: Anion Exchange Capacity
AFO	: Animal Feeding Operations
AHU	: Air Handling Unit
AI	: Artificial Intelligence
AISP	: Agricultural Input Subsidy Program
APEDA	: Agriculture and Processed Food Products Export Development Authority
ATMA	: Agricultural Technology Management Agency
BCA	: Bio-Control Agent
CA	: Conservational Agriculture
CAFO	: Concentrated Animal Feeding Operations
CDCP	: Centers for Disease Control and Prevention
cDNA	: Complementary DNA
CEC	: Cation Exchange Capacity
CFC	: Chlorofluorocarbons
CHN	: Carbon Hydrogen Nitrogen
COG	: Combined Crop Factors
CREAMS	: Chemicals, Runoff, and Erosion from Agricultural Management Systems
CSA	: Community-Supported Agriculture
CT	: Compost Tea
DMC	: Dairy Manure Compost
DNA	: Deoxyribonucleic acid
DNT	: Deoxy-Nucleoside Triphosphate
DPM	: Dust Production Model
EBI	: Ergosterol-Biosynthesis-Inhibiting
EF	: Erodible Fraction of soil
EFB	: Empty Fruit Bunch
EGEM	: Ephemeral Gully Erosion Model
ELISA	: Enzyme-Linked Immunosorbent Assay
ENSO	: El Nino-Southern Oscillation
EPA	: Environmental Protection Agency
EQIP	: Environmental Quality Incentives Programme
ET	: Evapo-transpiration

EUROSEM	: European Soil Erosion Model
FAO	: Food and Agriculture Organization
FYM	: Farm Yard Manure
GHG	: Green House Gas
GIS	: Geographical Information System
GMO	: Genetically Modified Organism
GPS	: Global Positioning System
HCFC	: Hydro chlorofluorocarbons
HDPE	: High-Density Polyethylene
ICAR	: Indian Council of Agricultural Research
ICT	: Information and Communication Technology
IDM	: Integrated Disease Management
IFAD	: International Fund for Agricultural Development
IFOAM	: International Federation of Organic Agriculture Movement
IFS	: Integrated Farming System
IITA	: International Institute of Tropical Agriculture
IK	: Indigenous Knowledge
IKS	: Indigenous Knowledge System
IoT	: Internet of Things
IPCC	: Intergovernmental Panel on Climate Change
IPM	: Integrated Pest Management
ISR	: Induced Systemic Resistance
IT	: Information Technology
ITK	: Indigenous Technical Knowledge
JA	: Jasmonic Acid
K’	: Soil Roughness Factor
KAU	: Kerala Agricultural University
KVK	: Krishi Vigyan Kendra
LAMP	: Loop-Mediated Isothermal Amplification
LDPE	: Low-Density Polyethylene
LED	: Light Emitting Diode
LER	: Land Equivalent Ration
ML	: Machine Learning
mRNA	: Messenger RNA
MSW	: Municipal Solid Waste
MUSLE	: Modified Universal Soil Loss Equation
MVD	: Modern Veterinary Drugs
NATP	: National Agricultural Technology Programme
NBAIR	: National Bureau of Agricultural Insect Resources
NE	: Natural Enemies
NeGP-A	: National e-Governance Plan in Agriculture

NFT : Nutrient Film Technique
NGO : Non-Governmental Organization
NHB : National Horticulture Board
NOPS : National Organic Program Standards
NPK : Nitrogen Phosphorus and Potassium
NPP : Net Primary Productivity
PAR : Participatory Action Research
PCR : Polymerase Chain Reaction
PDO : Pacific Decadal Oscillation
PE : Polyethylene
PLC : Poultry Litter Compost
PR : Pathogenesis Related
PUF : Polyurethane Foam
PVC : Polyvinyl Chloride
R&D : Research and Development
RAS : Re-circulatory Aquaculture System
RCA : Rolling Circle Amplification
RCR : Rolling Circle Replication
rDNA : Recombinant DNA
RFID : Radio-Frequency Identification
RNA : Ribonucleic Acid
RT : Reverse Transcriptase
RT-PCR : Real-time Reverse Transcriptase-Polymerase Chain Reaction
RUSLE : Revised Universal Soil Loss Equation
RWEQ : Revised Wind Erosion Equation
RWH : Rain Water Gathering
SA : Salicylic Acid
SAR : Systemic Acquired Resistance
SAU : State Agricultural University
SAVARS : Solar Assisted Vapor Absorption Refrigeration System
SCF : Soil Crust Factor
SDI : Surface Drip Irrigation
SETi : Soil Erosion by Tillage
SOC : Soil Organic Carbon
SOEL : Stiftung Ockologic and Lamdbau
SOM : Soil Organic Matter
SORET : Soil Redistribution by Tillage
SPEROS : Water and Tillage-Induced Soil Redistribution
SREP : Strategic Research and Tension Plans
SS : Scientific System
TEAM : Texas Tech Erosion Analysis Model

TEP	: Tillage Erosion Prediction
UAV	: Unmanned Aerial Vehicles
UHI	: Urban Heat Island
UK	: United Kingdom
UNDP	: The United Nations Development Programme
USD	: United States Dollar
USLE	: Universal Soil Loss Equation
VR	: Virtual Reality
VRT	: Variable Rate Technologies
WaTEM	: Water and Tillage Erosion Model
WEAM	: Wind Erosion Assessment Model
WEELS	: Wind Erosion and European Light Soils
WEPP	: Water Erosion Prediction Project
WEPS	: Wind Erosion Prediction System
WEQ	: Wind Erosion Equation
WESS	: Wind Erosion Stochastic Simulator
WF	: Weather Factor
WUE	: Water Use Efficiency
ZECC	: Zero Energy Cooling Chambers

1

Introduction to Regenerative Agriculture

***Haobijam James Watt*[1], *Devina Seram*[2], *and Y.S. Bagal*[3]**

[1,3]Department of Agricultural Economics and Extension, School of Agriculture Lovely Professional University, Phagwara, Punjab, Punjab, India

[2]Department of Entomology, School of Agriculture, Lovely Professional University, Phagwara, Punjab, India

Introduction

Regenerative agriculture is both a philosophy and a strategy for land management. It asks us to consider how all aspects of agriculture are connected through a web, which is a network of entities that grow, enhance exchange, and distribute, consumer products and services. Traditionally, supply chains have been organized linearly. It is about farming and ranching in a way that nourishes people as well as the earth, with specific practices differing from grower to grower and from region to region. This style of farming and ranching is also known as "permaculture." Although there is no definitive rule book, the holistic principles that underpin the dynamic system of regenerative agriculture are designed to address issues of inequity, reestablish the health of soil and ecosystems, and ensure that our land, waters, and climate are in better condition for future generations. It is a novel concept of uplifting the natural world rather than imposing an invasive control, regenerative agriculture intriguing environmental approach to agricultural production that enables the natural systems to renew themselves. Regenerative agriculture refers to farming practices that, in comparison to industrial agriculture, cause significantly less damage to the surrounding environment and the soil. When we give more to nature than what we receive from it, we bring nature closer to us and bring ourselves closer to nature at the same time. Farming that is in harmony with both the surrounding environment and the community fosters a feeling of integrity and togetherness that helps to strengthen the connection between the two. The concept of harmonious farming can be summed up in a few words: It involves engaging in agricultural practices and cropping systems that increase soil fertility return nutrients to the soil, increase soil water retention, remove carbon dioxide (CO_2) from the atmosphere, and, most importantly, adapt to the ecosystems in which they are grown.

Regenerative Agriculture: Way Forward

Regenerative agriculture is a way of using a more systems-based and holistic approach to land stewardship and employs several principles with the ultimate aim of increasing the land's productivity and biodiversity over time. In most cases, raising output and biodiversity is as simple as restoring soil health and function. Organic matter, which includes things like plant roots, earthworms, and microbes, is an essential part of good soil. Degenerative land management techniques, such as improper grazing management, have resulted in significantly lower soil organic matter levels in many of our agricultural lands.

Reducing or stopping soil erosion and boosting aggregate stability, water infiltration, water retention, nutrient cycling, plant health, crop yields, crop resilience, and biodiversity can all be accomplished by increasing the organic matter composition of our soils. Increasing the amount of organic matter in the soil also means less carbon is being released into the atmosphere as a greenhouse gas and more is being deposited where it can have a beneficial impact on the planet. There are no universally applicable guidelines for studying, adopting, or making the switch to regenerative agriculture; rather, growers and regions will need to develop their own methods and strategies. Soil health practices are not limited to farmland but can be implemented in any nonfarm setting, such as a backyard or a public park. The significance of soil to our ecosystem has recently been recognized by individuals, communities, and even major cities all over the world. However, when considering a farm or property, regenerative agriculture includes six essential components, where the first is crucial and specific to each individual; the remaining five are general concepts that can be applied in various situations:

1. The importance of knowing one's surroundings cannot be overstated.
2. Reduce the amount of chemical fertilizers, herbicides, and tillage used on the land.
3. During the growing season, prolong the "Living Root" effect by having a plant photosynthesize and pump carbon-based exudates into the soil to support the soil biology. Perennial plants are mainly adapted for this kind of growth.
4. Supply Soil Armor: Also known as "cover cropping," exposing bare soil to the elements is detrimental to soil health; therefore, it is advised that there is always some living or non-living debris covering the soil.
5. Animal Integration: If it is feasible, use planned grazing to have one or more species of animals travel across the fields.
6. Improve Biodiversity by introducing new species into our developing environment, whether through the introduction of bird perches, honeybee hives, honey, or multi-species cover crops.

Origin of Regenerative Agriculture

The concept of regenerative agriculture is founded on a variety of agricultural and ecological practices, specifically concentrating on the practice of composting and causing the least amount of disturbance to the soil as far as possible. Maynard Murray (2003) had similar ideas, using sea minerals. His work led to innovations in no-till practices, such as slash and mulch in tropical regions (Fukuoka & Masanobu, 2010). Sheet mulching is a regenerative agriculture practice that smothers weeds and adds nutrients to the soil below. Regenerative agriculture is based on various agricultural and ecological practices, with a particular emphasis on minimal soil disturbance and the practice of composting.

The word "regenerative agriculture" was first used by the Rodale Institute in the early 1980s (AFSIC). Rodale Publishing established the Regenerative Agriculture Association, which started publishing books on regenerative agriculture in 1987 and 1988 (TESA). While we continue to move forward under the banner of sustainability, we are, in essence, continuing to impede our own progress because we are not accepting a goal that is difficult enough. Robert Rodale quoted "I do not have a problem with the phrase 'sustainable'," but I do prefer the term "regenerative agriculture." However, the institute stopped using the term in the late 1980s, and it only appeared sporadically (Anonymous, 2005) until they released a white paper in 2014 titled "Regenerative Organic

Agriculture and Climate Change." The summary of the white paper states that "we could sequester more than 100% of current annual carbon dioxide (CO_2) emissions with a switch to common and inexpensive organic management practices, which we term 'regenerative organic agriculture.'" The white paper described agricultural practices that would allow for the sequestration. The residue that was left behind from a previous wheat harvest is now giving way to the emergence of newly planted soybean seedlings. This demonstrates how to rotate crops and plants without tilling the soil.

In his first book, *The Restoration Economy,* published in 2002, Storm Cunningham provided documentation of the commencement of what he referred to as "restorative agriculture." According to the definition provided by Cunningham, restorative agriculture is a method that not only rebuilds the quantity and quality of topsoil but also restores local biodiversity (particularly native pollinators) and the function of watersheds. In *The Restoration Economy,* "restorative agriculture" was one of the eight categories of "restorative development industries" and "restorative development disciplines" (Cunningham, 2002). The majority of people do not understand the effect that agriculture has on our planet. How the vast majority of our food, fiber, and fuel are produced at present is actually causing an alarming rate of damage to the ecosystem of our planet. This damage takes the form of a loss of topsoil, a loss of biodiversity, desertification, the devastation of habitat, as well as air and water pollution.

Agriculture is another industry that impacts significantly on climate change. Due to a lack of understanding of how soil and ecosystems work, this phenomenon has been going on for thousands of years. It is not a novel phenomenon, but it has been going on for thousands of years. The "conventional" agricultural practices that we use on a large scale today are harmful to the environment because they disrupt the natural cycles essential to our survival on this planet. The goal of regenerative agriculture is not only to stop causing damage to our ecosystem but also to actively improve it, all while maintaining our current levels of food, fiber, and fuel production. There is not one description of regenerative agriculture that is universally accepted by all. On the other hand, the majority of people would probably agree that its primary emphasis is on enhancing soil health by reintroducing carbon that has been removed from the atmosphere and placed in the soil through the application of a variety of agricultural management practices designed to operate in harmony with natural systems.

These are not "new" methods by any means. Since the beginning of time, indigenous people have engaged in activities that are mutually beneficial and restorative when they associate with the natural world. We can combine traditional knowledge with scientific research to bring about a rapid transformation in the way agriculture is practiced all over the world. This will allow us to create a future of plenty rather than one of scarcity. We must keep in mind that we are a part of nature and not distinct from it and that how we interact with nature has significant repercussions for both our civilization and the planet as a whole. The following are some instances of agriculture that promotes regeneration:

1. Permaculture
2. Crop growth with biological inputs on native grasslands
3. Forest management
4. Integrating livestock
5. Food forestry
6. Holistic grazing
7. Biodynamics
8. Diverse rotations
9. Cover crop

These techniques intend to improve the quality of the soil by establishing a self-sufficient ecosystem that also makes a positive contribution to the economy. This will result in a method of farming that both mitigates and reverts the negative effects on the environment. The practice of regenerative agriculture enlightens us to the fact that our planet is rich in resources, incredibly complex, and beautifully interconnected. Because of the interdependence of various factors of the system, if one mechanism is broken, then it will affect the others as well, and if we repair one, we will repair them all. In this sense, the health status of a system is determined by the state of health of all of the participants in that system. And it is precisely this procedure that will be responsible for the improvement in our future (Anonymous, 2022).

How Does Regenerative Agriculture Work

According to the Rodale Institute, one of the growing numbers of organizations advocating this strategy, regenerative agriculture is a method of farming that "improves the resources it uses, rather than destroying or depleting them." Regenerative agriculture is becoming increasingly popular. A wide range of approaches are utilized, and there is a significant amount of focus placed on taking an ecosystem-wide perspective when analyzing the agroecosystem. After each harvest, the land is typically tilled under and then left barren in conventional farming. Erosion of the soil, which deprives the soil of its nutrients and releases significant amounts of CO_2, is a side effect of tilling and plowing. Farmers can minimize the amount of physical disturbance caused to the soil by implementing no-till techniques, which helps the soil to keep its overall structure while also protecting it from erosion.

Planting cover crops between growing seasons (such as wheat), utilizing crop rotations, and applying compost and animal manures are all examples of practices that are utilized by regenerative agriculture systems. These practices help to restore the microbiome of the soil, which in turn helps to encourage nutrient growth. These microbial communities in the soil have been thrown out of equilibrium, both in terms of their structure and their function, as a result of using synthetic fertilizers in conventional farming. The part that livestock play is another important consideration. The conventional method of using cows on farms is not regenerative: If animals are confined to a single field for an extended period, they are more likely to overgraze the land, which will ultimately result in the soil becoming barren and compacted. A regenerative strategy, on the other hand, involves moving animals through the land in a pattern that is similar to traditional grazing patterns. This allows the animals to be free to roam over much larger areas before the risk of overgrazing arises. This strategy, also known as adaptive multi-paddock (AMP) grazing, places large numbers of animals in a limited space for only brief periods at a time, after which the animals are moved to a new confinement and the process is repeated. This enables each paddock to have a longer period during which it is not being grazed, which speeds up the development of the grass. In addition, the movement of cows' hooves helps to break up the soil, which results in the formation of a natural mulch that can assist in the soil's ability to hold onto moisture. However, an AMP strategy can also help to manage this, ensuring that this does not cause an excessive amount of disturbance and enabling soils to have periods of rest. Their wastes also contribute nutrients to the ground, which improves the soil fertility even further.

Taking a more systems-based and holistic approach to land stewardship, regenerative agriculture employs several principles with the ultimate aim of increasing the land's productivity and biodiversity over time. In most cases, raising output and biodiversity is as simple as restoring soil health and function. Organic matter, which includes things like plant roots, earthworms, and microbes, is an essential part of good soil. Degenerative land management techniques, such as improper grazing management, have resulted in significantly lower soil organic matter levels on many of our agricultural lands. Reducing or stopping soil erosion and boosting aggregate stability, water infiltration, water retention, nutrient cycling, plant health, crop yields, crop resilience, and biodiversity can all be accomplished by increasing the organic matter composition of our soils. Increasing the amount of organic matter in the soil also means less carbon is being released into the atmosphere as a greenhouse gas and more is being deposited where it can have a beneficial impact on the planet. There are no universally applicable guidelines for studying, adopting, or making the switch to regenerative agriculture; rather, growers and regions will need to develop their own methods and strategies. Soil health practices are not limited to farmland but can be implemented in any nonfarm setting, such as a backyard or a public park. The significance of soil to our ecosystem has recently been recognized by individuals, communities, and even major cities all over the world. However, when considering a farm or property, regenerative agriculture includes six essential components, where the first is crucial and specific to each individual; the remaining five are general concepts that can be applied in various situations:

1. Understand Context: Economic, personal, community, ecological, climate, bioregion, etc.
2. Minimize Disturbance: This refers to tillage, chemical fertilizers, pesticides, and more.
3. Establish a "Living Root": Have a plant photosynthesize and pump carbon-based exudates into the soil to feed the soil biology for as long as possible throughout the growing season. Perennials are able to do this well.
4. Provide Soil Armor: Also referred to as "cover cropping," bare soil exposed to the elements that harm soil health, so it's recommended to always have some living or dead debris covering the soil.
5. Integrate Animals: Have one or more types of animals move across fields if it can work in our context, otherwise known as planned grazing.
6. Enhance Biodiversity: Add diversity to whatever it is we are growing – this could be planting diverse hedgerows throughout the farm, installing owl boxes, integrating honeybees, or diverse multi-species cover crops.

Although these ideas are most powerful when used together, they need not be applied simultaneously.

Regenerative Agriculture and Its Effect on Climate

Even though different cultures all over the world have been practicing restorative land management for centuries, the dominant cultures have disregarded holistic practices that take into consideration the health of the soil and how ecosystems work. The rate at which soil was degraded accelerated during the Industrial Revolution as a result of the quick transformation of more landscapes using the use of machinery and artificial fertilizers. During both two world wars, massive wheat and maize monocultures were grown to satisfy the growing demand for food. Some of these developments were tremendously helpful in terms of producing a great deal of calories, but there were some concerning long-term effects associated with their use. Ecosystems that were both highly functional and complicated were stripped of their biodiversity, and the soil was also stripped of its organic matter. This paved the way for soil erosion, the loss of biodiversity, a decreased density of nutrients in our food, flooding, droughts, and an increase in the amount of carbon in our environment. The vast majority of land management practices implemented by humans, combined with our continuously increasing population, are leading to soil deterioration and desertification at an alarmingly rapid rate. A once-vibrant soil that is now devoid of life has been transformed into dirt. About 46% of the earth's surface is currently covered by drylands, and approximately 9% of those drylands are in the process of becoming severely deserted. During the past 40 years, we have suffered the loss of approximately one-third of our arable territory.

Benefits of Regenerative Agriculture to Farmers

Even in modern times, the typical farm in the United States still loses more than 4 tons of topsoil per acre per year. In addition to suffering historic crop losses as a result of flooding, drought, and other weather extremes, farmers are also accruing additional debt at a rate of about 4% per year. The vast majority of conventional farming practices have concentrated on raising crop yields through the application of numerous costly fertilizers, chemicals, and seeds; however, this method is not sustainable in the long term. A significant effect can be had on a farmer's bottom line by employing regenerative farming practices. These practices have the potential to cut down on soil erosion and even help build topsoil by increasing the amount of carbon that is stored in the soil. Farmers are aware that they can increase their profits by increasing yield, improving quality, raising prices, and lowering production costs; however, it is unclear how any or all of these goals can be achieved through the use of regenerative practices. To increase one's profitability, it is essential to have a more in-depth comprehension of the role that soil plays within the ecosystem.

By preventing soil erosion and increasing the amount of carbohydrates (food) that are pumped into the soil through plant exudates, regenerative agriculture practices aim to improve soil health and biodiversity. This is accomplished by preserving the soil from erosion. Farmers can still get yields from their crops even when there is a drought because healthy soil can infiltrate and store more water, which allows the farmers to save money on the expenses of irrigation. If the soil is healthy, then there will be more nutrients accessible to the plants, which will result in less money being spent on fertilizers. Because healthy soils result in healthier plants, producers are less likely to require the costly application of insecticides, herbicides, pesticides, fungicides, and other such chemicals. Producing food that is richer in nutrients should be easier for plants that are in better condition. A farmer may be able to command a higher price for their produce by cultivating plants that are in better health. This is because healthier plants have extended shelf lives, as well as improved appearance and flavor. While reducing tillage can improve the health of the soil, it can also reduce costs associated with machinery, labor, and fuel. The health of farmers' soils can be improved year after year, which is good for both their crops and their bottom line.

Benefits of Regenerative Agriculture to Nature

A beneficial environment is created throughout an ecosystem when the soil is healthy. Humans can create soil that is more resilient and produces fewer greenhouse gases like CO_2, methane, nitrous oxide, and water vapor if they provide support for the natural processes that occur in nature using traditional knowledge and established scientific principles. If large enough regions are regenerated, there is even the possibility that the climate could become more favorable.

Promotes Healthy Soil

A microbiota that is rich in variety is essential to the health of the soil. Agricultural systems are similar to dominoes in some ways; for example, removing animals from an agricultural system eliminates the ability of cows to graze the grass, which, if done correctly, would enable each grass stalk to grow to a level that is appropriate for the yearly renewal of soil. When animals are removed from an agricultural system, the cow's ability to graze the grass is also eliminated. The components of a regenerative system are not all of the same size and shape as the dominoes; rather, they are all different sizes, and their interactions with one another have various degrees of intensity. This is the primary distinction between restorative agriculture and dominoes. Due to the incredible complexity of the system, observing restorative agriculture can be both thrilling and amazing. There may not be a solution that works perfectly for everyone, but there is a guideline that must be followed: Encourage diversity. The methodology and philosophy of restoration agriculture are what are needed to re-establish, foster, and construct a microbiota that is rich in diversity in our soils.

Reducing Greenhouse Gas Emissions

The reduction of greenhouse gases is accomplished not only because regenerative agriculture generates fewer emissions, but also because the practices used in it simultaneously sequester and neutralize emissions. Livestock and agricultural production are the two primary contributors to the greenhouse gas emissions that are caused by agriculture. Animal feeding operations, also known as concentrated animal feeding operations, are what we're actually referring to when we say that livestock are releasing gas from animal feeding operations (AFOs), mainly concentrated animal feeding operations, also known as CAFOs. Currently, farmers and cattle in the United States live in deplorable circumstances, and greenhouse gases are released in massive quantities from the components used in CAFOs. Further, there is a significant amount of urine and feces on the ground inside CAFOs. Methane (CH_4) is also one of the principal gases released into the atmosphere by CAFOs. Methane gas is produced as a consequence of the digestive gas that cattle produce. However, all the responsibilities cannot be placed on the cows. There is a significant difference between a bovine that lives in a CAFO and a cow that forages for their food. Cows' stomachs become unsettled because of the food that they are fed in CAFOs, which consists of corn and soybeans, both of which cows have not evolved to digest. This causes cows to have digestive issues. Cattle have adapted their digestive systems to process grass, not genetically modified maize. The agriculture industry has covered up the problem by dosing animals with antibiotics rather than examining their diets. This practice causes further harm to the animals' gut biome and can be found in the flesh that consumers buy. Whether it is the livestock themselves or the manure pits, all of this activity results in a significant amount of production of CH_4. Dewormers and other medications that eliminate dung beetles are sometimes used by conventional farmers. Again, if we are in an environment that is natural and does not contain any medicine that kills the beetles, then we will be able to witness a beautiful process in which the beetles break down the feces, and then after a couple of days, the poop breaks down back into the soil, acting as a fertilizer. This process can only take place if the beetles are not killed by the medicine. It is remarkable what nature can teach us if we just give it a chance. Cows in a regenerative system move around the pasture more frequently, and if the people taking care of them are doing their jobs properly, the cows will engage in a practice known as "tall grazing." A heifer will only graze the grass up to a certain point when it is being managed to using the tall grazing method. This ensures that each section of grass has sufficient amounts of green matter remaining, which enables it to continue the process of photosynthesis throughout the year. This, in turn, causes it to develop more quickly and contributes to the retention of CH_4 in the soil. Another gas that is released from livestock feedlots is CO_2, which contributes to the greenhouse effect.

Photosynthesis is the process that drives the complete cycle of nutrients in the environment, which transfers CO_2 from the atmosphere into the soil and, *via* the plants, nourishes the biology that lives in the soil. Further, there are lifeforms in that soil that are based on carbon, and these lifeforms are eventually responsible for the creation of more microbes in that soil, which results in an increase in root mass and overall organic matter, all of which contain carbon.

Nitrogen is another potential problem. As a nutrient source, it is used as fertilizer. On the other hand, we use far too much of it, which results in the production of nitrogen oxide. Because nitrogen makes up 78% of the air we breathe, it is possible that nitrogen could be extracted from the environment and used to produce fertilizer. It turns out that nature is really excellent at extracting nitrogen from the air and converting it into a nutrient that plants can use. Microbes in the soil are responsible for converting the nitrogen in the atmosphere into a form that can be used by plants. This process results in the production of ammonium, nitrite, and nitrate, as well as amino acids, which are the fundamental components of proteins. Another piece of encouraging news is that there will never be a shortage of nitrogen. There are a great number of instances in which producers have been successful when using this kind of fertilizing resource. Farmers can save money and help keep nitrogen out of the atmosphere by limiting or eliminating their use of fertilizer altogether. This will also result in less nitrogen being released into the atmosphere. Because we can sequester CO_2 into the soil while simultaneously minimizing the amount of CH_4 and nitrogen that is released into the atmosphere, the methods mentioned above help reduce the amount of "bad gases" that are released into the atmosphere. The presence of water vapor in the atmosphere is a significant greenhouse element that is often overlooked by some individuals. Because of the warming trend, there is an increase in the amount of water vapor in the air. We probably already know that water vapor retains heat. When there is a greater concentration of water vapor in the atmosphere, that water vapor serves as an additional layer of greenhouse gas. At this moment, a significant quantity is involved. Similarly, the soil should be returned in its healthy state to the environment. Because it contains a greater amount of organic matter, healthy soil has a greater capacity to both store and absorb water. Humans can retain more water in the soil than in all of the dams, rivers, and lakes that we currently have, of course, we engage in regenerative agriculture and begin to increase the amount of organic matter in our soil.

Agricultural Methods that Promote Regenerative Agriculture

The practices of regenerative agriculture differ depending on the region, the context, and the history. The following practices are methods that farmers and agricultural laborers can implement that are best suited to the history of their territory.

No Till or Tillage

Tilling, or turning the soil over, is an essential part of conventional agricultural practices. Soil aggregates are broken up and organic matter is disturbed and exposed to air during tillage. When carbon is exposed to air, it undergoes oxidation and produces CO_2. Issues with tilling are possible. When considering tilling, context is everything. Many people advocate for "no till" farming, which may be the best-case scenario for minimal disturbance farming but rarely pans out in practice on land that has been abused for decades or centuries. Minimal tillage, also known as "low-till," is recommended for new growers, and not just for the apparent reasons. If a farmer wants a certain yield every year, she need only perform a single tillage exercise. Alternatively, a rancher may plow a single time to break up a tough hardpan and sow a cover crop. There is no universally applicable guideline regarding whether or not one should till the soil.

Compost

Compost is expensive, so farmers must consider the cost before using it. How can farms use compost cheaply and still benefit? Examine bacteria. Most compost users have soil microbes harmed. An extract of compost in water is a fine start. Farmers can inoculate their seeds by coating or dilution-dripping them, similar to tee seed probiotics. After sprouting, the seed does several things. Compost accelerates sprouting. Seeds amaze. They know when to germinate and exude, fueling soil microbes. Its seed microbiome starts growing. A grower can improve nutrient acquisition and plant disease resistance by diversifying the microbiome with compost extract and the lists continue. The Marine Carbon Project proves compost's longevity. One study found that a single application of composted organic waste increased net primary productivity (NPP) for three years without diminishing impacts. One application altered the yearly grassland ecosystem carbon balance and increased carbon storage. Compost has many proven benefits, so farmers just need to find the most cost-effective way to use it. Compost helps plants fight pests and diseases, photosynthesize better, and sequester more carbon into the soil. Composting rangelands improves soil and yields.

Crop Rotation

Instead of growing the same product year after year (which can become a familiar territory for pests and diseases), regenerative agriculture generally uses crop rotation to maintain healthy soil. Though crop rotation has been shown to improve soil health, there may be other methods to continue growing the same crop year after year. The condition of the earth at the outset determines the starting point for the grower. Growers can gain more assurance through soil testing. Crop rotation is a simple method to improve soil quality, however, it is a problem in most areas.

Cover Cropping

Cover cropping is an efficient method of improving soil quality. Cover crops serve an essential function in maintaining soil cover and feeding the soil with plant exudates, both of which are necessary if a farmer is to achieve the goal of having living roots in the soil for a live plant above ground to photosynthesize. Is there any bare ground in our mental image of nature (or an untamed place you're familiar with)? Unlikely. Natural soil is rarely completely barren. The soil is never bare except perhaps immediately following a particularly devastating fire or natural catastrophe. This is significant because soil contains a wide variety of organisms, from the microscopic to the macroscopic, and is thus a living entity in its own right. They need a safe place to live and sustenance to eat. Nothing will grow in barren earth, so they will not be fed. Without cover crops, we leave a big hole in the food web's web of life. Larger organisms that consume soil microbes starve when there are none to feed. A lack of microorganisms, which earthworms rely on for sustenance, means that farms with bare soil cannot sustain earthworm populations. When you are not growing your main product, planting cover crops is a smart move for soil maintenance. In a cover crop, what characteristics are most desired? It all comes down to the soil and the cultivator once more. According to Dr. Christine Jones, you should collect eight various plant species from a range of heights and leaf sizes representing at least four plant families. For instance, it is not ideal to rely solely on plants belonging to the bean family.

Strawberry fields

The variety of insects, birds, and other organisms that feed on them is possible due to the biodiversity. Soil microorganisms respond positively to increased diversity, and more plant species will produce exudates that nourish soil microbes. Maximum microbial biomass and CO_2 infusion are desired. Erosion can be avoided by using cover crops. The earth will be blown and washed away if it is bare. Farmers call rain that falls on a cover crop as "effective rainfall" because the water is retained in the soil, and a cover crop can prevent runoff (even when there's a breeze since the plants act as a buffer for the soil). When it rains, the water doesn't wash away along with the pesticides and harm our water supplies. A cover crop has the effect of actually cooling the atmosphere. The soil can dry out and reach temperatures of 130 degrees Fahrenheit or higher on a day when the air temperature is only 70 degrees, due to the "heat island effect" that occurs when the ground is unprotected from the weather. Heat is harmful for microbes, but it also prevents clouds from forming and makes the atmosphere more warm and humid. This means that if a cloud is moving through a region with no vegetation to absorb the heat, the cloud will dissipate and become water vapor instead of continuing on as a cloud. There would be numerous benefits to the environment if we covered our land instead of exposing it. That is why cover cropping is such a vital component of sustainable farming practices like regenerative agriculture

Reducing Pesticides

Large quantities of fertilizers and toxins are another problem in traditional agriculture, as they have an adverse effect on soil biology. There are many different labels for pesticides. Regenerative practices aim to reduce the use of harmful chemicals like insecticides, fungicides, pesticides, etc. Consider herbicides in the same light as antibiotics. They actively oppose any kind of existence on Earth. When you spray a pesticide on a plant, you risk wiping out more than just the plant itself. Fungicides have the same effect, killing off a wide variety of fungi that are essential for the proper functioning of many underground systems. Each of these pesticides has a microscopic degree of damage. Optimal soil is rich in microbial life and biological variety. Telling a working farmer to cease using pesticides is difficult. However, producers can begin to see signs of use if they keep a close eye on the soil and conduct tests. One sign of the presence of "cides" is the appearance of new parasites. A shrub that is being devoured by bugs has a message for you. Because a thriving plant is better able to withstand attacks from pests like insects and molds. In a community garden, for instance, you might see two identical plots, one of which is flourishing while the other is under damage from aphids. The only distinction is in the leadership. The insects are present, but they are picking the bad food to feed on. Typically, fertile soil provides those nutrients; the two interact mutually. Plants are the next logical step after microorganisms, and they rely on their exudates for survival. Whether or not bacteria, rather than the plant, communicate the plant's requirements for nutrients like phosphorus remains an open question. The bottom line is that plants grown in healthful soil tend to thrive.

Livestock Grazing

The livestock industry occupies a sizable part of farmland, and it requires special care and management. It is estimated that roughly 70% of the world's territory is devoted to raising livestock. Traditional grazing methods are used on the vast majority of farmland. Traditional grazing methods involve fencing off a sizable area of land and letting the livestock, sheep, or goats graze freely within that area. Because their produce matures and the grass regrows at the end of the season, many farmers and ranchers don't see a problem with this scenario. However, ungated grazing causes a shift in the grassland's species composition. Perennial plants, those that don't need to be replanted every year, will begin to disappear. In general, a perennial plant has a lifespan of two years, has extensive root systems, and can maintain its verdant coloration well into the dry season. Animals raised in a regenerative grazing system tend to be healthy overall because they are not given artificial growth promoters or antibiotics, and they feed on pesticide-free grass and forage. Consume healthily, and excrete healthily. Good excrement has the added benefit of attracting a wide variety of insects and dung beetles, which work to decompose it into usable fertilizer. When feces decompose to the point where they are no longer visible, you know they are healthy. There will be more perennials, which means greener during the dry season, healthier livestock, fertilized soil, more biodiversity, and a greater capacity to draw down carbon when a planned grazing system is implemented.

Silvopasture

The deliberate practice of integrating grazing livestock and trees into the same agricultural enterprise is known as silvopasture, which adds a new dimension to the traditional practice of livestock grazing. Imagine a tree crop, whether grown for timber, fruit, or nuts, that is incorporated with livestock, such as sheep, chickens, ducks, turkeys, or pigs, roaming underneath the trees. The trees could be grown for any of these purposes. Diversity is essential for anyone who is in charge of running a silvopasture enterprise. A symbiotic relationship exists where the animals and the plants are mutually beneficial to one another. The animals benefit from the trees' shelter, windbreak, and foliage, and the trees benefit from the nutrients that the animals' fertilizer provides.

Agroforestry

When it comes to combining different elements of restorative practices, agroforestry is comparable to silvopasture. In agroforestry, row crops and tree products are grown side by side. Other species benefit from the shade, windbreak, and habitat that the trees provide (which contributes to an increase in biodiversity), and the trees themselves gain from the fact that the crops maintain a degree of biodiversity in the soil. In addition, the trees help to prevent erosion and the loss of water due to weather, and they help to keep the temperature down. If you have successfully implemented agroforestry, you should see increased infiltration rates as well as enhanced nutrient cycling.

Strategies for Advancing Regenerative Agriculture

Provision of Education

Do you feel encouraged and energized to see the potential of regenerative agriculture become a reality? You have already invested the time and effort necessary to acquire a "topsoil" degree of comprehension regarding regenerative agriculture. Investing in your own personal growth by way of continued education and inspiration is a fantastic first move.

Participate in the Movement for Healthy Soils

The health of our communities depends on people like us taking an interest in environmental health and advocating for it. Start your search for answers on the property. Visit our town's farmer's market and listen to their stories; chances are good that they're dealing with interesting problems that might pique your interest. Getting our hands dirty is as simple as paying attention to the food we ingest before, during, and after we consume it. We may already be conscious consumers, but maybe there are ways we can return to regenerative best practices by beginning with a review of our purchasing habits. For instance, when we go grocery shopping, we can attempt to stock up on perennials, we can eat locally at least twice a week, or we can eat seasonally during our favorite time of year. These baby movements are the beginning of a new habit that will grow and flourish over time.

Begin Your Own Composting Program

It is not necessary to be a farmer or landscaper to engage in the practice of composting as it is another method for practicing regenerative living. Composting can be done by anyone, regardless of whether they live in an apartment, a house, or on a property. Compost pick-ups, drop-offs, and distribution sites can be found in many community garden centers across the country. If you reside on a property that has a backyard, your compost bin could be of great use to a garden project that is either just getting started or is already in progress.

Study Agricultural and Government Policies

The direction that farming methods develop in is heavily influenced by public policy. Governments typically craft agricultural policies with the goal of increasing agricultural output and yield within a set period while also improving farmers' financial standing and general quality of life. This strategy was created to foster extensive growth in the agricultural industry. The primary goals of agricultural policy in India are to address issues such as the dominance of low-value agriculture, a negative cost-benefit analysis of the sector's activities, and the slow development of co-ops and other forms of self-help.

Conclusion

The future of our food systems and the environment looks brighter and more sustainable thanks to regenerative agriculture. Regenerative agriculture not only secures the long-term profitability of farming but also mitigates climate change, conserves water resources, and improves food security by focusing on practices that restore soil health, encourage biodiversity, and reduce reliance on toxic chemicals. Adopting regenerative agriculture practices is essential to constructing a more resilient, healthier, and environmentally responsible agricultural system in the face of mounting problems posed by climate change and the need to feed a growing global population. We can pave the path for a more sustainable and prosperous future for farmers, consumers, and the planet by continuing to invest in research, education, and policy support for regenerative agriculture.

References

Anonymous (2005). Tracing the Evolution of Organic / Sustainable Agriculture (TESA1980) | Alternative Farming Systems Information Center| NAL | USDA.

Anonymous (2022). What Is Regenerative Agriculture? Introduction to Regenerative Agriculture. Retrieved Jan 25, 2023 from https://www.gallantintl.com/blogs/what-is-regenerative-agriculture.

AFSIC History Timeline. Alternative Farming Systems Information Center, United States National Agricultural Library, USDA. Retrieved 2017-03-09.

Cunningham, Storm. *The Restoration Economy*. Berrett-Koehler Publishers, 2002. p. 340.

Fukuoka, Masanobu. (2010). The one-straw revolution: An introduction to natural farming. New York Review Books. ISBN 978-1-59017-392-3. OCLC 681750905. and *Fukuoka,*

Masanobu Metreaud, Frederic P. (1993). The natural way of farming: The theory and practice of green philosophy. Bookventure. ISBN 978-81-85987-00-2. OCLC 870936183

Murray, Maynard. (2003). Sea energy agriculture. Acres U.S.A. ISBN 0-911311-70-X. OCLC 52379170. (originally published in 1976).

Open Food Network (NA). An Introduction to regenerative agriculture. Retrieved Jan 16 2023 from https://about.openfoodnetwork.org.au/project/an-introduction-to-regenerative-agriculture/.

The Third Pole (n.a.). Explainer: What is regenerative agriculture? Retrieved Jan 26, 2023 from https://www.thethirdpole.net/en/food/explainer-what-is-regenerative-agriculture/?gclid=Cj0KCQjwiZqhBhCJARIsACHHEH9AhdvzouD5wyJMTKHMAWG5sbB7G3ZsT0pIrTC49YqX33Y9OcqETywaAnmzEALw_wcB.

Tracing the Evolution of Organic / Sustainable Agriculture (TESA1980) | Alternative Farming Systems Information Center| NAL | USDA. Retrieved 2017-03-09.

2

Integrated Farming System (IFS) for the Future

Haobijam James Watt

Department of Agricultural Economics and Extension, School of Agriculture, Lovely Professional University, Phagwara, Punjab, India

Introduction

The term "Integrated Farming System (IFS)" (or simply "integrated farming") is a term used frequently and generically to describe a method of farming that is more integrated than monoculture methods. By virtue of the intensification of crops and related industries, an IFS offers the chance to enhance economic yield per unit area per unit time. This is accomplished through the system's use of several crop rotations. It improves the efficiency of fertilizers and natural resources, improves the recycling of nutrients, and contributes to increased food security. The IFS technique promotes ecological intensification by lowering the amount of anthropogenic inputs while simultaneously enhancing the performance of ecosystem services such as the recycling of nutrients, the creation of soil, and the soil's fertility. IFSs that are well-managed are seen to be safer than those that are not because they benefit from synergy among businesses, diversification of crops, and ecological stability (Behera et al., 2015). It requires paying attention to the minute details and constantly striving to improve in all facets of agricultural production through well-informed administrative practices. The IFS approach is widely regarded as the most important strategy for enhancing the nutritional status of families and ensuring the viability of their livelihoods. The research strategy that is centered on crops and cropping systems needs to give way to the research strategy that is based on agricultural systems, especially for smaller farms (Jha et al., 2011). Recycling of waste and increased effectiveness in terms of land utilization are two essential components of the IFS. The components and enterprises that make up the IFS might differ from region to region. This is because agroclimatic conditions, such as the kind of land, the availability of water, the socioeconomic standing of farmers, and market demand, all play a role. For effective holistic agricultural systems to be constructed, there must be strong connections and complementarities between the various components (Bell et al., 2014). Integrating land-based businesses into the biophysical and socio-economic contexts of farmers, such as aquaculture, poultry, duckeries, apiaries, as well as field and horticultural crops, is essential in the IFS to make farming highly profitable and reliable (Behera et al.,2004: Kashyap et al., 2015). Farmers who are limited in their access to resources can achieve sustainable output through the careful administration of agricultural residues and the distribution of limited resources.

Integrated Farming System: It is the integration of agriculture with livestock, poultry, and fish maintained at the same place to generate employment around the year and also get additional income.

Mono-cropping: It is the practice of growing a single crop year after year in the same place of land area.

Ecosystem: To put it simply, an ecosystem is any given region in which the interplay of climate, topography, plant and animal life, and other factors creates a self-sustaining community of life.

Livelihood security: It is the adequate and sustainable access to and control over resources, both material and social, to enable households to achieve their rights without undermining the natural resource base.

Farming Systems in India

Overexploitation of groundwater resources, soil degradation brought on by intensive tillage practices, decreased soil organic carbon (SOC), and shrinking trends in the size of landholdings are some of the factors that pose a risk to India's agricultural sector. Other factors are the requirement to preserve our natural resources, the adverse effects of climate change, declining factor productivity, falling factor productivity, and nutrient mining and multiple nutrient deficiencies. In terms of achieving sustainability, agriculture in our country is at a crucial crossroads for several reasons, the most important of which are as follows: (a) the region is having trouble generating sufficient income and employment for its large farming population; (b) failing to achieve environmental and energy security on the farm level; and (c) failing to confront or cope up with the effects of climate change (Behera and France, 2016). As a direct result of the various issues and concerns presented by modern agriculture, new concepts such as organic farming, natural farming, biodynamic agriculture, do-nothing agriculture, eco-farming, integrated agricultural systems, and so on have come into existence. Back to nature is what these types of farming practices are all about, to preserve the continuity of productivity between the soil, plants, and animals over the long term. Back-to-nature farming practices aim to preserve the continuity of productivity between the soil, plants, and animals. In light of these circumstances, it is necessary to investigate agricultural strategies that have the potential to boost productivity and provide adequate income and employment opportunities for smallholder farmers. In addition, it is necessary to investigate agricultural strategies that have the potential to generate renewable energy on the farm, put a stop to the loss of biodiversity, and reduce the amount of carbon emissions produced (Behera et al., 2015). All facets of agricultural food production must be sped up immediately, with adequate attention given to the maintenance and preservation of natural resources. Only through the management of resources in a sustainable manner and the implementation of farming practices that encourage participatory holistic thinking, this will be possible. All parts of agricultural food production must be sped up immediately to meet the ever-increasing demands for food posed by such a big population. Only then it will be possible to keep up with these demands. Specifically, for the smallholders with less than 2.0 ha of land who make up the vast majority of the farming community in developing countries, it is necessary to develop strategies and improved agricultural technologies that enable sufficient employment opportunities and income generation. This is especially true for the smallholders. This is because the amount of land that is available on a per capita basis is decreasing, which makes it more vital than it has ever been to find a solution to this problem. It is quite improbable that a single agricultural business, such as a typical monoculture system, will be able to provide financial support for all of the smallholder farmers. Because they take advantage of synergies between many businesses, a wider variety of food options, and more ecologically responsible farming practices, IFSs pose a lower risk to human health and the environment when they are properly managed. This is because IFS takes advantage of all of these factors. Based on this information, the International Fund for Agricultural Development (IFAD) has been advocating for the development of small and marginal farms, and researchers have developed a variety of solutions that have been of assistance to smallholder farmers by providing supplemental income and employment as well as reducing risk.

A "farming system" is a complicated network of interconnected components, such as land, flora, animals, tools, labor, financial resources, and others. The name "farming system" alludes to this complex network. Households involved in agriculture have some say over the inputs, which also have some sway on the outcome. To varying degrees, depending on the political, economic, institutional, and societal forces, all components are active on a variety of levels (Mahapatra, 1992). Sustainable production of high-quality food and other products using ecologically preferred technologies, maintenance of farm income, eradication or reduction of sources of present environment pollution generated by agriculture, and maintenance of mulberry orchards are all hallmarks of IFS. IFS is a method of agriculture that incorporates both the use of natural resources and various regulatory measures. Integration of two or more appropriate combinations of businesses, such as agricultural, dairy, piggery, fishery, poultry, beekeeping, etc., for each farm, in accordance with the availability of resources to support and fulfil the requirements of the farmer. Examples of these types of businesses include agricultural, dairy, piggery, fishery, poultry, etc. An IFS approach is not only a reliable method of obtaining relatively high productivity with a substantial economy of fertilizer, but it is also a concept of ecological soundness that can lead to sustainable agriculture. This is because an IFS approach reduces the amount of inputs needed to produce crops, which in turn reduces the amount of pollution caused by those inputs. This is due to the fact that IFSs are developed to reduce as much as possible the number of resources wasted (Swaminathan, 1987). The return on investment in integrated farming is higher, but so is the risk involved in achieving that return. Due to the enormous number of factors that have the potential to effect annual farm income, farming in India is a career that carries a high degree of financial risk. The labor for farming is physically demanding, and it occasionally even puts the employees in dangerous situations. When running a farm business, owners may experience mental strain as a result of having to deal with the regulatory framework as well as the fluctuating priorities that come with running the firm. On the other hand, threats to one's financial well-being are among the most significant. More than 80% of India's farmers cultivate their crops on parcels of land that are no larger than two hectares. Due to the scattered character of this little landholding, farming would be an endeavor that would be financially risky even under the most favorable conditions.

As a point of fact, the conditions are in no manner, shape, or form even remotely near to being ideal in any way. The concept of sustainable agriculture gained more support as the research community and farming communities became more aware of the need to strike a balance between productivity and the environmental and social repercussions of farming practices. This awareness led to an increase in the number of people who advocated for the concept of sustainable agriculture. According to Garibaldi et al., (2017), agricultural sustainability is a concept that takes into account the economic, environmental, and social aspects of farming, while simultaneously fostering the resilience and persistence of productive farming landscapes. This definition was provided in the article *Agricultural Sustainability*. Integrated farming is a method of agriculture that makes use of a number of different ideas and farming models in order to achieve the goal of growing more food (for the purpose of making a profit) while also providing benefits to both the environment and society. Integrated farming uses several different ideas and farming models to achieve the goal of growing more food (for the purpose of making a profit) (Garibaldi et al., 2017; Plumecocq et al., 2018; Pretty and Bharucha, 2014). Garibaldi et al., 2017, are the ones that came up with a new idea of farming method known as integrated farming. The following is a list of some of the ideas that have been proposed as a means of achieving agricultural sustainability; many of these concepts, such as integrated pest management, agroforestry, and organic agriculture, are now quite familiar, while others, such as precision farming and sustainable intensification, are becoming more common. Integrated pest management, Agroforestry, and organic agriculture are examples of concepts that are now quite familiar. All of these ideas have had an impact on the policy landscape at a variety of different scales. This is because policymakers are constantly looking for the most efficient method to encourage the adoption of an integrated agricultural system to put sustainable agriculture into practice. The fundamental impetus behind the adoption of a wide variety of agricultural practices was the desire to decrease dependency on resources that were brought in from the outside.

However, as a consequence of industrialization, farming shifted toward the production of commodities rather than food. The focus shifted depending on the agro-climatic conditions of the region as well as the proximity to industries such as sugar factories, soya processing plants, rice mills, oil mills, ginning mills, and daal mills, amongst other types of mills. Simultaneously, agricultural practices such as poultry farming, dairy farming, piggery, beekeeping, fish cultivation, vegetable farming, fruit farming, floriculture, and mushroom cultivation are examples of farming practices (Meena et al., 2018). However, as time went on, the viability of farming based on a single product became questionable as a result of the ever-changing patterns of the market and the reliance on inputs from outside sources. This was because farming requires inputs from outside sources. The so-called green revolution further expanded the crop-based farming system through the uncontrolled use of fertilizers, which ultimately led to a situation in which the land became overworked and the crop yield decreased.

Integrated Farming System in Agriculture

IFS has four primary goals, which are the following: The maximization of the yield of all component enterprises in order to provide a consistent and stable income; the rejuvenation or amelioration of the system's productivity in order to achieve agro-ecological equilibrium; the avoidance of the build-up of insect-pest, disease, and weed populations through natural cropping system management in order to keep them at a low level of intensity; and the reduction of the use of chemicals (fertilizers). The implementation of an IFS ultimately leads to a more reliable and long-term source of income for the farm as a whole. This is achieved by launching several businesses with the goal of deriving the maximum benefit from the utilization of easily accessible natural resources in order to satisfy the needs of the family. It is hoped that this will result in a minimal level of farm income required for the farm family to maintain a sustained interest in farming, hence preventing people from abandoning the agricultural industry altogether. IFS models have to be developed based on the agro-climatic situations, holding size, availability of resources like land, water, labor, marketing facilities, risk factors, family size, the ability of the farm family members to participate in the farming activity, their knowledge and skill level, and other such factors. It is obvious that the profit margin changes depending on the ecosystem (rain-fed irrigated), managerial abilities, and the socioeconomic situations that affect a person's ability to make a living (Rathore et al., 2019). The IFS utilizes a network of interconnected businesses so that the waste from one component of the system can be used as input for another component of the system. This contributes to the reduction of costs as well as the improvement of either the production or the income of farmers. IFS makes it possible for the by-products of one kind of agricultural production to serve as a resource for another kind of agricultural output. Because it recycles trash into usable resources, it not only helps us get rid of the garbage but also ensures an overall increase in the production that can be achieved through agricultural systems. This is a situation in which everyone benefits (Anonymous, 2010).

The IFS approach is recognized as a resource management strategy to achieve economic and sustained productivity that satisfies the varied needs of farm households while simultaneously conserving the resource base and preserving a high level of environmental quality. This goal can be accomplished by meeting the diverse needs of farm households through the use of IFSs (Lal and Miller, 1990). IFS refers to an entire complex of activities, including the development, management, and allocation of resources, as well as the decisions and activities, that take place within an operational farm unit, or combinations of units, and ultimately result in agricultural production, processing, and marketing of the products. It is a total farm management method that blends the ecological concern of a diverse and healthy environment with the economic demands of agriculture to ensure a continued supply of wholesome food at a price that is reasonable to the consumer. On the other hand, the integrated agricultural system is a dynamic idea that needs to have the flexibility to be applicable on any farm, in any country, and it must always be sensitive to change and technological advancements. Above all else, the IFS is a viable path forward for agriculture that will benefit society as a whole, not just those who work in the industry. IFS can be defined as a positive interaction between two or more components of different nature, such as field and horticultural crops, livestock, aquaculture or fishery, poultry, duckery, apiculture, sericulture, mushroom cultivation, biogas production, silviculture, etc., within the biophysical and socio-economic environment of the farmers in order to increase productivity and profitability in a manner that is both environmentally friendly and sustainable. (Behera et al., 2004; Rautaray et al., 2005). A well-planned combination of two or more of these farm businesses, each of which makes use of innovative agronomic management technologies, has the potential to increase farm income while also contributing to the recycling of farm residues. The selection of firms must be based on the cardinal principles of limiting the competition and maximizing the complementarities between the enterprises. The rising number of people all over the world led to an increase in the demand for agricultural products. The technologies that will be used in the North Eastern region include IFSs for the Hilly region of the North Eastern region. To such an extent that, ultimately, it calls for a creative new approach to agriculture that makes use of sophisticated special configurations (Rizzo et al., 2013). The realization of a climate-smart adaptation of a new farming system that generated an increasing amount of biomass for energy and maintained equilibrium in the ecosystem became a fundamental criterion as a result of these points of view.

On the other hand, because there is a finite amount of land available, increasing agriculture farming any further is neither practical nor desirable. This is because of the pressing need to lessen the negative effects on biodiversity and natural resources (Foley et al., 2011; Lemaire et al., 2013). There was a need for farmers to practice integrated land-based farming of field crops, horticultural crops, poultry, fisheries, and apiary within the biophysical and socioeconomic condition of the farm in order to gain more products and profit from it as the landholding is continuously decreasing (Behera et al., 2004). In a developing country like India, where the primary source of income for rural residents is agriculture, there is a significant risk that arises as a result of the shrinking size of agricultural land. This creates a significant worry regarding the farmers' ability to maintain their businesses and maintain their profits. Small and marginal farmers made up 85% of India's overall farming community but only owned 44% of the country's total operating land. Despite this disparity, they were the driving force behind India's rural economy. It was discovered that 50% of the operating land holdings shrunk from 2.28 hectares (ha) in 1970–1971 to 1.15 hectares (ha) in 2010–2011. By 2020–2021, it is anticipated that it will be 0.95 ha (−12.04% change from the previous average landholding in 2015–2016).

The amount of land that was available per person fell from 0.5 ha in 1950–1951 to only 0.15 ha at the turn of the century, and it is anticipated that this number will fall to approximately 0.95 ha by 2020. Therefore, it was vital to build procedures and rural advancements that enable satisfactory employment and the generation of money, in particular for small and marginal farmers who made up a maximum of over 80% of the agricultural community's population (Behera et al., 2013). Under the auspices of the National Agricultural Technology Programme (NATP), the nation was comprehensively portrayed into five agroecosystems. These agroecosystems are irrigated, rainfed, arid, coastal, hilly, and mountainous. This was done to appropriately address the farmer's particular requirements through an interdisciplinary and between-institutional approach.

Types of Integrated Farming System

Rao (1996) suggested different types of IFSs as follows:

Integrated Fish-Paddy Farming

Paddy farmers all over the world practice this sort of farming, which is considered to be a traditional method of farming. The breeding of fish in the paddy field is part of an on-farm production and value-adding combination. There is a tremendous amount of potential in integrated fish and paddy farming. It has the potential to improve the farmers' standard of living by raising the average income of their families and by giving them employment opportunities (Haobijam et al., 2016). This farming utilized two distinct types of paddies, where one type was used throughout each season. The wet season was used to cultivate a tall variety, while the dry season was used to cultivate a high-yielding variety. This method of farming was well suited for the paddy fields that were located in low-lying locations that received an average of a lot of rain each year. The farmers typically dug their fields using one of three distinct trench designs (Sharma et al., n.d.). They are as follows:

Perimeter Type: Paddy was planted in the middle of this layout, and the soil gently slopes out from the center in all directions toward the surrounding trench.

Central Pond Type: In this configuration, there was a pond amid the field that was surrounded by paddy that grew all around the region and drained into the center pond.

Lateral Trench Type: In this configuration, trenches were excavated on one end of a paddy field that had a gentle slope, or two or more trenches were excavated on the other side of the paddy growing area. The total amount of water utilized by this sort of integrated farming should be approximately equal to one-third of the total land area. During the rainy season, the water level ought to be kept at between 2.5 and 3 cm. The stocking rates per acre of fish range from 4000 to 6000 fingerlings with various fish species, including a proportion of Rohu (30%), Catla (25%), and Mrigal (45%), as well as tiny amounts of Carp and Prawn. Paddy output was from 3500 to 4500 kg/ha, while fish production for 3 to 4 months was in the range of 200 to 300 kg/ha, and for 5 to 6 months was in the range of 400 to 500 kg/ha.

Integrated Fish-Pig Farming

In India, this procedure was utilized because it was both an economically viable and scientifically established practice. The fish pond was built using this approach with the necessary depth of 1 to 1.5 m and a side slope of 1.5 m. The pig shed was built primarily on the dyke of the pond or near it with a system for draining all of the biomass to the pond, where it can then turn into outstanding pond manure, thereby increasing the production or biological activities performed in the pond water, and therefore increasing fish production (Agrifarming, n.d.). The pig shed needs to be set up so that each pig has a lying space between 1.1 and 1.5 m^2, and the dung area needs to be 1.5 m^2 per pig. This will ensure that the pigs are able to grow up healthy and strong (Vikaspedia, n.d.). Excreta from the pig, which are easily digestible, were added to the fish's diet as a supplementary food source. The digestibility rate for the fish was approximately 70%. It was proposed that a stocking rate of 8000–8500 fingerlings per hectare be used in conjunction with a ratio of surface feeders at 40%, column feeders at 20%, bottom feeders at 30%, and macro vegetation feeders at 10% in order to achieve a greater fish output. Fixed cultures of Indian main carps can also be taken with a species ratio of 40% surface feeders, 30% column feeders, and 30% bottom feeders. In an integrated agricultural system that also utilized fish, the manure from 30 to 40 pigs was adequate to fertilize a one-hectare water area, as was indicated. The production of manure from a pig was approximately 500–600 kg of biomass per pig per year. The live weight of the fish produced by this method of farming was approximately 6000–7000 kg/ha/year, while the amount of hog meat produced was approximately 70–80 kg per pig/6–7 months. Through the integration of pig and fish farming, farmers stand to gain an annual profit of 73,2923 yen per hectare of land (Haobijam and Souvik, 2018).

Integrated Fish-Duck Farming: This method of farming was an excellent blend of many approaches. The droppings of ducks were utilized in the production of feed and fertilizer, and they also assisted in the loosening of the pond bottom, which led to the release of nutrients and boosted the fish productivity in the pond. The fish pond was built with the desired depth of 1 to 1.5 m and a side slope of 1.5 m. The majority of the duck houses were constructed on the dyke of the pond and near to the fish pond. It is recommended that a distance of 0.3–0.5 m be kept between each bird, and that there be one male duck for every 5–6 female ducks. The selection of fish species was very crucial, and the maximum number of species should have been composed of 60% plankton feeders and 40% omnivorous fish. The stocking rates of fish in this kind ranged between 8000 and 8500 fingerlings per hectare, with a species ratio of 20% Catla, 20% Silver carp, 20% Rohu, 15% Mrigal, 10% Grass carp, and 15% Common Carp. Grass carp made up the remaining 15% of the fish population (Abhipedia, n.d.). In integrated farming with fish, it was sufficient to supply a one-hectare water area with 200–300 ducklings/ha, as was advised, in order to fertilize the water area. The amount of manure produced by ducks was approximately 45–55 kg of excrement per duck per year. This method of farming resulted in a production of approximately 3500–4000 kg of live weight per hectare per year from the fish, 18,000–18,500 eggs, and 500–600 kg of meat from the duck each year.

Integrated Fish-Chicken Farming

In this system of agriculture, the cost of feed and fertilizer for fish cultivation was decreased thanks to the use of meat or eggs from chickens, which resulted in the greatest possible return on the fewest possible resources invested. To a large extent, the chicken coop was constructed on top of or near the fish ponds. Additionally, the chicken manure was used as a source of recycled fertilizer and added to the ponds. In order to make the most efficient use of space and cut down on the amount of manual labor required to transport manure into the ponds, an efficient method was developed for erecting a chicken shelter above the pond. In integrated farming with fish, it was sufficient to stock approximately 500 to 600 poultry chicks per hectare in order to fertilize a one-hectare water area. This was the recommendation (Vikaspedia, n.d.).

Due to the fact that the amount of manure produced by chickens was approximately 50 kg per day (0.10–0.08 kg of excreta per bird per year every day, 36.5–29.2 kg per bird per year). The amount of fish produced by this method of farming was approximately 4500 to 5000 kg/ha/year, while the amount of eggs produced was approximately 70,000, and the amount of chicken meat produced was approximately 1000 kg/year.

Integrated Fish-Cattle Farming

It is a method of farming that is widely used in many different parts of the world. Countries such as Hong Kong, Taiwan, and the Philippines have commercialized this type of farming to a high degree and earned the highest return on their investment (Celkau, n.d.). In comparison to the other types of livestock excrement, the manorial value of cattle excrement, as well as dung, urine, and the act of cleaning the cattle shed, was exceptional. Cow manure used to sink at a rate that was 6 cm per minute slower than the sinking rate of any other type of livestock manure. This allowed the fish to have enough time to consume it. The cattle shed was constructed at the dyke of the ponds or near the pond so that the waste and washing from both the shed and the animals could be directly drained into the fish pond, where it could be used as fertilizers and feed. To fertilize a one-hectare water area that was used for integrated farming with fish, the stocking of cows should be between five and six cows per hectare (ha), as was advised. The production of manure from the cow was about 4000–5000 kg dung and 3500–4000 liters of urine on an annual basis per cow. As a result, the production of fish and milk from this type of farming was approximately 3000–4000 kg per hectare per year and 9000 kg per year, respectively.

Integrated Fish-Rabbit Farming

The fish and the rabbits that are raised in this style of farming both come out ahead. The rabbit's shelter was constructed along the dike of the ponds so that the shed's wastes and washing could be directly drained into the pond. This was done in order to keep the ponds clean. The byproducts of the rabbits made an excellent organic fish food and manure when they were composted (Krishisewa, n.d.). In integrated farming with fish, it was sufficient to use 400–500 rabbits, as was indicated, to fertilize an area of one hectare that contained water. The amount of manure produced by rabbits was approximately 60–70 kg of excreta per year, which equated to approximately 140–150 g each day. The yield from this farming was approximately 3500–4000 kg of fish per hectare per year in addition to approximately 9600 kg of rabbit meat per year.

Integrated Fish-Goat Farming

In this style of farming, raising goats and fish was done in conjunction with one another. The goat shelter was constructed on the dyke of the pond or in close proximity to the pond so that any waste produced by the goats, as well as any washing done in the shed, could be drained directly into the pond. The orientation of the shelter must be such that it receives the appropriate amount of sunshine and ventilation. The needed floor spacing was 10 square feet per goat, and the required distance of 3 feet above the ground should be maintained (Sharma et al., n.d.). It took some time for the droppings of the goats to decay, so there was sufficient time to compost them before utilizing them in the pond. According to the recommendations, a flock of 50 to 60 goats was sufficient to fertilize a one-hectare water area that was being farmed integrated with fish. The optimal depth for the fish pond is between 2 and 2.5 m.

The recommended species for stocking rates per hectare of fish in this type are grass carp (10%), silver carp (20%), common carp (15%), catla (20%), rohu (20%), and mrigal. The range of stocking rates per hectare of fish in this type was between 7000 and 8000 fingerling/ha (15%). This type of integrated farming resulted in an output of approximately 750–800 kg of goat meat (Chevon/year) and approximately 3500–4000 kg of fish per hectare per year.

Integrated Fish-Horticulture/Agriculture Farming

Here, horticultural and agricultural products were grown on the dyke of the pond in order to make the most efficient use of the available space while still achieving a high rate of return on investment. In general, a space of 200 cm^2 was required for a family of four to five individuals in order to satisfy the daily requirement of two portions of fruits and vegetables (Vikaspedia, n.d.). Growing fruits, vegetables, pulses, and oilseeds on the dykes, which account for around 25–33% of the total farmland, allows a family to not only satisfy their own needs for fruits and vegetables but also make additional revenue by selling their harvest. In this method of farming, some of the fish species that are recommended include Channamarulius, C. strialus, Common carp, and tilapia. Some of the fruit crops that are recommended include bananas, papayas, coconuts, lemons, guavas, etc., and some of the vegetables that are recommended include bitter guards, lady's finger, bottle gourd, brinjal, lobia, cucumbers, French beans, pumpkins, and tomatoes, The cultivation of crops within the dikes of ponds contributed to the preservation of the pond embankments' strength and stability against soil erosion, as well as to the prevention of the pond water from becoming unnecessarily cloudy. There were a total of 12 distinct forms of fish farming found in Manipur, according to the data from the integrated fish farming system (Haobijam & Souvik, 2020). In addition, agricultural innovations such as paira and relay cropping contributed to an increase in the farmer's annual income as well as their net productivity. It was reflected that paira cropping of lentil in the rice field, also known as the rice-lentil cropping system, provided a tremendous enhancement in lentil production, as was the case in multiple studies conducted by Bandyopadhyay et al., 2018, and other researchers (Nandi et al., 2021; Saha et al., 2021). These researchers conducted their research in 2018. Outstanding and lucrative results were achieved through the integration of fish farming with other agricultural pursuits, such as horticulture or livestock (Haobijam & Souvik, 2020).

Advantages of IFS

Pooling and sharing of resources and inputs, efficient use of family labor, conservation, preservation, and utilization of farm biomass including nonconventional feed and fodder resources, effective use of manure and animal waste, regulation of soil fertility and health, improved space utilization, diversified products, income, and employment are some of the benefits of IFS, which is a strategy to ensure sustainable benefiting current and future generations via the responsible utilization of natural resources. The following is a list of the significant benefits that can be gained from applying IFS.

Productivity, A Rise in Food Supply, and Assurance Of Adequate Nutrition

The importance of the IFS approach lies in the fact that it has the potential to offer a chance to increase the system's productivity or economic yield per unit area per unit time. This is accomplished by the intensification of crop and allied businesses in order to satisfy the ever-increasing demand for food, feed, and fuel to support the ever-growing human and animal populations. When grown on the same amount of land, horticultural and vegetable crops can produce two to three times the number of calories than cereal crops do. Incorporating activities such as beekeeping, fishing, mushroom growing, and sericulture into a two- or three-tiered IFS results in the production of a significant amount of additional high-energy food while having little impact on the production of food grains.

Profitability

IFS increases land productivity and profitability by lowering production costs through the recycling of wastes and by-products from one business as inputs to other businesses. This allows IFS to improve land productivity and profitability.

Sustainability

IFS places a strong emphasis on the efficient and effective utilization of wastes and by-products generated by linked components, as well as the achievement of agroecological equilibrium through the reduction of the buildup of diseases and pests.

Balanced Food

There are a variety of interconnected parts that, when combined, result in the production of a wide range of products. These products, in turn, help the farm family maintain a healthy diet.

Environmental Safety

The IFS approach recycles waste materials as well as nutrients in an effective manner, makes farming less dependent on external inputs, and maximizes resource utilization by linking appropriate enterprises and components. As a result, environmental pollution caused by the intensive use of external inputs is reduced.

Resource Recycling

IFS is able to execute efficient recycling of waste materials and by-products such as crop residues and livestock wastes. Because of this, there is less dependence on external inputs (such as fertilizers, agrochemicals, feeds, and energy), which results in a production system that is more stable. It is possible to restore soil fertility by using organic manure, recycling biomass, incorporating legumes into cropping systems, and other similar practices. It is common practice for other types of businesses to make use of crop residue and plant biomass as a form of input. Some examples of this practice include the cultivation of mushrooms, and the use of feed blocks, mulch, and substrates in vermicomposting.

Year-Round Income

Through the sale of various farm produce such as milk, eggs, mushrooms, vegetables, fruits, and food grains, amongst other things, an IFS ensures that farmers maintain a steady stream of income all through the year.

Risk Minimization

IFS makes available a production system that is both stable and sustainable using a diverse array of crops and businesses. This contributes to the reduction of potential risks and increased resistance to the effects of climate change.

Use of marginal and wastelands

It is possible to combine crop production with forestry, fishing, poultry, dairying, mushroom growing, and beekeeping; additionally, all of these activities are capable of being carried out on marginal and wastelands as well.

Increased Employment

Farm families are seeing a rise in their level of life as a result of a 200–400% increase in the number of farm families with productive jobs and increased income. The integration of agricultural businesses is contingent on a variety of factors, including the availability of resources, the geological and climatic characteristics of the chosen region, the current level of resource utilization, land, labor, capital, and skills, the economics of the proposed IFS, returns from the existing farming system, and the managerial ability of the farmer.

Conclusion

In light of the current circumstances, adopting a strategy for an IFS is of utmost significance and will save time thanks to its effective management of the natural resources that are currently available. In addition to this, it enables low-income persons living in rural regions to sustainably improve their standard of living by generating sufficient money, finding jobs, and benefiting from increased employment opportunities. The synergistic effect of the various businesses that are a part of the IFS needs to be exploited further in order to increase the effectiveness of the recycling of agricultural by-products and the utilization of resources. Because it utilized the same land resource to produce both carbohydrates and animal protein at the same time, the rice-based IFS offers tremendous opportunities for food security and poverty mitigation in rural regions. This was especially true in places where rice was the primary crop. On the other hand, vegetables and fruits grown on the dykes and bunds meet the demand for vitamins and minerals that help in implementing a balanced diet, decreasing hunger and malnutrition of the farm family. Bunds and dykes can be used to grow vegetables and fruits.

References

Abhipedia. (n.d.). Fish-duck integrated farming system in Assam. Retrieved August 22, 2022, from https://abhipedia.abhimanu.com/Article/State/OTM1MD AEEQQVV/Fish---Duck-Integrated-Farming-System-in-Assam-Assam-State.

Agrifarming. (n.d.). Pig farm with fish farm, pig manure as fish feed. Retrieved November 22, 2020, from https://www.agrifarming.in/pig-farm-with-fish-farm-pigmanure-as-fish-feed.

Anonymous. A manual on integrated farming system. Caribbean Agricultural Research and Development Institute, (Ministry of Economic Development, Belize), 2010: 1-58.

Bandyopadhyay, P.K., Halder, S., Monda, K., Singh, K.C., Nandi, R., Ghosh, P.K. Response of lentil (Lensculinaries) to post-rice residual soil moisture under contrasting tillage practices. *Agricultural Research*, 7, no. 4, (2018): 463-79.

Behera, U. K., Babu, A., Kaechele, H. and France, J. Energy self-sufficient sustainable integrated farming systems for livelihood security under a changing climate scenario in an Indian context: A case-study approach. *CAB Reviews*, 10, no. 19 (2015): 11.

Behera, U.K., Dass, A., Rautaray, S.K., Choudhary, A.K., Rana, D.S. 2013. Integrated farming system research in India: An overview. In U. K. Behera, A. Dass, S.K. Rautaray, A. K. Choudhary, & D.S. Rana, (Eds.), Integrated farming systems for livelihood security of small and marginal farmers (pp. 40-78). Division of Agronomy, IARI, New Delhi.

Behera, U. K., France, J. Integrated farming systems for the livelihood security of small and marginal farmers in India and other developing countries. *Advances in Agronomy*, 138, (2016): 235-82.

Behera, U.K., Kaechele, H., France, J. Integrated animal and cropping systems in single and multi-objective frameworks for enhancing the livelihood security of farmers and agricultural sustainability in Northern India. *Anim. Prod. Sci.* 55, (2015): 1338-46.

Behera, U.K., Jha, K.P., Mahapatra, I.C. Integrated management of available resources of the small and marginal farmers for generation of income and employment in eastern India. *Crop Res.* 27, (2004): 83-9.

Bell, L.W., Moore, A.D., Kirkegaard, J.A. Evolution in crop–livestock integration systems that improve farm productivity and environmental performance in Australia. *Eur. J. Agron.* 57, no. 9, (2014): 10-20.

Celkau (n.d.). Duck-fish culture. Kerala Agricultural University, Thrissur, Retrieved August 21, 2022, from http://www.celkau.in/Fisheries/CultureFisheries/Intergrated %20Fish%20 Farming/duck_fish_sys.aspx

Foley, J.A., Ramankutty, N., Brauman, K.A., Cassidy, E.S., Gerber, J.S., Johnston, M., et al. Solutions for a cultivated 10 planets. *Nature*, 478, (2011): 337-42.

Garibaldi, L.A., Gemmill-Herren, B., D'Annolfo, R., Graeub, B.E., Cunningham, S.A., Breeze, T.D. Farming approaches for greater biodiversity, livelihoods, and food security. *Trends in Ecology and Evolution,* 32, no.1, (1017): 68-80.

Gupta, V., Rai, P.K., Risam, K.S. Integrated crop-livestock farming systems: A strategy for resource conservation and environmental sustainability. *Indian Research Journal of Extension Education,* 2, (2012): 49-54.

Haobijam, J.W, Balkho, R.M, Rede, G.D., Mandal, T.K. Socio-economic characteristics of the farmers practising paddy-cum-fish cultivation in the low lying paddy field of Manipur. *Journal of Crop and Weed*, 12, no. 3, (2016): 94-9.

Haobijam, J.W., Ghosh, S. Integrated pig-fish farming: A case study in Imphal west district of Manipur. *The Pharma Innovation Journal*, 7, no. 1, (2018): 495-9.

Haobijam, J.W., Ghosh, S. Strengths, weaknesses, opportunities and threats (SWOT) of integrated fish farming system (IFFS) as perceived by the farmers in Manipur. *Journal of Community Mobilization and Sustainable Development*, 15, no. 2, (2020): 448-54.

Jha, S.; Bacon, C.M., Philpott, S.M., Rice, R.A., Méndez, V.E., Läderach, P. A review of ecosystem services, farmer livelihoods, and value chains in shade coffee agroecosystems. In Integrating Agriculture, Conservation and Ecotourism: Examples from the Field; Springer: Dordrecht, The Netherlands, 2011; pp. 141-208.

Kashyap, P., Kansal, A., Prusty, A.K., Singh, J.P. Evaluation of horticulture based IFS models for providing nutritional security to small and marginal farmers of western plain zone of Uttar Pradesh, India. *Int. J. Econ. Plants*. 2, (2015): 15-17.

Krishisewa. (n.d.). Fish cum rabbit farming: An innovative idea for entrepreneurs in rural areas. Krishisewa. Retrieved August 15, 2022, from https://www.krishisewa.com/articles/livestock/1111-fish-cum-rabbit-farming-is-aninnovative-idea-for-entrepreneurs-in-rural-areas.html

Lal, R., Miller, F.P., Sustainable farming for tropics. In: Singh, R.P. (Ed.), Sustainable agriculture: Issues and Prospective., Indian Society of Agronomy, IARI, New Delhi (1990)1, 69-89.

Lemaire, G., Franzluebbers, A., Carvalho, P.C.F., Dedieu, B. 2013. Integrated crop-livestock systems: Strategies to achieve synergy between agricultural production and environmental quality. Agriculture, Ecosystems and Environment (in press). http://doi: 10.1016/j.agee.2013.08.009.

Mahapatra, I.C., Farming systems research challenges and opportunities. Eastern Indian Farming System Research and Extension. *Newsletter,* 6(4) (1992)3–10.

Meena, L.R., Kumar, D., Meena, L.K., Singh, K., Singh, S.P., Panwar, A.S., Integrated farming system model for doubling farmers' income in western Uttar Pradesh. *Indian Farming*, 68, no. 3, (2018): 17-19.

Nandi, R., Mondal, K., Singh, K.C., Saha, M., Bandyopadhyay, P.K., Ghosh, P.K. (2021). Yield-water relationships of lentil grown under different rice establishments in the lower plain of India. *Agricultural Water Management*, 246, (2016): 1066-75.

Pretty, J., Bharucha, Z.P.,. Sustainable intensification in agricultural systems. *Annals of Botany,* 114, no. 8, (2014): 1571-96.

Plumecocq, G., Debril, T., Duru, M., Magrini, M.B., Sarthou, J., Therond, O. The plurality of values in sustainable agriculture models: Diverse lock-in and coevolution patterns. *Ecology and Society,* 23, no. 1, (2018): 21.

Rao, D.R. 1996. Integrated fish farming. In environmentally sound technologies for women in agriculture (pp. 213). Rao., D.R. (Ed.) International Institute for Rural Reconstruction. http://www.nzdl.org/cgi-bin/library?e=d-00000-00---off-0cdl--00-0----0-10-0---0---0direct-10---4-------0-11--11-en-50---20-about---00-0-1-00-0--4----0-0-11-10-0utfZz-8-1 0 & c l = C L 1 . 1 9 3 & d = H A S H c 8 8 330c33115c69e2f6724.10.1 &gt=1%252020-11-2020

Rathore, V.S., Tanwar, S.P.S., Kumar, P., Yadav, O.P. Integrated farming system: Key to sustainability in arid and semi-arid regions. *Indian Journal of Agricultural Sciences,* 89, no. 2, (2019): 181-92.

Rautaray, S. K., Das, P. C. and Sinhababu, D. P. Increasing farm income through a rice-fish based integrated farming system in rainfed lowlands of Assam. *Indian Journal of Agricultural Sciences*, 75(2), (2005) 79–82.

Rizzo, D., Marraccini, E., Lardon, S., Rapey, H., Debolini, M., Benoît, M., & Thenail, C. (2013). Farming systems designing landscapes: Land management units at the interface between agronomy and geography, Geografisk Tidsskrift-Danish Journal of Geography, 113(2), 71-86.

Saha, M., Sarkar, A., Bandyopadhyay, P.K., Nandi, R., Singh, K C. Tillage and potassium management for improving yield, physiological and biochemical responses of rainfed lentil under moisture stressed rice-fallow. *Journal of Soil Science and Plant Nutrition*, 21, (2021): 637-54.

Sharma, A., Sarma, M., Sharma, S., Sharma., S., Das, M.G., Tripathi, A.K. (n.d.). Integrated paddy-cum-fish culture [Brochure]. Krishi Vigyan Kendra Ri-Bhoi, ICAR Research Complex for NEH Region, Meghalaya. http://www.nicraicar.in/nicrarevised/images/publications/TDC/Integrated%20 paddy - cum - fish % 20 culture _ K V K _ Ri -Bhoi_ Meghalaya%20(English).pdf

Swaminathan, M.S., 1987. Inaugural address. In: International symposium on sustainable agriculture. International Rice Research Institute, Las Banos, Philippines.

Vikaspedia. (n.d.). Technologies for Northeastern region: Integrated farming systems for North Eastern Hilly region. Vikaspedia. Retrieved August 22, 202, from https://vikaspedia.in/agriculture/livestock/technologiesfor- north-eastern region/integrated-farming-systems-for north-eastern-hilly-region.

3

Climate Change and Sustainable Agriculture

Y.S. Bagal

School of Agriculture, Lovely Professional University, Phagwara, Punjab India

Introduction

The problem of dealing with climate change, a critical global issue, has come to define our era. Human activities are rapidly altering Earth's climate, with far-reaching repercussions for many industries, including agriculture. The necessity for environmentally responsible food production is pressing as the global population rises. One possible way to lessen the effects of global warming on food production is to adopt the sustainable agriculture practice of striking a balance between environmental protection, social justice, and economic viability. The impacts of climate change on farming may be seen all across the world right now. Global food security is threatened by climate change, which is causing higher temperatures, changing precipitation patterns, and increasing extreme weather events. Greenhouse gas emissions, deforestation, and unsustainable farming practices all have a hand in accelerating global warming. Therefore, it is crucial to tackle the double problem of lowering agriculture's environmental imprint while also adapting it to climate change. Sustainable agriculture works toward a compromise by encouraging robust and flexible agricultural practices. It aims to ensure the sustainability of food production in the future while reducing environmental effects by incorporating climate change adaptation and mitigation techniques into agricultural practices. Agricultural systems can be made more resistant to climate change by the use of sustainable farming practices such as agroforestry, crop diversification, soil conservation, and water management. Sustainable farming practices help communities adapt to climate change, boost local economies, and secure food supplies for future generations.

Climate on Agricultural Sustainability

When it comes to growing crops and raising livestock, the weather is the most crucial natural resource. The dominating cause of the total inter-annual variability in productivity in many places, climatic fluctuations continue to damage ecosystem services and are thus a major concern for the agricultural sector (Howden et al., 2007). People living in rural areas are more susceptible to the effects of bad weather, climatic fluctuation, and change because they rely so heavily on agriculture for their survival and economic well-being (Molnar 2010). A large number of the world's small-scale farmers have always felt the effects of climatic instability since they tend to labor on marginal land in the tropics. The suitability of farmland for various crops and pastures is being negatively impacted by the increasing weather and climatic changes. By 2050, the world's population is projected to exceed 9.1 billion, necessitating a 70% increase in food production (FAO 2009a). Poverty and food insecurity are anticipated to persist at high rates in Sub-Saharan Africa, Asia, and Latin America due to their rapid population expansion and depletion of natural resources. Efforts to preserve sustainability are in jeopardy due to rising population growth and shrinking arable land, especially in less developed nations (Sivakumar et al., 2000). The already difficult task of feeding the world while preventing the agricultural system from collapsing under the weight of climate change has become much more daunting. This highlights the need for agricultural policies, programs, and practices that are "climate savvy" for future food security (FAO, 2010c). Regardless of location or food system, climate science plays a crucial part in this process and in the development and implementation of effective sustainable agriculture techniques. The rural population, especially those who rely on agriculture for subsistence, suffers greatly from food insecurity and is highly susceptible to the direct effects of extreme weather. Despite the increasing population and food demand, agriculture faces a variety of hazards and problems that might seriously hinder its capacity to attain global food security.

The United Nations Development Programme, UNDP (2006) reports that water consumption has increased more than sevenfold between 1900 and 2000 and that this trend is expected to continue. Increasing precipitation is predicted by climate models for high latitudes and certain sections of the tropics while decreasing precipitation is predicted for some sub-tropical and lower mid-latitude regions (Bates et al., 2008). The last 50 years have seen a roughly three-fold rise in world average cereal output because of high-yielding cultivars, intensive inputs, and irrigation. However, this trend is now challenged by flat or even falling per capita grain production. There are more than 277 million hectares of land that is irrigated, and agriculture uses around 70% of the world's freshwater and as much as 95% in several poor nations (FAO, 2007). Irrigated agriculture accounts for a large portion of the world's water consumption, thus sustainable farming strategies that maximize water production are especially important. In addition to the current difficulties posed by the effects of climate, the danger of flooding is expected to rise as the 21st century progresses because of an expected increase in the severity of tropical cyclones (Knutson et al., 2010) and the frequency of heavy precipitation occurrences. Furthermore, it is expected that the percentage of arid land will rise and that summers will get drier overall, particularly in the sub-tropics, low, and mid-latitudes (Bates et al., 2008). Despite continuous efforts (Mills, 2005) in numerous sectors, including agriculture, the susceptibility of these vulnerable areas to the effects of hydro-meteorological catastrophes has grown considerably over the last several decades.

Agricultural biodiversity is in danger from climate change and rising climatic unpredictability, although it serves numerous essential purposes in ensuring the long-term security of food and livelihoods. According to the IPCC's Fourth Assessment Report (IPCC 2007a), the rate of species extinction is expected to speed up as the planet warms. About 20–30% of plant and animal species evaluated so far are expected to be in greater danger of extinction if the global average temperature rises to surpass 1.5–2.5°C above 1980–1999 levels, according to the paper. Key services of the agroecosystem, including nutrient and water cycle maintenance, pest and disease regulation, pollination, and land-erosion control, may be lost if climate-related consequences are not included in the management of agricultural biodiversity. Additional crises in local agricultural output are anticipated to arise as a result of climate change-induced shifts in the occurrence, distribution, and severity of pests and diseases. In particular, the important parameters influencing the amount of damage from pests are recognized to be CO_2-temperature and CO_2-precipitation interactions (Easterling et al., 2007).

Communities in desert areas that rely heavily on their food systems may be impacted hard by climate change. Large swaths of the most fragile ecosystems may be found in coastal regions, which are especially susceptible to hazards including flooding, storms, erosion, and saltwater intrusion. Communities that rely on fisheries are especially susceptible to climate change impacts such as rising sea levels, changes in ocean salinity, cyclones, and declining fish supplies as a result of rising water temperatures (Hall-Spencer et al., 2008). Climate variability related to the El Nio-Southern Oscillation (ENSO) and Pacific Decadal Oscillation (PDO) affects the majority of the world's largest marine fisheries (Lehodey et al., 2003). Coastal areas are especially vulnerable due to several factors, including the effects of anticipated warming, the relative significance of fisheries to many national economies, and a limited capability to adapt to the consequences of climate change (Allison et al., 2009). About 20% of the Earth's continental areas are covered by mountain ranges, and many of the world's cultural minorities call these places home. Ecosystems in mountainous areas are particularly vulnerable to storms and severe rains, which may cause avalanches, landslides, and floods on already unstable terrain. Isolation, marginalization, and a lack of sophisticated early warning systems for climate-related dangers all contribute to their extreme vulnerability. Unsustainable natural resource management is a major problem in mountain areas due to rising populations, high rates of agricultural and animal production, and excessive land degradation (FAO, 2009b). The very irregular and scant rainfall, together with overexploitation and degradation of natural resources, make dry and semi-arid regions especially susceptible to the effects of climate change. The danger of starvation rises in arid and tropical locations as temperatures rise (IPCC, 2007a). Frequent droughts in arid and semiarid regions make land more susceptible to desertification or speed up the process of desertification.

Food Security

All four of food security's pillars availability, accessibility, use, and stability will be impacted by climate change and rising climatic unpredictability (FAO, 2008b). The growing frequency and severity of severe weather events, as well as the accompanying unpredictability in rainy season patterns, are having serious consequences for rural and urban food security, distribution systems, livelihood assets, human health, and emergency food supplies (FAO, 2008a). Battisti & Naylor (2009) found that increased growing season temperatures, as well as other expected changes in the frequency and intensity of weather and climatic events, have serious ramifications for agricultural output, farm revenue, and food security. To mitigate the effects of climate, its unpredictability, and predicted change, the whole food chain must use sustainable agricultural practices and adaptive modifications to food system operations.

Sustainable agriculture may be understood in several ways. Basically, sustainable agriculture incorporates three primary goals—environmental health, economic profitability, and social and economic equality. The foundation of sustainability is the recognition that we cannot provide for the requirements of today's population if we also expect future generations to be self-sufficient. Sustainable agriculture was defined in several ways, and Sivakumar et al. (2000) emphasized the need to set long-term objectives that may be attained by methods conditioned by appropriate agroclimatic knowledge. To appreciate how agroclimatic data contributes to long-term sustainability, it is helpful to use a systems approach. We consider the whole system, from the scale of an individual farm to that of the surrounding environment, and from there to the communities both near and distant that are impacted by agricultural practices and the weather. By using a systems-based approach, we are equipped with knowledge and methodologies to probe the links between agriculture and weather. The FAO's official definition of sustainable agriculture (FAO, 1995) encompasses all facets of sustainability in the following ways: When it comes to agriculture, sustainable development means prioritizing the needs of future generations while also protecting the needs of the present. This is accomplished through careful management of the land's natural resources and the strategic implementation of technological and institutional advancements. Conservation of plant and animal genetic resources land, water; suitable technology; non-degradation of the environment; economic viability and social acceptability are all hallmarks of sustainable development in the agricultural, forestry, and fishing industries.

Sustainability initiatives aim to incorporate environmental, social, and economic considerations into policymaking at the local, national, and international levels (Dovers and Handmer, 1993). To be clear, everyone involved in the agricultural system—farmers, service providers, policymakers, academics, merchants, and consumers—has some duty to play in ensuring that agriculture is practiced sustainably. Everyone has a special function to fulfill, and together we can all do more to support the sustainable agricultural movement. One of the conditions for developing climate-responsive and resilient agricultural policies, programs, and practices is the systematic description of agro-climatic resources, which climate science is already helping with. However, a fresh perspective is required that recognizes climate as a resource that can help with land, water, plant, and animal genetic resource base planning and management, uses weather and climate information on seasonal and intra-seasonal time scales to mitigate climate risks, and employs medium to long-term climate change scenarios to improve resilient adaptation in the service of achieving sustainable agriculture.

Resource Management

Because it influences almost every biological activity, from the functioning of the ecosystem to the growth of individual plants, climate is an essential resource for the agricultural sector. Photosynthesis, remobilization, crop growth, crop development, assimilate partitioning, leaf area development, water and nutrient uptake, transpiration, evaporation, and stress from a deficit or excess of water are just some of the plant physiological processes that are influenced by climate and weather (deVries et al., 1989). Climate science has long been recognized for its role in providing the necessary information for understanding these interactions, the time has come to revitalize and expand this role by shifting the focus toward on-the-ground actions to achieve sustainable management of food system resources.

Farmers' access to climatic data based on their own requirements may aid in the efficient administration of vital agricultural resources (land, water, and genetic resources). Recognizing climate as a resource, having climate and agricultural services quantitatively measure it, then communicating the results to farmers and having them appropriately manage it are all essential parts of resource management. Farmers often manage their natural resources based on their knowledge of the local climate, but these methods may not be enough in the face of global climate change. As a result of the shifting conditions, it is crucial to have a deeper comprehension of the meteorological, ecological, and socioeconomic factors at play. Sustainable management of all natural resources, including climate, is essential for ensuring enough food and water for everyone and attaining sustainable agricultural development for present and future generations. To mitigate climate change's negative effects, sustainable natural resource management is a crucial first step (Tilman et al., 2002). A natural resources management strategy for agriculture should include the susceptibility of different agricultural systems to climate change, as well as specific regions of concern and climate hotspots. Glantz et al., (2009) suggest that before a climate hotspot develops, key areas of concern should be identified and prioritized in the management of natural resources via a climate window. With the recent enhancements in the availability of advanced climatic information, these areas of concern may be identified sooner, which helps with resource management. To mitigate the effects of climate change on natural resources, it is important to have a thorough knowledge of the climatic resources in a given area (WRI, 2009). To comprehend these differences, use data on this resource for a variety of purposes, such as agro-ecological zoning, land-use planning, and sustainable land management; forest management that takes into account climatic conditions and underground and surface water levels;

rainfall expectations and infiltration rates; river basin management, including pre-release of water from dams to store anticipated excess water; and so on. Sustainable management methods may help mitigate the negative consequences of future hazards by adjusting operational procedures in response to actual weather conditions. Soil and slope protection techniques, run-off reduction, water storage potential enhancement in the upper soil layers via simple tillage practices and water conservation techniques (Selvaraju et al., 1999), and increased groundwater recharge are all examples of sustainable soil and water management practices that would benefit researching the connections between weather, soil, and vegetation is essential for smart water management. Predicting the reliability of the water supply may be made easier with the use of water-harvesting systems, methods to raise the moisture content of soil, and irrigation reservoirs.

According to Pretty et al., (2006), these methods also increase water production, which is particularly helpful for rain-fed farms. Water run-off, water infiltration, and soil erosion are all lowered when no-till techniques are used (Blanco-Canqui and Lal 2008). Soil moisture can be preserved, water infiltration can be maximized, nutrient runoff can be reduced, and crop yields can be increased with the help of conservation tillage, residue management, mulching, and proper fertilizer use. To better manage water resources, Bates et al., (2008) suggested including information regarding present climatic variability in water management. Studies on the use of climate data and information for water resource management, water demand, and irrigation optimization under drought conditions, and crop area decision strategies conditioned on climate information all attest to the significance of climate science in water management and increasing water productivity (Selvaraju et al., 2007). Several of the pilot projects looked at ways to enhance the accuracy of seasonal forecasts and their practicality on farms. Scaling up these activities is possible, but only with careful planning and consistent backing from climate scientists.

The protection and administration of genetic resources are both affected by climate change. The understanding of abiotic and biotic interactions in the context of climatic fluctuation and change is helpful for the preservation of plant and animal genetic resources in general. The coordinated variations in temperature, precipitation, CO_2 fertilization, and pest and disease dynamics highlight the importance of climate research in supporting biodiversity conservation and breeding efforts. To facilitate agricultural adaptation to climate change, researchers working on crop improvement must have a thorough understanding of the future climate to develop short-duration genotypes that are tolerant to temperature and drought stress, resistant to pests and disease, and capable of understanding genotype-environment interactions. Investments in breeding tend to follow a 12 to 20-year cycle, so knowing how that can be affected by climate change is crucial for picking the right breeding strategies.

The effects of climate change on agriculture and the possibility of reduced inputs were studied by Romay et al., (2010), who concluded that maintaining yields in different environments is a primary breeding objective and highlighted the importance of climate knowledge in developing climate-resilient maize genotypes. Crop-breeding programs, which aim to choose plants with stable genotypes across circumstances, rely heavily on climate services (Hammer et al., 2010). Therefore, various climatic factors, particularly those associated with high temperatures, should be considered if yield under stress circumstances is an objective of breeding. Climate data should also be considered when choosing locations for in-situ conservation of genetic resources. (FAO, 2008c).

Risk Management

Recent improvements in climate prediction, analysis, and synthesis of climate information may expand livelihood opportunities in agriculture by allowing for better management of climate risk. Soil moisture conditions and the status of crops, in quality and quantity, may be forecasted in advance via the monitoring and analysis of meteorological and climatic data, unlocking the proactive potential of controlling climate risk. Forecasting crop yields has historically been used in climate risk management, but doing so needs a broad variety of data, including but not limited to meteorological, agrometeorological, soil, and remotely sensed information, as well as agricultural statistics (Gommes, 1998). Several value-added indices may be constructed from meteorological and agronomic data, useful for monitoring soil and agricultural conditions and predicting crop output for climate risk reduction, emergency preparedness, and proactive food security management (Tadesse et al., 2008). Decisions about climate risk may benefit from the use of weather and climate predictions and early warning systems (Challinor, 2009). Hansen et al., (2006) stress the need for decision-makers to be adaptable and use risk management measures to mitigate the destructive consequences of climatic extremes and make the best use of limited resources. Climatic and Agriculture initiative case studies have shown the efficacy of risk mitigation and yield enhancement via the use of locally tailored management solutions based on reliable climate data (Sivakumar and Hansen, 2007). To ensure climate-based guidance is integrated into the planning and management of crops, animals, forests, and fisheries, national extension services and national meteorological agencies must work together as part of agricultural policy. As food systems expand into more marginal, fragile, and climate risk-prone regions, the role of climate science in the climate risk management of food and agricultural resources is becoming more crucial (FAO, 2008a).

The consequences of intra-seasonal to inter-annual climate variability highlight the need for high-quality and trustworthy climate projections to inform many agricultural choices. Increased intra-seasonal accuracy in predicting summer monsoon rainfall variability is a need for agricultural growth in southern Asia (Webster et al., 1998). According to Donald et al., (2004), intra-seasonal predictions are crucial for determining when and how to implement changes to risk-mitigation measures for commercial and subsistence farmers. When it comes to explaining the intra-seasonal climatic variability that is useful for agricultural applications, many indices and approaches are used. Patterns of suppressed and/or enhanced precipitation beyond the synoptic scale have been attributed to the Madden-Julian Oscillation's passage (Bond and Vecchi, 2003). Indicators of active and break monsoons in southern Asia may be gleaned from the propagation features of the Monsoon Intra-Seasonal Oscillation (Hoyos and Webster, 2007). Using intra-seasonal forecasting skills, farmers may evaluate the probability and timing of rainfall events and temperature swings on time scales of up to 6 weeks, therefore improving tactical risk management.

On the other hand, seasonal climate estimates are crucial for making risk-averse strategic decisions in agriculture. It's not only farmers who may benefit from seasonal weather forecasts in making calculated decisions about crop and animal management; policymakers, nonprofits, and companies can also use this information to plan ahead. Using seasonal to inter-annual climate forecasts and sources of predictability, such as ENSO-related indicators, may have more potential for risk reduction and opportunity management (Goddard et al., 2001). Small-scale farmers and rural communities may benefit economically from these changes. Corporations, including those involved in the grain, sugar, and cotton gins (Hansen et al., 2006; Meinke et al., 2006). To achieve the potential advantages of such predictions in reality, further study is required on climate predictability sources, significant underlying climate processes, and efforts for integration into the operational forecasting scheme. According to a report published by the National Academy of Sciences in 2010, the value of forecasts to decision-makers and researchers could be improved if operational intra-seasonal and inter-annual forecast centers made adjustments to their methods for obtaining and disseminating forecast information and strengthened their collaborations with the external research community. Future improvements in observational capacity, statistical and dynamical models, and data assimilation systems will allow these forecast systems to better correctly portray the variables and processes that serve as inputs.

Agricultural index insurance is a promising new tool for mitigating climate and weather-related risks (Skees, 2008). Insurance premiums for agriculture and forestry in 2001 were estimated to be around US$6.5 billion annually, according to the FAO (2005). This was only about 0.4% of the estimated total farm gate value of agricultural production worldwide. Strengthening weather observation networks, monitoring severe climate events, standardizing indices, data sharing, early warning systems, and capacity development are all necessary for providing localized and needs-based climate information to promote index-based weather insurance.

In Malawi, a crop-specific water balance is used to create value-added crop-weather variables that may be integrated with other data in an insurance method based on an FAO-evaluated index (Chavula and Gommes, 2006). By using gridded data, this method may compensate for individual missing stations. Whatever the situation may be, a large quantity of monitoring infrastructure and data points is required to provide a plausible explanation of spatial fluctuation. Index-based weather insurance systems need the support of climate science and national hydro-meteorological services relevant to a specific agroecological zone so that they can produce accurate and timely crop production indices. When the microflora associated with plants undergo changes due to climate change, new disease and pest patterns may arise. To effectively manage and control newly emerged illnesses and pests, we need new indices of economic threshold levels and their connections to climate. Early detection of disease and pest outbreaks might be aided by weather and climate forecasts, leading to more effective control measures and lessened management costs. Keeping an eye on pest thresholds may be one way in which climate information systems contribute to the development of integrated pest management methods. There has long been consensus that the most efficient approach to controlling pests and diseases is via a mix of continual operational monitoring and early warning systems based on weather patterns. Models of these dangers and weather forecasts resulting from climate simulations may be used to set up early warning systems for pests and illnesses (Shaw, 2009).

Farmers need to use climate risk management methods to effectively respond to climatic unpredictability and change. Among these tactics are a wide range of approaches and procedures designed to help farmers recognize, assess, and prioritize climate change-related risks and vulnerabilities. Using a combination of historical climate data and weather and climate prediction information, the technique analyses potential outcomes and generates alternatives for agricultural decision-making. The localized climate-risk management system relies heavily on real-time weather, crop information, weather and climate projections, prices of inputs, and pricing of farm output to provide needs-based and location-specific farm advisories. Agro-meteorological techniques and methodologies should use not just up-to-the-minute data on weather and crops, but also historical climatic data, location-specific socio-economic factors, and local coping mechanisms when assessing exposures and risks and developing mitigation plans at the farm level. Methods for managing climate-related risks at the community level are achieved by:

- Examining and compiling data on regional circumstances, such as those pertaining to farmers' decision-making challenges, social and physical risks, and environmental conditions, the current environment and the data and information products that are already accessible on the subject;
- Using a wide variety of techniques and tools (e.g., crop models) to analyze susceptibility and climate risks and optimize management choice alternatives in past, near-real-time, and forecast products;
- Creating individualized information products for at-risk populations and deciding on suitable management choices based on the local context and its requirements;
- Easing the way for people to take action on the ground by spreading knowledge about the climate, available solutions, and how the feedback loop is being kept tabs on and assessed.

The four main features of localized climate risk management systems are adaptive management techniques; farmer agro-advisor training; climatic, agricultural, and socio-economic data collection and analysis/prediction; and hazard mapping. The method seeks to provide a comprehensive range of advice regarding crops to be planted, time and quantity of inputs that might be used, and management practices to be followed to prevent and/or reduce risks to be prepared to implement management decisions on short notice based on the anticipated weather and climate. Local advisories will detail the information on input availability with agricultural support services, input suppliers, local cooperatives, and farmer service-based community organizations to facilitate actionable decision-making based on both climate information and the availability of the necessary inputs. The development of ICT has allowed for the systematic collection of real-time data from local communities, its analysis by technical institutes providing climatic and agricultural services, and its subsequent delivery to farmers in the form of management recommendations.

In order to better prepare for and respond to disasters and food shortages, combining climate data with analyses of livelihoods and market information might provide better early action plans. For analyses of food security and humanitarian responses, climate monitoring and forecasting using cutting-edge technology is a crucial input because of the reliance of large, scattered populations on rain-fed agriculture and pastoralism. Drought index maps and agricultural water balance models rely heavily on satellite rainfall estimates for climate monitoring, as emphasized by Verdin et al., (2005). Satellite-based rainfall estimations and ensemble forecasting models are very helpful, as are remote sensing approaches (De Groeve and Riva, 2009).

Effective flood forecasting methods (Hopson & Webster, 2010) have substantial prospects for safeguarding agricultural livelihoods and preventing food-related calamities. Opportunities to improve national capacities for climate risk management and emergency responses can be found, for instance, in the FAO's Rainfall Estimate Routine, an independent method to estimate the amount of rainfall, especially for certain regions in Africa. The climate risk management strategy is widely recognized as an effective method for mitigating the growing severity of climate-related disasters. From 14% of all-natural disasters in the 1980s to 20% in the 1990s and 27% since 2000, according to data from the FAO's global information and early warning system on food and agriculture, sudden-onset catastrophes, notably floods, have grown (FAO, 2008d). These tendencies have significant ramifications since sudden-onset situations of vast and catastrophic intensity leave far less time for preparation and response than less severe slow-onset catastrophes to safeguard agriculture-based livelihood systems via the strategic preparation of risk reduction measures and the mobilization of resources for responding to catastrophes. Climate information for the evaluation of climate risks on regional and temporal scales is crucial for planning suitable actions since they rely on the intensity and frequency of hazards. Humanitarian assistance is necessary to help families deal with calamities and make the most of their limited resources when neither a safety net nor enough financial support mechanisms are in place (Haile 2005). Accurate and efficient humanitarian assistance will allow families to take advantage of possibilities to adopt livelihood methods that are both sustainable and productive. Investments in poverty reduction efforts would be more effective if accompanied by response mechanisms that would protect livelihoods during times of crisis due to extreme weather events. Weather variables' effects on rural communities' food security and vulnerability can now be predicted and monitored more accurately thanks to our better understanding of climate variability, including ENSO effects. There is potential for consumers to make more informed decisions about where and how to store food and seeds by using data offered by climate and meteorological services. Rainfall, air moisture, and temperature all have a role in how well grain stores, while the drying effects of sun and wind are welcome. If the farmer makes decisions based on a reliable estimate for the following five to seven days, he or she may be able to reduce the risk of deterioration during drying. The usefulness of weather reports has increased with the development of short-season-variety crops, which may be harvested and stored long before the traditional dry season starts.

Adaptation Measures

There is an immediate need for resilient adaptation measures, particularly in the poorest and most at-risk nations. Resilient adaptation is a multi-pronged approach that combines the more traditional tactics of adaptation, mitigation, and prevention (Glantz et al., 2009). To facilitate the creation of robust adaptation plans, the climate research community supplies the most crucial input: Climatic scenarios. Using climate scenarios, we may assess how likely various future climates are to be, how the Earth's ecosystem will react to our actions, what kind of havoc they'll wreak, and what kind of solutions we'll need to adapt or mitigate (Moss et al., 2010). Advice for policymakers and farmers on how to effectively adapt to climate change; provision of needed assistance such as information, technology, alternative sources of income and employment, post-harvest facilities, credit facilities, and insurance mechanisms; With the use of data on climate scenarios and probable alterations in seasonal rainfall pattern, new strategies for grazing and livestock management may be mapped out. Herders and others who rely on livestock for a living need access to climatic information to relocate to new grazing areas and count stock, two forms of early adaptation to climate change (Thornton et al.,2009). Indigenous peoples and forest-dependent communities may benefit from climate and meteorological services since they allow for the calculation of bushfire hazards based on scenarios of temperature, humidity, and wind.

The capacity to evaluate risks and identify impacts, prioritize local strategies for dealing with climate change, put those strategies into action, get political support for them, and eventually scale them up all benefit enormously from climate knowledge. An evaluation of the consequences of climatic fluctuation and change on agricultural systems, taking into account the biophysical and socioeconomic environment and the possible adaptation strategies. However, agricultural production systems are dynamic in their own right, therefore it's important to develop more effective adaptation strategies to lessen the impact of global warming in specific regions. In this regard, Moss et al., (2010) emphasized the importance of climate science in delivering the next generation of scenarios for climate change and evaluation, which might provide the foundation for the creation of locally tailored adaption plans by identifying susceptible zones and hot spots of risk.

If we can better understand the vulnerabilities that are specific to various regions, we may gain insight into the aspects of system vulnerability and adaptation effectiveness, such as resilience and coping mechanisms, and enhance our ability to adapt to climatic fluctuation and change. The Fourth Assessment Report of the Intergovernmental Panel on Climate Change (IPCC, 2007b) states that increases in soil carbon sinks, decreases in greenhouse gas emissions, and the provision of biomass feedstock for energy use are all potential outcomes of adopting sustainable agriculture methods. Increasing soil carbon sequestration via better agricultural and grazing land management forestry and agroforestry projects are all influenced by air and soil temperature and soil moisture. (Palm et al., 2005), enhancing the effectiveness of nutrient management, and restoring organic soils (Lal and Follett, 2009), and conserving agricultural and degraded lands. Further, improvements in climate modeling and forecasting technology have improved the potential for sustainable livestock and pasture management in grassland ecosystems (Luseno et al., 2003).

Data Tools and Methods For Achieving Sustainable Agriculture

Planning sustainable agricultural systems and production techniques requires compiling a database of climatological, meteorological, phenological, soil, and agronomic information. Doraiswamy et al., (2000) cited the availability of an accurate meteorological and agro-meteorological database as a crucial first step in the investigation and administration of agricultural productivity. Impact evaluations, planning for adaptation and mitigation techniques all need data on climate change scenarios in addition to standard agro-meteorological data for planning (Moss et al., 2010). Complex impact assessment techniques and procedures are available for a variety of industries, including agriculture. However, many poor nations lack access to the full range of climatic variables needed to properly use these tools (Vega, 2008).

Organizations and agencies in the fields of agriculture and climate services will need to collaborate closely in order to develop a platform for sharing climate and agricultural data for the purpose of impact assessment. It is useful to use crop growth models to accurately predict climate risk associated with agricultural output and to create crop and soil management approaches that decrease risk to acceptable levels. Quantitative methods for the evaluation of climate hazards in crop production were described by Muchow and Bellamy (1991). Methods such as crop modeling for calculating climate risk, and optimizing agronomic management techniques and decision-making to reduce risk, were among those used. Crop growth simulation models paired with GIS data are often used to study crop suitability, geographical, and temporal yield variability, and water productivity at river basin and regional scales (Ines et al., 2002).

It is possible to predict future crop yields with the help of several different methodologies that use a combination of observed meteorological variables, historical agricultural yields, and climate change scenarios to account for the CO_2 fertilization effect, technological trends, and potential adaptation options (UNFCCC, 2008). Using process-based crop growth models may help with decision-making, but these models need data on the environment, soil, and the eco-physiology of each crop in question. The strategy's efficacy depends on our level of knowledge of the physical and physiological systems that govern crop growth in response to climate. In addition, there are simple algorithms that can be used to convert climate data into actionable insights, such as the quantity of water that is available to crops in the root zone throughout the growth season and whether or not they are experiencing water stress or surplus (FAO 2010b). Advantage in the agricultural sector need easy access to the tools used to create synthetic (incremental), analog, and model-based climate change scenarios in order to evaluate their potential effects. Many national hydro-meteorological services are currently working to enhance their meteorological monitoring infrastructure, build a climate database, advocate for policies at the national and regional levels, expand their educational and training opportunities, and create innovative, user-friendly tools and methods to better assist policymakers (WMO, 2010a).

Research and Capacity Building

It is crucial to be able to locate, gather, and distribute data, as well as apply information and applicable methodologies for data analysis and construct knowledge pertinent to climate and weather information and food security. The most pressing issues are the dissemination of cutting-edge information in the field of climate science, the improvement of agro-meteorological observation techniques, the creation of tailored forecasting products, the organization of data, and the development of models for assessing the effects of climate change on agricultural practices. The link between policymakers and the agricultural community may be improved via the enhancement of agriculture extension services by addressing climate risks and planning for adaptation and mitigation. Strengthening the technical skills of agro-meteorological and agricultural extension services in interpreting climate information, producing impact outlooks, and devising adaptation alternatives is essential for coping with climate change and increasing climatic variability. Combining agricultural study with climate analysis needs to provide several novel adaptation and mitigation choices tailored to individual regions. To manage climate hazards in the past and the present requires a different kind of study than that required to adapt to quickly changing climatic circumstances.

There has to be fresh funding for Research and Development and the infrastructure that supports it. Improving automated data collecting, distribution, and analysis; creating needs-based early warning and forecasting systems; and creating communication protocols and dissemination channels are all top research priorities in the field of climate science. To reap the benefits of new technology and practices in agriculture, substantial governmental and private expenditures in R&D are necessary (FAO, 2009a). Based on their study, Meinke et al., (2006) concluded that in order to make good use of weather and climate information, institutions need to work together to facilitate experiential learning and action research. Groups of individuals may learn from their own experiences and share those lessons with others via a process called Participatory Action Research (PAR) (Wadsworth, 1998). In order to effectively apply climate information in agricultural decision making, the PAR approach can be adopted. This is done by involving climate services, agriculture support services, and users in a joint examination of current climate risks and actions to enhance climate risk and opportunity management based on the local context. The World Meteorological Organization (2010b) developed a worldwide framework for climate services to bolster the development, availability, delivery, and use of science-based climate prediction and services, and it integrates all prerequisites for doing so.

Enabling Policies

Advance knowledge of weather, climate, and value-added secondary indicators has been shown to enhance farmers' decision-making, resulting in higher farm earnings and more effective risk management (Sivakumar and Hansen, 2007). Such weather and climate information services must be maintained, enhanced, and appropriately institutionalized to guarantee their ongoing viability and continual government funding. Several obstacles must be overcome to maintain the flow of climate knowledge for policymaking at all levels of government. Challenges include a lack of resources, shifting priorities within institutions, and the inability to implement new policies (Subbiah and Selvaraju, 2007). To guarantee institutionalization and bridge the gap between research, policy, and users in the production of climate projections, Maria et al., (2002) emphasized the necessity to bridge this divide. There are several developing countries where, within the confines of the current institutional frameworks, a system for the efficient flow of information from information providers to agriculture support services and farmers is feasible. Establishing novel channels of communication and feedback is crucial for improving the efficiency of weather and climate applications within current institutional frameworks. The environment is conducive to the maintenance of communication and feedback systems of this kind.

Institutional frameworks at the national and sub-national levels that facilitate the collaboration of specialists from a variety of disciplines in order to produce, translate, transmit, and use climate information in agricultural decision making. There is a way to further guarantee sustainability by advising government agencies and local institutions to prioritize the need for climate services in their development plans. This will allow them to obtain the climate information required for the management of natural resources and climate-related risks and the planning of resilient adaptation strategies. To guarantee that choices in agriculture and food security are based on the most up-to-date and accurate meteorological data available, a prioritization along these lines may open the door to getting greater financial and technical help. To have an impact, enabling local institutions and services such as input suppliers, agricultural cooperatives, regulated markets, etc., must actively promote needs-based advice and new technologies. Sustainable agriculture and food security rely on the continuous use of weather and climate services, which may be advanced by establishing enabling institutions that are independent, self-reliant, and democratic. More effective collaboration between governmental bodies, as well as active backing from the climate research community, would guarantee the required steps toward institutionalization would increase productivity as at-risk populations better use weather and climate data in making informed decisions.

Conclusion

Sustainable agriculture is all about minimizing damage to the natural environment while maximizing the utilization of most of the resources available. The first step in developing methods that contribute to sustainable agriculture is to have an appreciation for the interplay between plant and animal species, as well as climate, land, soil, and water. By using a resource-based approach, climate research may help characterize agro-climatic resources and pinpoint farming strategies that are good for the planet, the economy, and society. Climate science may help by advising on how to best manage water resources and increase water production, as well as how to best adapt businesses and livelihoods to changing climates. Also, particular contributions from climate science are needed for integrated watershed management and land use and land-use planning, all of which contribute to the sustainable use of natural resources for agriculture and are reliant on climatic conditions.

Long-term resource preservation and harvesting are at the heart of the notion of sustainable agriculture. This is a radical departure from the emphasis placed on resource extraction and output maximization during and after the green revolution of the 1960s. A deeper knowledge of the constraints and production potentials of crops, animals, forests, and fisheries is necessary for developing sustainable food production methods. Challenges posed by climate change, rising energy demands, and shifting socioeconomic conditions should be included in any such analysis and characterization.

Coastal, mountainous, arid, and semi-arid regions are especially vulnerable to climate change because of the increased frequency and severity of weather and climate-related events like drought, dry and wet spells, water scarcity, cyclonic storms, floods, landslides, erosion, seasonal flooding, and salinization. These weather events are not unprecedented, but they are becoming more problematic for agricultural systems as population growth and industrialization drive food production into formerly uncultivated areas. Therefore, it is important to adopt a strategy for integrated climate risk management that encourages the use of weather and climate data to improve resource utilization efficiency, reduce exposure to climate-related hazards, and speed up the creation of long-term, climate-resilient responses to climate change. The potential economic value of climate research has been proven, and there is widespread agreement that it is crucial within a risk management framework. Nonetheless, a renewed emphasis on cooperation, collaboration, and policy support is necessary to institutionalize efforts related to early warning systems for extreme events; the application of weather and climate information products; the promotion of weather-crop insurance; the monitoring of crop or livestock pests and diseases; and the forecasting of crop yields and information on food security. Collectively, these efforts might help attain food security and promote sustainable agriculture by making better use of existing resources. The frequency, intensity, and geographical and temporal changes of climatic hazards are all changing as a result of climate change, and it's possible that business as usual won't be enough to address these new threats. Areas of concern in terms of shifting biophysical and socioeconomic factors must be systematically identified, and localized adaptation and mitigation strategies must be planned. Agriculturalists have access to and are actively contributing to the evolution of a wide range of specialized resources for making use of climate and weather data in order to lessen their own exposure to risk, better manage climate-related hazards, and better prepare for the effects of climate change. The field of climate science must be specifically targeted if any of these initiatives are to succeed.

References

Allison EH, Perry AL, Badjeck MC, et al. Vulnerability of national economies to the impacts of climate change on fisheries. Fish Fisheries 10, no. 2, (2009):173-96.

Bates BC, Kundzewicz ZW, Wu S, Palutikof JP (eds). 2008. Climate change and water. Technical Paper of the Intergovernmental Panel on Climate Change, IPCC Secretariat, Geneva.

Battisti DS, Naylor RL Historical warnings of future food security with unprecedented seasonal heat. *Science,* 323, (2009): 240-4.

Blanco-Canqui H, Lal R. No-tillage and soil-profile carbon sequestration: An on-farm assessment. *Soil Sci Soc Am J* , 72 (2008): 693-701.

Bond N, Vecchi G. The influence of the Madden-Julian Oscillation on precipitation in Washington and Oregon. *Weather Forecast,* 18, (2003): 600–13.

Challinor A. Towards the development of adaptation options using climate and crop yield forecasting at seasonal to multi-decadal timescales. *Environ Sci Policy 12,* (2009): 453-65.

Chavula A, Gommes R. 2006. Development of a weather yield index (WYX) for maize crop insurance in Malawi. Food and Agriculture Organization of the United Nations, Rome.

De Groeve T, Riva P. 2009. Global real-time detection of major floods using passive microwave remote sensing. In: Proc 33rd Int Sym on remote sens environ, May 2009, Stresa, Italy. International Center for Remote Sensing of Environment, Tucson, AZ, p 587-90.

deVries PFWT, Jansen DM, ten Berge HFM, Bakema A (1989) Simulation of ecophysiological processes of growth in several annual crops. Simulation Monograph 29, IRRI, Los Banos, and Pudoc, Wageningen

Donald A, Meinke H, Power B, Wheeler M, Ribbe J .2004. Forecasting with the Madden-Julian Oscillation and the applications for risk management, new directions for a diverse planet. In: Fischer T, Turner N, Angus J, McIntyre L, Robertson M, Borrell A, Lloyd D (eds). Proc 4th Int Crop Sci Congr. The Regional Institute, Gosford, Australia.

Doraiswamy PC, Pasteris PA, Jones KC, Motha RP, Nejedlik P. Techniques for methods of collection, database management and distribution of agrometeorological data. *Agric For Meteorol* 103, (2000): 83-97.

Dovers SR, Handmer JW. Contradictions in sustainability. *Environ Conserv*, 20, (1993): 217-22.

Easterling WE, Aggarwal PK, Batima P, et al. 2007. Food, fiber and forest products. In: Parry ML ,Canziani OF, Palutikof JP, vander Linden PJ, Hanson CE (eds) Climate change 2007: Impacts, adaptation and vulnerability. Contribution of Working Group II to the 4th Assessment Report of the Intergovernmental Panel on Climate Chang, p 273-313, Cambridge: Cambridge University Press.

FAO (Food and Agriculture Organization). 1995. Dimensions of need: An atlas of food and agriculture. UNFAO, Rome.

FAO. 2005. Insurance of crops in developing countries. FAO Agricultural Services Bulletin 159, FAO of the United Nations, Rome.

FAO. 2007. The state of food and agriculture—paying farmers for environmental services. FAO of the United Nations, Rome

FAO. 2008a. Climate change adaptation and mitigation: Challenges and opportunities for food security. Information document prepared for the high level conference on World Food Security: The Challenges of Climate Change and Bioenergy, 3–5 June 2008, UNFAO, Rome.

FAO. 2008b. Climate change and food security: A framework document. Interdepartmental working group on climate change, FAO of the United Nations, Rome.

FAO. 2008c. Report of the FAO expert meeting on climate change and biodiversity for food and agriculture. FAO of the United Nations, Rome.

FAO. 2008d. The state of food insecurity in the world 2008. High food prices and food security threats and opportunities. FAO of the United Nations, Rome.

FAO. 2009a. How to feed the world in 2050. FAO of the United Nations, Rome.

FAO. 2009b. Sustainable management of mountains. International mountain day—information note on disaster risk management in mountains. FAO of the United Nations, Rome.

FAO. 2010a. FAOSTAT, FAO statistical databases. FAO of the United Nations, Rome.

FAO. 2010b. Impact of climate change on agricultural yields in Morocco. FAO of the United Nations, Rome.

FAO. 2010c. "Climate-Smart" agriculture: Policies, practices and financing for food security, adaptation and mitigation. UNFAO, Rome.

Glantz MH, Gommes R, Selvaraju R. 2009. Coping with a changing climate: Considerations for adaptation and mitigation in agriculture. Food and Agriculture Organization of the United Nations, Rome.

Goddard L, Mason SJ, Zebiak SE, Repelewski CF, Basher R, Cane MA. Current approaches to seasonal to interannual climate predictions. *Int J Climatol,* 21, (2001): 1111-52.

Gommes R. 1998. Agrometeorological crop yield forecasting methods. Food and Agriculture Organization of the United Nations, Rome.

Haile M. Weather patterns, food security and humanitarian response in sub-Saharan Africa. Philos Trans R Soc Lond, 360, (2005): 2169-82.

Hall-Spencer JM, Rodolfo-Metalpa R, Martin S. Volcanic carbon dioxide vents show ecosystem effects of ocean acidification. *Nature,* 454 (2008): 96-99.

Hammer GL, Oosterom EV, McLean G, Chapman SC, Broad I, Harland P, Muchow RC. Adapting APSIM to model the physiology and genetics of complex adaptive traits in field crops. *J Exp Bot,* 61 (2010): 2185-202.

Hansen JW, Sivakumar MVK, Bates BC. Advances in applying climate prediction to agriculture. *Clim Res (Spec Issue),* 33, (2006): 1-2.

Hopson T, Webster PJ. A 1–10 day ensemble forecasting scheme for the major river basins of Bangladesh: forecasting severe floods of 2003–2007. *J Hydrometeorol,* 11 (2010): 618-41.

Howden SM, Soussana JF, Tubiello FN, Chhetri N, Dunlop M, Meinke M. Adapting agriculture to climate change. *Proc Natl Acad Sci* 104, no. 50, (2007):19691-96.

Hoyos CD, Webster PJ. The role of intraseasonal variability in the nature of Asian monsoon precipitation. *J Clim,* 20 (2007): 4402-24.

Ines AVM, Gupta AD, Loof R. Application of GIS and crop growth models in estimating water productivity. *Agric Water Manag,* 54 (2002): 205-25.

IPCC (Intergovernmental Panel on Climate Change) (2007a) Impacts, adaptation and vulnerability. In: Parry ML, Canziani OF, Palutikof JP, van der Linden PJ, Hanson CE (eds) Climate Change. 2007. contribution of Working Group II to the 4th assessment report of the Intergovernmental Panel on Climate Change, Cambridge: Cambridge University Press.

IPCC (Intergovernmental Panel on Climate Change). 2007b. Mitigation. In: Metz B, Davidson OR, Bosch PR, Dave R, Meyer LA (eds). Climate Change 2007: contribution of Working Group III to the 4th assessment report of the Intergovernmental Panel on Climate Change, Cambridge: Cambridge University Press.

Knutson TR, McBride JL, Chan J, et al. Tropical cyclones and climate change. *Nat Geosci,* 3 (2010): 157-63.

Lal R, Follett RF. 2009. Soil carbon sequestration and greenhouse effect, 2nd edn. Madison, WI : Soil Science Society of America.

Lehodey P, Chai F, Hampton J. Modelling climate related variability of tuna populations from a coupled ocean biogeochemical–populations dynamics model. *Fish Oceanogr,* 12, (2003): 483-94.

Lemos MC, Finan TJ, Fox RW, Nelson DR, Tucker J. The use of seasonal climate forecasting in policy making: lessons from northeast Brazil. *Clim Change*, 55 (2002): 479–507.

Luseno WK, Mcpeak JG, Barrett CB, Little PD, Gebru G (2003) Assessing the value of climate forecast information for pastoralists: evidence from southern Ethiopia and northern Kenya. World Dev 31:1477–1494.

Meinke H, Nelson R, Kokic P, Stone R, Selvaraju R, Baethgen W. Actionable climate knowledge: from analysis to synthesis. *Clim Res,* 33, (2006):101-110.

Mills E. Insurance in a climate of change. *Science*, (2005):1040-44.

Milly PCD, Dunne KA, Vecchia AV. Global pattern of trends in streamflow and water availability in a changing climate. *Nature,* 438, (2005): 347-50.

Molnar JJ. Climate change and societal response: livelihoods, communities, and the environment. *Rural Sociol*, 75, (2010): 1-16.

Moss RH, Edmonds JA, Hibbard KA, et al. The next generation of scenarios for climate change research and assessment. *Nature*, 463, (2010): 747-56.

Muchow RC, Bellamy JA. 1991. Climatic risk in crop production: models and management for the semi-arid tropics and subtropics. Wallingford: CAB International.

National Academy of Sciences. 2010. Assessment of intraseasonal to interannual climate prediction and predictability. Committee on assessment of intraseasonal to interannual prediction and predictability, Board on Atmospheric Sciences and Climate, Division of Earth and Life Sciences, National Research Council, Washington, DC: The National Academies Press.

Palm CA, van Noordwijk M, Woomer PL, et al. 2005. Carbon losses and sequestration following land use change in the humid tropics. In: Palm CA, Vosti SA, Sanchez PA, Ericksen PJ (eds). Slash and burn: the search for alternatives, p 41-63, New York: Columbia University Press,.

Pretty JN, Noble AD, Bossio D, Dixon J, Hine RE, Penning de Vries FWT, Morison JTL. Resource-conserving agriculture increases yields in developing countries. Environ Sci Technol 40, (2016): 1114-19.

Romay MC, Malvara RA, Campob L, Alvarezc A, MorenoGonzálezb J, Ordása A, Revillaa P. Climatic and genotypic effects for grain yield in maize under stress conditions. *Crop Sci,* 50, (2010): 51-58.

Selvaraju R, Subbian P, Balasubramanian A, Lal R. Land configuration and soil nutrient management options for sustainable crop production on Alfisols and Vertisols of southern peninsular India. *Soil Tillage Res*, 52 (1999): 203-16.

Selvaraju R, Meinke H, Hansen J (2007) Climate forecast for better water management in agriculture: a case study for southern India. In: Sivakumar MVK, Hansen J (eds). Climate prediction and agriculture: advances and challenges, p 143-155 Heidelberg: Springer-Verlag, .

Shaw MW (2009) Preparing for changes in plant disease due to climate change. *Plant Prot Sci,* 45 (2009): 3-10.

Sivakumar MVK, Hansen J. Climate prediction and agriculture: advances and challenges. Heidelberg: Springer-Verlag.

Sivakumar MVK, Gommes R, Baer W. Agrometeorology and sustainable agriculture. *Agric For Meteorol*, 103, (2000): 11-26.

Skees JR. Challenges for use of index-based weather insurance in lower income countries. *Agricult Finance Rev*, (2008): 197-17.

Subbiah AR, Selvaraju R. 2007. Institutionalizing climate forecast applications for agriculture. In: Sivakumar MVK, Hansen J (eds). Climate prediction and agriculture: advances and challenges. p 57–61, Heidelberg : Springer-Verlag.

Tadesse T, Haile M, Senay G, Wardlow BD, Knutson CL (2008). The need for integration of drought monitoring tools for proactive food security management in subSaharan Africa. *Nat Resour Forum*, 32, (2008): 265-79.

Thornton PK, van de Steeg J, Notenbaert A, Herrero M The impacts of climate change on livestock and livestock systems in developing countries: a review of what we know and what we need to know. *Agric Syst*, 101, (2009): 113-27.

Tilman D, Cassman KG, Matson PA, Naylor R, Polasky S Agriculture sustainability and intensive production practices. *Nature*, 418, (2002): 671-77.

UNDP (United Nations Development Programme). 2006. Human Development Report 2006, beyond scarcity: power, poverty and the global water crisis. UNDP, New York.

UNFCCC (United Nations Framework Convention on Climate Change). 2008. Compendium on methods and tools to evaluate impacts of, and vulnerability and adaptation to, climate change. UNFCCC.

Vega RER. 2008. Workbook on climate change impact assessment in agriculture—basic knowledge, methodologies and tools. Caribbean community climate change centre, Belmopan, Belize.

Verdin J, Funk C, Senay G, Choularton R. Climate science and famine early warning. *Philos Trans R Soc Lond B Biol Sci.*, 360, (2005):2155-168.

Wadsworth Y. 1998. What is Participatory Action Research? Action Research International, Paper 2. Institute of Workplace Research, Learning and Development, and Southern Cross University Press, Australia.

Webster PJ, Palmer T, Yanai M, Tomas R, Magaña V, Shukla J, Yasunari A. Monsoons: processes, predictability and the prospects for prediction. *J Geophys Res*, 103, (1998): 14451-510.

WMO (World Meteorological Organization). 2010a. Building capacity around the world. WMO Bulletin 59, World Meteorological Organization, Geneva, p 46-52.

WMO (World Meteorological Organization). 2010b. Global framework for climate services (GFCS). World Meteorological Organization, Geneva.

WRI (World Resources Institute). 2009. The Bellagio framework of adaptation assessment and prioritization. Working Paper, Washington, DC: World Resources Institute.

4

Technology-led Development in Horticulture: Food for Thought

***Chongtham Tania*[1] *and Haobijam James Watt*[2]**

[1]*ICAR RC NEHR Manipur Centre, Umiam, Meghalaya, India*

[2]*Department of Agricultural Economics and Extension, School of Agriculture, Lovely Professional University, Phagwara, Punjab, India*

Introduction

The traditional methods of food production are coming under increasing strain due to the ever-increasing population of the world and the ever-decreasing availability of land and water resources. The good news is that there are currently concrete solutions to some of these difficulties that can be provided by technology. The fact that climate change is happening while the global population is growing poses the question of how we can ensure that the food we produce is reliable and of consistent quality. Is it possible to grow food in settings that can be regulated, allowing for production that is not dependent on the weather or temperature? It is just a computer enthusiast who is thrilled about how the industry is evolving, but maybe that makes it sound like a visionary. Over the course, gathering a significant amount of expertise working on initiatives that include the use of technology towards horticulture concerns is important. In order to keep newly emerging pests and diseases under control, horticultural crops require a higher degree of personal attention and plant monitoring than arable crops. The use of this intensive but integrated management strategy cleared the path for technology advancements that automated and optimized the farming processes that were previously handled manually, hence introducing increased precision and dependability.

The Importance of Adoption of Technologies

Recently, the necessity to improve production, earnings, and productivity was the primary driving force behind the choice of technology that was made available to farmers. The primary obstacles included a lack of access to cash, a lack of familiarity with how to use the technology, and exposure to market risks, the latter of which was mitigated by the policies of the government in many countries. In the past, "good policy practices" were therefore rather straightforward, relating primarily to increasing output, and the goal of agricultural and horticulture policies was to increase productivity in agriculture and horticulture. This was the case because in the past, "good policy practices" were related to increasing output. It is possible that agricultural research and extension services could center their efforts, for instance, on increasing the overall productivity of smaller farms.

Now, horticulture must fulfill a variety of aims, including being internationally competitive, producing horticulture products of high quality, and satisfying goals related to sustainability. Horticulture producers want quick access to new technologies to maintain their position in the industry's competitive landscape. Farmers today face a far broader range of challenges, but also a wider range of opportunities. They need to be profitable, comply with environmental standards and regulations, and deal with both direct and indirect pressure from customers and lobby groups. These pressures come from both individuals and organizations. They might also be inundated with information from a wide variety of government and industry sources, which makes it more difficult for them to choose the most appropriate technologies. Farmers also need to change their production and management practices in response to agricultural policies that consider environmental conditions, and the farmers can do so if well-directed through government support and incentives. There is a possibility that there will be ambiguity over the future policy on the environment, particularly concerning support, trade, and the pressures coming from the agro-food industry. Farmers that take advantage of new technologies are investing. However, it takes some time for the rewards to start coming in, and farmers may be hesitant to spend in a climate that is unclear and has more limits, especially in cases when some of the benefits are for society. Who should bear the financial burden: Farmers or society?

The evolution of technology has been the driving force behind increases in horticultural output as well as the advancement of the horticulture industry. Research affects the productivity of agricultural systems because it leads to the development of new technologies, which, if they are suitable to the conditions in which farmers operate, are quickly implemented. Throughout agricultural and horticultural innovation's history, the primary responsibility for recognizing and incorporating economic and environmental elements into the process of product development and distribution has traditionally been borne by researchers and extension workers. Researchers design the invention, extension workers promote its use, and farmers either adopt or reject the innovation based on the characteristics that are relevant to them. This is generally described as a top-down process.

Characteristics of Required Technologies

Technologies that More Precisely Target Pests and Diseases

It is highly unlikely that there will be an end to the demand for pharmaceuticals and chemical pesticides in the agricultural industry any time soon. It is anticipated that technological advances in the science of pest control will continue to produce chemical control agents that, over time, are at least as effective in controlling pests as the ones they replace, but which are also less toxic, less persistent, and less mobile through the soil. This will be the result of continued research and development. Farmers should be able to be more economical in their use of pest control agents, especially insecticides, thanks to the increased application of monitoring and knowledge-based systems, which are aided by reductions in the costs of electronic sensors and computers. This should enable farmers to apply these agents only when and where they are necessary, as opposed to applying them according to predetermined dosages and schedules.

Technologies that Administer Nutrients More Efficiently

Manure application and crop burning have historically been the two primary methods that farmers have relied on to give nutrients to root zones. The use of inorganic fertilizers made it possible to separate crop production from animal husbandry; they also helped restore fertility to soils that had been depleted; and they led to the development of livestock production that was based on grain and other feed ingredients. Over the years, research into the specialized requirements of various crop-soil combinations and animals has led to the development of fertilizers and feeds that are more scientifically designed. It is anticipated that the widespread adoption of technology that applies fertilizers only at the times and in the amounts required would result in further increases in crop yields while simultaneously minimizing the quantity of nutrients lost to leaching and runoff.

Technologies that Reduce Wastage Following Harvesting

A significant number of the techniques that are currently utilized for irrigating crops date back to the dawn of civilization. The difficulty, which existed in ancient Mesopotamia as well as in modern times, is that transporting water through open channels and furrows is inefficient since a significant amount of water is lost to evaporation before it reaches the root zone. The majority of the water used in farming is transported to fields by pipes; however, the technical efficiency of farming could still be improved through the increased application of technologies such as precision fertilization, which combine more accurate measurement of actual crop needs with means to deliver the water more accurately and in more precise dosages. Pipes transport the majority of the water used in farming.

Technologies that Reduce Wastage Following Harvesting

The demand for primary horticultural commodities is a derived demand, which is in part defined by the amount of product that is wasted between the producer and the final customer. There is already a high level of efficiency in the technologies used for harvesting, transporting, storing, processing, and distributing agricultural products; as a result, there is a significantly lower level of waste than in countries where the necessary capital and infrastructure are in a much shorter supply. The vast majority of plants and animals may have virtually every aspect of their bodies repurposed for some kind of commercial use, even if it's just for the production of feed, fertilizer, or electricity. It is possible to achieve a greater reduction in post-harvest losses, although the point of final consumption accounts for the greatest amount of waste (concerning the quantity purchased).

Technologies that Disseminate Knowledge

When determining what constituted "excellent farming practices," farmers traditionally relied on their own experiences as well as those of their neighbors. The environmental implications are becoming an increasingly important focus of both the advice and information provided by publicly sponsored bodies and the agricultural and food industry. The spread of information regarding environmentally friendly technologies has been given a boost by the proliferation of the internet. When compared to arable crops, horticultural plants need more frequent and in-depth checks for signs of pests and diseases. With the help of automation and optimization tools, previously labor-intensive farm tasks have become more precise and reliable thanks to a management strategy that emphasizes both intensity and integration.

Some Effective Technologies for Long-Term Sustainability for Horticultural Crops

Automation in the Garden

The scouting of crops is an essential stage in the management of diseases and pests. It provides the farmer with the ability to make timely judgments and take precautionary actions. It saves time and money, identifies patterns, and forecasts future dangers and potential issues. A magnifying glass, a plastic bag to collect plant samples, a clipboard, a marker, and a camera are the standard pieces of plant monitoring equipment (Schumann, 2010). These days, sticky traps and smartphone apps may provide a far more precise picture of the diseases and pests that are affecting a crop; in fact, some mobile apps can even make suggestions for possible treatments (Villordon, 2013). The recent development of digital scouting applications for mobile devices that offer biological solutions for horticulture and give a monitoring platform for the registration of illnesses is an example of recent technological progress. For instance, a scanner can count the insects that are on sticky cards with the use of accurate picture recognition. We can scan all of the sticky traps in a crop using this method, and the dashboard on a desktop or laptop computer will provide a direct schematic overview of the results. After that, the data can be combined with those from other measurements, and a comprehensive analysis can offer recommendations for biological remedies. Additionally, the application may be equipped with the capacity for machine or deep learning, which enables it to increase its predictive potential over time.

Use of Drones for Pest and Disease Identification and Management

Drones, also known as unmanned aerial vehicles (UAV), are currently being utilized in agriculture to locate "hotspots" in crops that are infested with pests and diseases and then accurately disperse helpful insects from the sky. In the coming years, drones equipped with sensors and imaging capabilities will play an increasingly important role in diagnosing crop damage and limiting the amount of damage that occurs (Psirofonia et al., 2017). It is now possible to link a novel release mechanism that was developed for the airborne release of beneficial insects (biological organisms) to a drone. This enables the biological solution to be distributed precisely where it is required the most in the crop. It is also possible to employ drones that have been fitted with scouting gear to discover and diagnose diseases and pests from remote locations (Papa et al., 2018). The highly modified drones, which may have a wing span of 1.5 meters, can lift 15 kg, fly over a strawberry field that is 20 hectares in size in an hour, and distribute helpful biological organisms. These drones can target specific locations, sometimes known as "hot spots," where diseases or pests are prevalent. They are more accurate than equipment that is operated by humans. What was once a tool used by the military for observing and hitting specific places held by the enemy is now being put to use as an instrument for sustainable agriculture.

Alternatives to Traditional Growth in Soil

The demand for natural resources like land and water is undergoing a fundamental transformation as a result of technological advancements. Plants are grown via hydroponics in a relatively tiny amount of solid matter, which is often comprised of sand and gravel. A solution that can be recycled supplies the plant with the precise quantity of nutrients it needs, cuts water usage by 90% and leads to fewer issues with pests (Barbosa et al., 2015). Leaf lettuce, tomatoes, peppers, strawberries, and herbs are some of the most common plants that are cultivated using hydroponic methods (Aatif et al., 2014). The plant is grown inside a humid atmosphere that is used for aeroponics, and then the roots are sprayed with a nutritional solution as the plant grows (Hayden, 2006). We can cultivate using these approaches even in the most challenging situations, such as deep underground or in highly polluted urban areas. Together, the practices of vertical farming (growing low crops in multiple layers, most of the time inside buildings) and urban farming (the growing of plants within and around cities) in conjunction with technologies such as hydroponics enable us to make more efficient use of space and decrease the distance that the food must travel to reach consumers. Another fresh and fascinating innovation coming out of Japan is the use of a biopolymer in the process of "film farming." The filmmakers claim that their method requires 90% less water than traditional farming practices and presents a workable alternative to resource-intensive agricultural practices.

Symbiotic organisms and plant defenses: Biocontrol

The application of microorganisms that stimulate plant growth at the beginning stages of the cultivation process is the primary focus of current research. These have the potential to improve the microbial makeup of the surrounding environment around the roots of the plants. Microbes that are beneficial to plants can stimulate the immune systems of the plants and cause them to develop more robustly (Glick, 2012). When they dwell on or near the root system, they engage in competition with pathogens for space as well as nutrition (these are microorganisms that can cause diseases). For example, Trichoderma has a direct effect on the pathogen both by its ability to predate on the pathogen and by the antibiotic compounds that it secretes in the root zone. Microbial insecticides that are designed to work on foliage pests infiltrate the harmful organism (often while it is in the larval stage) and feed on the tissue of the host plant. Beneficial microorganisms can be applied to crops in a timely and effective manner through the use of chemicals that are applied via a seed coating (video hosted on YouTube). Additionally, biopesticides have been generated in this manner as a means of providing support for the plant. Microbial biopesticides can be broken down into three distinct types such as bacteria, fungi, oomycetes, and even viruses. Biopesticides offer a wide variety of tools for the nonchemical management of pests and diseases, but they are subject to stringent regulations. Every product needs to be registered in the specific country where it will be sold.

Precision agriculture for weed control, soil health, and better yields

The requirement to increase food production in a manner that is not detrimental to the environment has resulted in the widespread implementation of "precision agriculture." This phrase refers to a variety of agricultural approaches that make use of technology to supply crops with exactly (neither more nor less than is necessary) the required number of nutrients and treatments for a high yield (Cassman, 1999). Automation helps farmers save money by preventing the unnecessary overapplication of agrochemicals, which in turn reduces negative impacts on the wider environment (such as nutrient leaching). Additionally, predictive technologies can assist farmers in customizing crop varieties for improved yield (Chen, et al., 2011). Here are two illustrations of this: The Blue River Technology, which is a precise weed management machine, as well as methods to improve soil health and increase pepper harvest on a farm that has problems with salinity.

Ensuring Constant Crop Quality

Take a step back and look at the bigger picture: What are the possible outcomes of using that technology? What kind of implications could this technology have on the way that we produce food in the future? It is currently possible for robots to sow seeds, and tend and harvest plants. In addition, sensors can better record data, which may then be utilized to set up closed-loop controls, which are what govern the behaviors of a robot. As a consequence of this, there is now a level of reliability, replicability, and quality control in the manufacture of food that did not previously exist. When it comes to planning, both customers and suppliers can benefit from a system that is dependable and highly optimized. This is of particular benefit to manufacturers of medicinal herbs or plant-based medications, such as those, who rely on maintaining a consistent quality of their products. To draw further insights from the data obtained on plant growth in these controlled conditions, machine learning and the handling of large amounts of data are essential components. What we discover could be applied to the process of optimizing light, soil conditions, plant spacing, or other variables to further accelerate the production cycle or assist us in producing food that is richer in nutrients. These kinds of data could be the cornerstone of a company's long-term success, helping to mold algorithmic processes and propelling further improvement.

Growing Food in Controlled Environments

It is not as simple as going to the hardware store and hanging some bulbs in order to provide your plants with optimal light conditions. This indicates that there is a demand for manufacturing LEDs that emit light at a particular wavelength that encourages photosynthesis. Sensors that utilize cutting-edge technology are yet another connected technology that will advance in tandem with the transformation of food production. These sensors will have to gather a substantial amount of information regarding, among other things, the nutritive value and moisture content of the soil, the growth of plants, and the amount of light. What if these automated gardens—or farms, since that is what we are truly discussing at this point—were located in urban areas? The advancement of technology may one day make it possible to cultivate plants not only on the ground but also on walls or on raised platforms in addition to growing them on the ground. This type of farming is known as urban or vertical farming. We might have modest vertical farms as a local complement within cities in addition to the conventional farms that are located outside of the cities. We might be able to cut down on both the expenses of transportation and logistics and the amount of carbon dioxide that is emitted if we take advantage of these new technologies and bring them into the cities. You are undoubtedly aware that there are many new ways and ideas on how to produce and transport food, resulting in the creation of a new digitalized segment of the food industry that is growing more willing to embrace high technology.

The United States of America, Europe, and Asia are among the regions where vertical farming business models have already been shown to be profitable for several enterprises. The reduced cost of logistics is because they are located in metropolitan regions. They can do this by utilizing technology and optimizing the growth conditions, which results in them consuming up to 95% less water than conventional farms and, in some cases, producing zero agricultural runoff. In addition, given that they are built indoors and in a vertical position, they call for a significantly reduced amount of land (Marz 2020).

Rootstock Exploitation

The protection of food, health, and livelihoods can all be strengthened through biodiversity's provision of essential components. It is the raw material for the biotechnology industry and a farming system that is resistant to the effects of climate change. To save rare, endangered, and threatened species through education, social mobilization, and regulation, as well as their utilization for future use, every nation faces the difficult task of formulating a plan that can be put into action to do so. Since the beginning of human history, wild fruit plants have provided a significant amount of the world's food supply. The usefulness of these fruits increased throughout history alongside the progression of man. For instance, there has been a gradual shift toward a declining trend in the productivity of temperate fruit crops. The most pressing requirement of the present day is to improve the production, productivity, and quality of nutrient-dense food while minimizing negative impacts on the environment. This should be done to raise the income of farmers and thereby ensure the food and nutritional security of households. It is necessary to transition standard low-density plantations into high-density plantations in order to shorten the period spent as a juvenile and increase productivity per unit area. To enhance fruit production, now is the ideal time to search for new rootstocks and pollinators. Breeding programs for disease resistance or as pollinizers have been identified to enhance the production of horticultural fruits for human consumption. A perfect rootstock would have uniformity in vigor, resistance to diseases and pests, and high quality and productivity, all without compromising either of those factors. Wild sources are a significant source of resistant and elite traits, which are utilized in crop improvement programs. The evolution of fruit crop plants was significantly influenced by biodiversity, which also serves as a genetic reservoir for breeding purposes. Therefore, the diversity of crops, species, and varieties is currently of the utmost importance to cope with the change in climate as well as the failure of monocrops as a result of unavoidable circumstances (Pramanick et al. 2017).

Quality Planting Material

It is absolutely necessary to enact and strictly enforce nursery laws and regulations in order to guarantee a high level of quality planting materials that are produced by nurseries. Appendix 1 contains a list that serves as an example of an indicative list of quantitative standards for seedlings to be used in plantations. This list was issued by the Forest Department of Madhya Pradesh. After the establishment of laws and regulations, the subsequent step may involve the process of registering all nurseries with the designated line departments of the state government.

The National Horticulture Board (NHB) is responsible for implementing a system that recognizes horticulture nurseries as legitimate businesses. It is necessary to obtain registration as a nursery if the nursery will be producing plants for distribution or sale. Appendix 2 contains a template that has been provided by the NHB for use in applying for registration. It is required that anyone who buys planting material from a registered nursery intending to resell it becomes registered as a Nursery Stock Dealer. The National Home Builders' Association (NHB) has also implemented a new system that is distinct from the licensing of Horticulture Nurseries and makes use of a five-grade rating based on the number of stars, with one star denoting satisfactory performance and five stars denoting outstanding performance. The information presented here is readily adaptable for use in agroforestry nurseries. Another system (Appendix 3) that is currently in use in a variety of countries is the Nursery Accreditation system. The quality of the planting material may be significantly improved as a result of these actions. After a nursery has been registered, recognized, or accredited, it ought to be given the authority to produce planting material of high quality by adhering to a protocol that is outlined in this document. Nurseries are required to comply with standard operating procedures and must provide a purity guarantee and a phytosanitary certificate for all of the material that is produced and sold (Handa et al., 2019).

Aeroponics Production

The term "aeroponics" comes from the Greek words "aero-" and "pnos," which translate into "air" and "labor," respectively. It refers to a method of cultivating plants that does not involve the utilization of soil or water as a growing medium. All of the essential growing parameters are maintained while the plant roots are kept suspended in air or mist and the plant is allowed to continue growing. The cultivation of plants using this method is distinct from hydroponics, aquaponics, and plant tissue culture. In hydroponics, rather than using soil or compost as a growing medium, a nutrient solution that contains all of the necessary minerals for healthy plant growth is used. Aquaponics makes use of water and the waste products of fish as the growth medium, whereas aeroponics cultivation is done without the use of a growth medium (Lakhiar et al., 2018). A plant growth support structure keeps the plant in place while keeping it separated from its roots when it is being grown using aeroponics. In the process of cultivating plants using the aeroponics method, the roots of hanging plants are irrigated with a nutrient-rich water solution that is sprayed onto them in the form of a mist using sprinklers. The roots of the plant are responsible for the absorption of water and nutrients, while the plant growth conditioning environment is responsible for the maintenance and control of growing parameters such as temperature, humidity, pH, and the electrical conductivity of the nutrient solution (Muthoni and Kabira 2014). It has been demonstrated that the vegetables that are grown using the aeroponic system have a high quality, that they are hygienic, healthy, fresh, crunchy, flavorful, and that they are accompanied by a high flavor. The opportunity needs of society's middle class and upper class can be met by the production of vegetables using aeroponics. Because of this reason, numerous aeroponic systems have been developed in the country of Indonesia. The oxygenation of each grain of the fine nutrient solution mist is one of the most significant benefits that can be gained from using aeroponics. This provides the roots with a plentiful supply of oxygen, which makes root respiration easier and leads to an increase in the amount of photosynthetic activity. According to (Mbiyu et al., 2012), the following are some of the advantages of aeroponics in comparison to other cultivation systems:

- Uses a significantly smaller amount of water
- Offers a consistent supply of water to plants at all times of the year
- Allows for easier observation as well as more comfortable and hygienic harvesting of corpses
- Aeration of the roots should be optimized
- Permits intensive production to take place within a constrained area

When trying to maximize production, the quality of the plant, its upkeep, and supervision, as well as the timing of the harvest, are all important factors. As a result, dependable and accurate aeroponics environmental control is absolutely necessary in order to accomplish this objective. The cultivation of plants through the use of aeroponics is a technique that was developed on the theory that in order to achieve the plant's full potential for production, it must be provided with the best possible conditions for growth. When compared to hydroponics, aeroponics systems achieve optimal root aeration, which is one of the most important factors in increasing yield. When compared to traditional planting methods that involve using soil as the medium, the advantages of the aeroponic system are clear. When grown in soil media, conventional methods produce between three and six mini tubers per plant, whereas aeroponic methods produce more than 40 mini tubers per plant (Chindi et al., 2014). Plants grown using aeroponics showed a greater increase in vegetative growth but a slower formation of tubers. However, this resulted in a 70% higher total tuber production per plant and 2.5 times the number of tubers compared to plants grown using hydroponic systems. The high production that can be achieved through the use of aeroponics can be attributed, in large part, to the high nutrient absorption efficiency, the high stolon development, the relative absence of pests and diseases, and the ease with which plants can be managed (Mbiyu et al., 2018).

Other Supportive Technologies

In addition to this, technology is laying the groundwork for the agricultural and horticulture sectors of the world's future. The technology of drones comes to mind at this moment. It has produced possibly the most high-profile technological innovation in the world of agriculture and land management to this point so far. In Denmark, a deadly invasive plant species known as the Giant Hogweed was the focus of an intriguing study that we carried out a year ago. This project used drones, remote sensing, and AI to help identify and eliminate the Giant Hogweed. Other businesses, such as Dendra, have also automated the planting of seeds with the help of drone technology. These drones are helping to rebuild ecosystems by spreading the seeds of native plants and trees across a wide area. Both studies demonstrate how farmers' routines for planting crops and tending to vegetation are being disrupted by the rise of digital technology. Precision farming is a term that has been used to describe the process of managing a farm with the assistance of remote-controlled aircraft and other technological advancements (Marz, 2020).

Conclusion

In the near future, it is anticipated that there will be even more technologies like UAVs, robots, and computer-controlled farming, which will be used in all levels of food production.

References

Aatif, A., Iqbal,K., Aziem, S., Mahato, P., Negi, A. K. A review on the science of growing crops without soil (soilless culture)- a novel alternative for growing crops." *International Journal of Agriculture and Crop Sciences*, 7, no. 11, (2014): 833.

Barbosa, G. L., Gadelha, F. D. A., Kublik, N., Proctor, A., Reichelm, L., Weissinger, E., Wohlleb, G. M. and Halden, R. U. Comparison of land, water, and energy requirements of lettuce grown using hydroponic vs. conventional agricultural methods. *International Journal of Environmental Research and Public Health*, 12, no. 6, (2015): 6879-91. https://doi:org/10.3390/ijerph120606879

Cassman, K. G. Ecological intensification of cereal production systems: Yield potential, soil quality, and precision agriculture. *Proceedings of the National Academy of Sciences*, 96, no. 11, (1999): 5952-59. https://doi.org/10.1073/pnas.96.11.5952

Chen, X.-P., Cui, Z.-L., Vitousek, P., M., Cassman, K. G., Matson, P. A., Bai, J.-S., et al. Integrated soil-crop system management for food security. *Proceedings of the National Academy of Sciences*, 108, no. 16, (2011): 6399-404. https://doi.org/10.1073/pnas.1101419108

Chindi, A., Giorgis, G.W., Solomon, A., Tessama, L., Negash, K. (2014). Rapid Multiplication Techniques (RMTs): A tool for the production of quality seed potato (Solanum tuberosum L.) in Ethiopia. *Asian J. Crop Sci.* 6, no. 9, (2014): 176-85.

Glick, B. R. Plant growth-promoting bacteria: Mechanisms and applications. *Scientifica*, (2012): 1-15. https://doi.org/10.6064/2012/963401

Handa, A. K., Dhyani, S. K. and Rizvi. J. 2019. Guidelines to Produce Quality Planting Material of Agro-forestry Species. https://doi.org/10.13140/RG.2.2.31253.14566

Hayde, N, Anita. L. Aeroponic and hydroponic systems for medicinal herb, rhizome, and root crops. *HortScience* 41, no, 3, (2016): 536-38. https://doi.org/10.21273/HORTSCI.41.3.536

Lakhiar, I. A. Gao, J., Syed, T. N., Chandio, F.A., Buttar, N.A. Modern plant cultivation technologies in agriculture under controlled environment: A review on aeroponics. *J. Plant Interact.* 13, (2018): 338-52.

Marz. 2020. Technology in Horticulture for Reliable and Optimized Food Production. Retrieved Oct 15, 2022 from https://nttdata-solutions.com/ch/blog/technology-in-horticulture-for-reliable-and-optimized-food-production/

Mbiyu, M. W., Muthoni, J., Kabira, J., Elmar, G., Muchira, C., Pwaipwai, P., et al. Use of aeroponics technique for potato (Solanum tuberosum) minitubers production in Kenya. *J. Hortic.*, 4, (2012): 172-77.

Mbiyu, M. W., Lung'aho, C., Otieno, S. A., Nyongesa, M.W., Muchui, M. N., Ogemma, J.N. Performance of five potato varieties with regards to growth and production of minitubers under an aeroponic system in central highlands of Kenya. *Afr. J. Agric. Res.*, 13, (2018): 366-78.

Muthoni, J., Kabira. Multiplication of seed potatoes in a conventional potato breeding programme: A case of Kenya's national potato programme. *Austral. J. Crop Sci.*, 8, (2014): 1195-99.

Papa, R., Manetto, G., Cerruto, E., Failla, S. "Mechanical distribution of beneficial arthropods in greenhouse and open field: A review", *Journal of Agricultural Engineering*, 49, no. 2. (2018): 81-91.https://doi.org/10.4081/jae.2018.785

Pramanick, K. K., Shukla, A. K., Watpade, S., Kumar, J. Rootstocks and their exploitation in improving production and quality of apple crops. *International Journal of Tropical Agriculture,*. 35, no. 3, (2017): 721-729.

Psirofonia, P., Samaritakis, V., Eliopoulos, P., Potamitis, I. Use of unmanned aerial vehicles for agricultural applications with emphasis on crop protection: Three novel case studies. *Int. J. Agric. Sci. Technol*, 5 (207): 30-39. https://doi:10.12783/ijast.2017.0501.03

Schumann, A. W. Precise placement and variable rate fertilizer application technologies for horticultural crops. *HortTechnology,* 20, no. 1, (2010): 34-40. https://doi.org/10.21273/HORTTECH.20.1.34

Villordon, A. "I Have an App for That: Introduction to Mobile Application Development in Horticulture." *HortTechnology,* 23, no. 4, (2013): 398. https://doi.org/10.21273/HORTTECH.23.4.398

5

Horticultural Crop Production Through Intercropping

***Rahat Ashraf*[1] *and Y. S. Bagal*[2]**

[1]*School of Agricultural Sciences and Technology, RIMT University, Mandi Gobindgarh, Punjab, India*

[2]*School of Agriculture, Lovely Professional University, Phagwara, Punjab, India*

Introduction

The simultaneous cultivation of two or more crops in the same field is known as intercropping. Sustainable agriculture aims to at least build agricultural systems that are inspired by nature. Efficiency and the idea that there are no waste products in nature are key components of sustainable agriculture since nature incorporates plants and animals into a variety of landscapes. When domesticating crops replaced hunting and gathering, the landscape became altered. Due to humankind's limited ability to produce a wide variety of agricultural plants and livestock, the amount of biological diversity throughout a large portion of the planet has significantly decreased. In nature, cooperation predominates over competition. Mutually beneficial interactions between species within groups are one example of cooperation. A few types of creatures may initially inhabit an abandoned agricultural field if it is left unattended and unplanted, but over several years, a complex community made up of numerous wild species emerges. When a community reaches a high level of variety, it becomes stable. Diverse communities are stable and have fewer swings in species abundance. Enterprise diversification, crop rotation, the use of windbreaks, the provision of more habitats for microorganisms, intercropping, and the integration of crop farming with animal production are methods that increase diversity and stability on the farm (Reddy and Willey,1981; Reddy et al., 1992). Each crop should have enough room while growing with another one or more to enhance cooperation and minimize competition. This is made possible by the elements of spatial layout, plant density, crop maturity dates, and plant architecture. The spatial arrangements are Row intercropping—Growing two or more crops together at the same time with at least one crop planted in rows; Strip intercropping—Growing two or more crops together in strips wide enough to separate crop production using machines, but close enough to interact; Mixed cropping—Growing two or more crops together in no distinct row arrangement; Relay intercropping—Plant a second crop into a standing crop at a time when the standing crop is at its reproductive stage but before harvesting.

The seedling rate of each crop on the combination is reduced from the maximum rate to maximize plant density and lessen competition from crowding. In the blend, the crops will produce well (Hiesbick, 1980; Magaguda et al., 1979; Prabhakar et al., 1983). Planning intercrop with staggered maturity dates or developmental stages takes the use of changes in peak nutrient, water, and sunshine resource needs. As a result, it is easier to harvest the grain, separate it, and lessen competition (Akyempong et al., 1999). One intercrop can collect sunlight that would not otherwise be available to others thanks to the structure of the plant. According to Reddy and Willey (1981) and Gardiner and Craker (1981), this is crucial for the development and production of grain and legume crops. Using the Land Equivalent Ratio (L. E. R.) approach, the intercrop components' performance is assessed. It assesses the benefits of intercropping systems on the total yields of the two crops. It offers a uniform framework for contrasting systems in various contexts and with various crop combinations. Timeliness of planting, correct fertilizing, pest and disease management, and expected profitability of intercrops are just a few of the many factors that need to be considered in order to ensure a successful harvest (Prabhakar et al., 1983; Reddy and Willey, 1981).

Benefits of Intercropping in Horticulture

Growers are becoming more interested in intercropping as a tool for sustainable long-term land management because it has the potential to lessen the burden on the farm's resources by making more effective use of, and even replacing, the natural resources used during crop production. Intercropping is a farming method that has been used for centuries in developing nations across the globe, including those in Central America, Asia, and Africa (Allen and Obura, 1983; Altier and Liebman, 1994; Bekunda and Woomer, 1996; 1999; Grossman and Qualles; 1993; Henzel and Valdis; 1977; Jodha; 1979; Prabhakar et al., 1983). Food security, erosion management, and higher crop yield are only some of the many benefits of intercropping (Bekunda and Woomer, 1996, 1999; Jodha, 1979; Owuor et al., 2002). Increased distance between plants of the same species, thanks to the planting of other crops between them, results in a decrease in plant diseases, an increase in total farm production and profitability, a decrease in the weed population through allelopathy and efficient crop rotation, and a decrease in insect and mite pest populations (Maguguda et al.,1979). Further, soil warmth is kept at bay when an intercrop acts as a suitable cover. Therefore, soil organic matter is protected from combustion and nutrient loss is avoided. Additionally, it creates a microclimate that may be ideal for growing complementary crops. Beans, for instance, climbing beans, can exploit the maize stalks in an intercrop to help them grow (Maguguda et al.,1979).

Potential Problems of Intercropping

(i) **Competition for water, light, and nutrients between intercropped crops can reduce harvests:** Reddy and Willey (1981) and Roger and Dennis (1993) both found that making adjustments to the physical layout of the crops helped to lessen the intensity of the harvest inherent competition. Plants that are further apart from one another have less competition for water. Mechanization challenges and rising labor demands are two more issues.

(ii) **Crop rotation and Intercropping:** Crop rotation (temporal diversification) tactics may significantly reduce weed population density and biomass output. As compared to monoculture, crop rotation reduced the number of weeds that germinated and sprouted in the test crops. Weed seed density was often lower in crop rotations than in monocultures of component crops. Weed biomass was often lower in the intercrop when a major crop was inter-sown with a "smother" crop species (Jodha, 1979). Weed biomass was lower in intercrops made up of two or more major crops than in any of the constituent single crops (Jodha, 1979).

It appears that crop sequences that generate different patterns of resource competition, allelopathic interference, soil disturbance, and mechanical damage are key to the success of rotation systems for weed suppression. These factors combine to make the environment unstable and frequently inhospitable, which in turn discourages the growth of weeds. There are two potential areas in which intercropping can outperform single crops in terms of weed management. First, if intercrops are more successful than solitary crops in usurping resources from weeds or suppressing weed development through allelopathy, then better crop production and less weed growth may be attained. Alternately, if intercrops employ resources that are not exploitable by weeds or convert resources to harvestable material more effectively than solitary crops, then intercrops may give yield advantages without lowering weed development below levels found in component sole crops (Jodha, 1979).

Different Intercropping Systems

Intercropping of Fruit Crops

Fruiting crops are frequently intercropped with annual crops in East Africa. Banana, for instance, is intercropped with food and/or fodder crops to boost and utilize efficiency for smallholder farms (Clark and Francis,1985). The wood volume of Grevillarobusta was found to be greatest after 31/2 years when a banana was intercropped at three densities of 208, 313, and 0.25 trees per hectare, whereas yields of banana and beans in the intercrop system were unaffected (Roger and Dennis (1992) and Reddy et al. 1992). Banana, when intercropped with sweet potatoes and beans, has been shown to decrease the prevalence of weevils and nematodes (Roger and Dennis, 1993; Lima and Lopez, 1979; Reddy et al., 1992; Lima and Lopez, 1993). Reduced populations of Striga hermontheca (Striga) and Maize stemborer (Chilopartellu) have resulted from intercropping Maize with desmodium, legume, and Napier grass in Kenya (Ahmed et al., 2008). There has been a lot of success with intercropping bananas and potatoes in India. In India, mustard has also been intercropped with bananas. Many different kinds of fruit trees have been planted among each other. Intercropping citrus mandarin seedlings with cucumber, for instance, resulted in a high yield of cucumber fruit with minimal impact on the growth of the former, presumably because citrus seedlings are rather slow-growing (Nataraja and Nairk,1992). In Kenya, juvenile fruit trees are intercropped with a wide variety of short-term crops such as beans, peas, potatoes, maize, millet, and exotic and indigenous vegetables to ensure that families have access to nutritious food and a steady income while they wait for the trees to mature.

Intercropping of Vegetable Crops

In an intercrop system with cereals, beans, and peas, the yield has been shown to increase because of the transfer of biologically fixed nitrogen from the roots of legumes to the root zone of the companion crop (Akyempong et al., 1999; Gold et al., 1999; Ndungu et al., 2003; Reddy and Willey, 1981; Nanecke, 1989). The use of commercial fertilizers may be reduced if this naturally fixed nitrogen is utilized instead (Lima and Lopez, 1979; Hiesbick, 1980). Intercropped between rows of field maize grown on elevated soil beds are vegetables including cabbage, cucumber, radish, snap bean, and broccoli. Beets and okra, pepper and onion, and other intercropped vegetable combinations with distinct architectural and nutritional profiles are common in tropical Asia. Vegetable cropping systems, such as those based on maize and cowpea, are common in certain parts of Africa. Plant height (Spiejer et al., 2006), leaf size, and leaf orientation and distribution within the plant canopy are all elements that affect whether or not two crops may be intercropped effectively (Sogbedji et al., 2006). These factors modify the light transmission through the canopies (Grossman and Quales, 1993), which in turn may impact the photosynthetic rates of the leaves inside the canopy. For plants to thrive in Intercropping systems when the canopies are at various heights, the understory crop must be able to tolerate some shadow (Sogbedji et al., 2006).

Land Equivalent Ratio (LER)

The efficiency of intercropping systems may be evaluated using this technique. It is the most used method for calculating the impact of intercropping on total crop output. Mandal and Roy (1986) Intercropping yields are attained in a smaller fraction of the land compared to monoculture yields (Trenbath, 1976). Intercroppin"s monetary benefit was determined using a cost-benefit analysis (Trenbath, 1979). The LER puts more weight on total land area than on the amount of time a field is used for harvesting. To address this shortcoming, the LER technique was revised (Liebman and Elizabeth, 1993) to account for the full growing season (ATER). Using LER as a metric for productivity evaluation, ATER accounts for the amount of time the crop was present on the ground (Liebman and Elizabeth, 1993). When planting maize in between cowpeas and soybeans, yields improved by 22–32% on an LER basis. The productivity of both wheat and chickpeas increased when grown in an intercropping method. The rise was 29% (Mandal and Roy, 1986) when the financial benefits of intercropping barley and lentil with a single irrigation and a straw mulch were taken into account, the increased yield was 28% (Ndungu, 1987). To ensure the viability of an intercropping system for vegetable crops in a given area, it is necessary to identify the most appropriate cultural methods for that population of plants. It has been found that the number of sweet corn ears produced by a plant in an intercropping with southern peas did not change, but that yields rose with plant population density. Similarly to the research on sweet corn and southern peas, it has been documented by other investigators (Mashingaidze et al., 2000; Trenbath, 1976, Reddy and Willey, 1981) that increasing the population of maize in a bean and maize combination led to a higher maize production (Reddy and Willey, 1981). In a 1981 study conducted by Reddy and Willey, the density of the plants and the amount of nitrogen fertilizer used did not affect the yield per plant. In a cereal-legume intercrop, light interception has been shown to significantly impact plant development and production (Grossman and Quales, 1993). Intercropping southern pea and sweet corn reduces the amount of light absorbed by each crop individually compared to growing each crop as a monoculture system of those crops likely contributed to the decrease in yields for those crops individually (Reddy and Willey, 1981). Reduced light interception had a greater effect on intercropped southern pea than on intercropped sweet corn, resulting in reduced yield (Reddy and Willey, 1981). The development of maize and beans in an intercrop was shown to be enhanced in a prior study (Clark and Francis, 1985) by staggered planting of the crops at different times. Putting off planting either southern peas or sweet corn resulted in lower yields from the intercropped crops (Reddy and Willey 1981).

Many African countries have used the intercropping of maize and legumes for decades to ensure their citizen" access to affordable, nutritious food. Due to a lack of nutrients, agricultural output has generally decreased. According to the literature (Clark and Francis, 1985; Grossman and Quales, 1993; Hiesbick, 1980), this is a problem. When beans are included in the rotation, maize benefits from the nitrogen (N) that is fixed by the legume. Staggering the maize rows and intercropping with legumes like groundnuts and cowpeas in crop rotation systems have both been shown to increase harvests. In Kenya and elsewhere, farmers have found success with crop rotations that pair vegetables with grains. Specifically, in the Machakos area of Kenya, maize and pigeon peas are grown together as an intercrop. Rows of corn and pigeon peas are planted as soon as it starts to rain. About the time the pigeon pea begins to blossom and produce fruit, (after 5 months), the maize is harvested. Due to their deep roots, pigeon peas can fix nitrogen into the soil, increasing the lan"s fertility. It is also utilized for food security, providing shade, and as a source of fuel. In South Africa, maize and pigeon peas are often planted at the same time as an intercrop, although the former takes a little longer to develop, thus it is typically harvested before the latter can produce any significant biomass due to its long growing season. Pigeon pea thrives for several additional months on leftover soil moisture after maize is harvested, resulting in a full canopy cover and high yields. It has been shown that yields from intercrop Maize are comparable to those from mono-crop Maize when planted at the same spacing. Growing corn and pigeon pea together decreases the amount of nitrogen and phosphorus fertilizer required in subsequent harvests (Spiejer et al., 1993). One study indicated that using pigeon leaves to add 75 to 90 kg of nitrogen per hectare significantly improved the timing of the maize harvest (Sánchez et al., 1996). Pigeon peas have the potential to make excellent use of the unused soluble P that is present in the soil after fertilizing maize (Bahl and Pasricha,1998). There is a lot of pest and disease harm from pigeon usage (Allen and Obura,1983, Sogbedji et al., 2006). Aside from South Malawi, parts of Mozambique and Tanzania, Benin, and southern Nigeria are also using maize-pigeon pea intercropping (Sileshi and Mafongoya, 2003). In Kenya, maize and bananas are often grown along with sweet potato as an intercrop. Although the tubers are consumed, the leaves and stems of the sweet potato are utilized as a feed source for cattle. When the soil has been prepared for planting by weeding and fertilizing, sweet potatoes are planted in between rows of corn. This intercropping of sweet potatoes, maize, and bananas has the dual benefits of increasing crop diversity and reducing the land's environmental impact. The sweet potato's ability to provide long-term soil cover helps cut down on weed control expenses and soil erosion.

Improvements in maize-legume intercropping have made it possible for farmers to plant a broad variety of edible legumes in the understory of their fields. It is possible to grow maize at its optimal population while still keeping equal yields by shifting every other row to provide a larger inter-row for the legume or strip crop. Because of the latter problem, intercropping with Maize is uncommon, but this method makes it possible to reap the benefits of growing Maize alongside high-value food legumes like groundnuts, green gram, soybean, etc. (Woomer et al., 2004). Better interactions for legume intercrops, or MBILI, have been documented (Sileshi and Mafongoya, 2003). The goal of this system is to boost legume productivity, and it does so by applying Rock Phosphate (Buresh et al., 1997) and fortified manure compost (Ndungu et al., 2003) to the Maize legume intercrop. However, the system has had a positive effect on Maize yields as well. The standard advice for planting maize seed in Western Kenya is 75 cm between rows and 30 cm inside each row, yielding around 44,000 plants per hectare from 75 kg of seed. In the spaces between the maize rows of 75 cm, beans are sown. MBILI allows for a new row configuration. The distance (the gap) between maize rows is 100 cm. Within the space, two rows of legumes (with 33 cm row spacing) are planted. Staggering the rows of maize into a two-by-two pattern allows the legumes to better compete with the maize while yet maintaining the same plant populations as conventional intercrops. When possible, the rows should be laid out in an east-west orientation on flat ground. The cost-benefit analysis has shown a high rate of return. Similarly, positive outcomes have been seen with sorghum-legume intercropping in Uganda (Owuor et al., 2002). The breeding work at IITA (International Institute of Tropical Agriculture) is focused on increasing cowpea biomass to produce reasonable quantities of both grain and fodder with minimal application of pesticides. Sorghum and Millet have also been intercropped with cowpea in West Africa, where cowpea is used for human food and fodder (Sileshi and Mafongoya, 2003). Finally, in Zimbabwe, combining maize and pumpkin into a single crop has been shown to minimize weeding needs compared to maize alone (Mashingaidze et al., 2000). High rates of intercropping decreased Pumpkin plant height and usable leaf area per plant. This is consistent with the recommendation made by (Olaniyan et al., 2006) that Pumpkin plant spacing be increased to promote optimal vining behavior. However, leaf yields and partial Leaf Area Index increased with high levels of Pumpkin intercropping populations, demonstrating the importance of intercropping populations in determining yields (Mashingaidze et al., 2000). Pumpkin has a prostrate growth habit and requires a lot of free space, both of which were lacking in the intercropping between it and Maize, resulting in shorter vines. Groundnut output was also drastically decreased when intercropped with Pumpkin populations of 0.5%, 1%, and 2% of the groundnut population. All pumpkin-groundnut intercropping populations outperformed groundnut monoculture in terms of yield and weed suppression (Mashingaidze et al., 2000).

Intercropping for More Effective Use of Resources

Intercropping systems have gained the attention of agricultural planners because of the role they play in reducing inputs in sustainable agriculture by making more efficient use of available resources (Ouma and Jeruto, 2010). Since intercropping makes better use of soil nutrients than monoculture, it has been linked to higher yields in general (Eskandari, 2011; Mobasser et al., 2014). Overall productivity may be increased in a given region by increasing the efficiency with which available resources are put to use, which is one of the main motivations for intercropping. Component plant species in these systems have varied roots and canopy qualities, allowing them to utilize resources in a complementary manner, resulting in more effective crop production (Lithourgidis et al., 2011; Moradi et al., 2014). When compared to monoculture, intercropping maximizes the utilization of solar radiation, water, and nutritional components by producing two crops that mutually benefit one another (Eskandari, 2011). By capitalizing on variances in canopy growth rate, canopy, root structure, and time and space, good intercropping applications optimize resource usage across the life of the crop. Intercropping has beneficial impacts by lowering rivalry across crop species and allowing for more efficient use of scarce resources through complementary resource utilization and niche differentiation over area and time (Li et al., 2014). Intercropping systems interact with one another in both direct and indirect ways to promote growth.

Intercropping systems can improve resource efficiency by supplying symbiotic nitrogen from legumes or by making accessible inorganic phosphorus fixed in soil due to a decrease in pH caused by nitrogen-fixing legumes (Jensen, 1996a; Aminifar and Ghanbari, 2014). In intercropping systems, nitrogen fixation by a legume crop may be the most cost-effective and convenient means of delivering nitrogen to the non-legume. Several studies have suggested that legume plants might supply non-legume plants with biologically fixed nitrogen (Karlidag and Yildirim, 2009b). Increased efficiency in the use of soil nutrient components by intercropping systems compared to that of pure stands allows for less nitrate to be lost to runoff (Zhang and Li, 2003). The research conducted by Whitmore and Schröder (2007) indicated that intercropping systems may be utilized to lessen nutrient contamination while maintaining yields and that yields and profits are inversely connected with residual nitrate after harvest. In their study, Sobkowicz and Sniady (2004) found that the intercropping system of triticale and beans increased nitrogen absorption more than the total output. The fact that nitrogen intake is greater in intercropping than in monoculture suggests that the two systems use nitrogen in a complementary manner because the two crops use distinct nitrogen sources.

Mycorrhizal fungi and other soil-plant associations facilitate nitrogen transfer from legume to non-legume crops, while root interactions in intercropping systems maximize soil utilization. Exudation of amino acids and plant growth-stimulating materials, extracellular enzymes, acidification, and competition-induced alteration of root characteristics are all examples of soil-plant interactions that significantly alter resource mobilization (Hauggaard-Nielsen and Jensen, 2005). It has been claimed that certain legume crops can increase overall yields by fixing atmospheric nitrogen and increasing the availability of phosphorus (Snapp and Silim, 2002). Intercropping, according to research by Mobesser et al., (2014), increases the absorption of nutrients like phosphorus. Some kinds of legumes are able to solubilize fixed phosphorus by altering the chemistry of their rhizosphere (Hauggaard- Nielsen and Jensen 2005).

In order to maximize the uptake of soil nutrients like nitrogen and phosphorus, cereals-legume intercropping strategies can be quite useful. For example, root-induced acidification from nitrogen fixation in legumes might lead to significant fluctuations in phosphorus availability in the rhizosphere, leading to the concept that various intercrop species can exploit distinct phosphorus pools (Betencourt et al., 2010). Because the roots of the many plants in an intercropping system explore a greater area of soil, the plants in these systems can absorb more nutrients than those grown in isolation. The depth of an intercropping system's root system has a direct bearing on the plant nutrients that may be used. Crops with contrasting root properties can make more efficient use of scarce resources. Selecting crops with dissimilar rooting properties, nutrient requirements, and peak demand times might alleviate nutrient competition in intercropping systems. Reduced nutritional concentration in plant sections is a direct result of nutrient competition (Varghese, 2000; Morris and Garrity, 1993). Much previous research (Santos et al., 2002; Yildirim and Guvenc, 2005; Guvenc and Yildirim, 2006; Yildirim and Turan, 2013) noted that plant nutrient concentrations were comparable between intercropping and solo cropping, showing effective utilization of available resources by intercrops. Recommendations were made for improved water efficiency in intercropping systems (Moradi et al., 2014). Intercrops, according to Kanton and Dennet (2004), generated more dry matter than individually. By decreasing wind speed, providing shade, and increasing infiltration, intercropping systems help to save water in the soil and improve the structure of the soil. Plants' capacity to absorb water and access it is affected by the root architecture of intercrops. When comparing maize monoculture to maize monoculture with an intercropping system, the maize-cowpea system consistently exhibited superior water usage efficiency because of increased light interception, lower water evaporation, and soil water conservation (Mobesser et al., 2014). Many physiological indicators for interspecific facilitation have been found via research on rhizosphere processes and nutrient usage in intercropping systems (Li et al., 2014).

Soil water is conserved, water evaporation is reduced, and plant nutrients are better used in intercropped systems compared to mono-cropping systems (Ghanbari et al., 2010). It has been suggested that arbuscular mycorrhizal fungi can form connections between the roots of different crops in intercropping systems (He et al., 2003). According to Jensen (1996b), arbuscular mycorrhizal hyphae have a faster turnover rate and produce more phosphorus-rich root materials, both of which contribute to an increase in the plant nutrients phosphate and nitrogen that emerge from the hyphae over time. In addition, He et al., (2003) demonstrated a form of two-way transfer strategy by demonstrating nitrogen transfer from nonnitrogen-fixing crops to nitrogen-fixing mycorrhizal crops. Short-season plants can be grown in the spaces left between rows of long-season plants while the former take their time to mature and form a complete canopy (Fukai and Trenbath, 1993). Between the rows of long-season plants, you may cultivate a component plant that can make optimal use of the sunlight, water, and nutrients that are available. The productivity, profitability, and utilization of resources like water and nutrient elements are all said to be higher in intercropping systems of crops having different under and above-ground characteristics and growth spans than in monocropping (Vargeshe, 2000; Yildirim and Guvenc, 2005; Karlidag and Yildirim, 2009b).

Intercropping for More Stable System

When a producer's available resources are extremely limited, sustaining total yield and economic return should take precedence over maximizing productivity or economic return. Providing food and income year-round, lowering the risk of failure at all times, keeping financial expenditures low, and participating in community activities are all examples of family objectives (Fukai and Widmore, 1993). Intercropping systems have been widely hailed as one of the most reliable strategies to ensure the long-term viability of crop production (Coolman and Hoyt, 1993). This is because they provide farmers with a backup plan in the event of crop failure in monocropping. Some plants have the power to alter unfavorable environmental conditions and increase the availability of resources for others (Lambers et al., 1998). When subjected to the same set of adverse conditions, one species of plant may do badly while another thrives (Horwith, 1985). Since the benefits of intercropping methods are restricted to low-input and small-scale agriculture operations (Baumann et al., 2001), there is a growing interest in intercropping. The technique of intercropping has been proposed as a viable alternative to solitary cropping, especially for small farmers with limited resources, by Itulya et al., (1997). While increasing total yields, intercropping strategies may also be used to mitigate risk by ensuring that, even if one crop fails, another can supply enough sustenance to last until the next harvest (Whitmore and Schröder, 2007). For low-input and small-scale farms, intercropping strategies may be more stable than mono cropping, as shown by Guvenc and Yildirim (2006). This might be because the production from one intercrop compensates for the loss of yield in a different crop (Zimmermann, 1996).

Increasing Productivity with Intercropping Systems

Growing many crops at once is often done to increase overall production per acre. By making better use of available natural resources, total production may be boosted through intercropping strategies. Simple natural laws govern the management of component crops to maximize their synergistic benefits and minimize their competitiveness, and the only real limitation on their use is the creativity of farmers and agronomists. Financial gain from evaluating the viability of various cropping patterns is an important factor in determining the longevity of any production system. Researchers have shown that intercropping increases agricultural output, productivity, and profitability per acre of land (Emuh et al., 2006; Karlidag and Yildirim, 2007; Yildirim and Turan, 2013). The efficient use of resources per acre of land for intercropping may be an indicator of this phenomenon (Sharaiha and Hattar, 1993). Profits and food production on fewer acres may both benefit from intercropping. Yildirim and Guvenc (2005), Karlidag and Yildirim (2007), Yildirim and Karlidag (2009b), and Mahant et al., (2012) all found that intercropping methods were suitable for use on a commercial scale because the higher yields they provided. Rehman et al., (2010) found that when maize was intercropped with cowpea, the net farm and cost-benefit ratio was much higher than when either crop was grown alone. The LER has been consistently established as a higher one in previous investigations. Yildirim and Guvenc (2004), Yildirim and Guvenc (2005), Karlidag and Yildirim (2007), and Yildirim and Turan (2013) all agree that LER values are a good indicator of the biological efficiency of intercropping systems. Intercropping is more effective than monocropping if the LER is greater than one (Vandermeer, 1989). The most successful intercropping strategies involve planting crops with varying maturation dates, such that peak demands on soil, water, and sunlight all occur at various times (Fukai and Trenbath, 1993). Intercropping systems have been found to be more beneficial than single-crop systems (Karlidag and Yildirim, 2009a; Yildirim and Turan, 2013). As long as the plants receive water and nutrients in sufficient amounts and at the appropriate timing, intercropping systems can make efficient use of existing resources (Abusuwarand Al-Solimani, 2013). Competition between crops can be reduced by the use of intercrops by making use of their complementing impact (Sharaiha and Hattar, 1993).

Intercropping for Pest, Disease, and Weed Control

In order to lower insect population densities in field-grown crops, intercropping is one of the IPM strategies used. The field of integrated pest management has expanded its focus to incorporate agroecosystem management as a result of the growing interest in cultural control strategies. The beneficial effects of intercropping on pest, disease, and weed management are well documented (Baumann et al., 2000). There are many different kinds of herbivores, and it has been found that intercropping systems may lower their populations in a lot of different situations. Some of the other crops' pests may find refuge in the component crop (Vandermeer, 1989). Disease incidence was shown to be lower in barley-grain legumes (pea, faba bean, and lupine bean) intercrops than in similar mono crops by as much as 20–40%, according to research by Hauggaard-Nielsen et al., (2008). Keeping a dense cover crop on the ground kills weed seeds before they have a chance to germinate. Insect or disease situations and their natural enemies are influenced by factors such as spatial and temporal controls, physical and olfactory barriers, microclimate alteration, odor effects, and color and trapping effects between intercrops. It's possible that a disease pandemic may be contained by a more diverse agricultural ecology.

Natural ecosystems benefit from increased biodiversity when intercropping is used. Insects and illnesses can have a lessening influence on populations when variety is increased (Anil et al., 1998). There are two hypotheses for reducing the detrimental effects of insects and diseases: (a) Limiter crop theory states that the presence of even a single intercrop can significantly reduce a pest's capacity to attack its host; (b) Trap crop theory states that by planting one intercrop species so that a pest or disease that typically attacks another crop will be drawn to this intercrop, or "Natural enemies" (predators and parasites) may be more prevalent in intercropping than in solitary cropping (Aminifar and Ghanbari, 2014).

Weed population density and biomass output can be reduced by the use of crop rotation and intercropping strategies. Weeds are easier to manage in intercropping systems than in mono-cropping systems. When intercropping approaches are superior to monocropping in extracting resources from weeds or inhibiting weed development via allelopathy, this goal can be fulfilled. In addition, if intercrops use sources that are not exploitable by weeds or convert resources to harvestable portions more effectively than solitary crops, intercropping systems can have certain benefits without lowering weed growth below levels specified in component crops (Liebman and Dyck., 1993). If only one type of crop is grown, those few resources will likely go to the weeds instead of the intended produce. Altieri (1995) and Aminifar and Ghanbari (2014) both found that compared to single-crop farming, intercropping strategies that make greater use of ecological resources and fill the resulting void are more successful in suppressing weed growth. Some growth-regulating substances secreted by intercrops can significantly contribute to competitive interactions amongst intercrops, including weeds, and hence change the ultimate intercrop composition (Hauggaard-Nielsen and Jensen, 2005). Corn and soybean intercropping, as demonstrated by Aghaalikhani et al., (2009), has a synergistic impact that allows for more efficient solar radiation consumption, leading to reduced weed growth and sustained crop yields.

Conclusion

Plant-plant interactions are a defining feature of intercropping systems since these interactions can have a significant impact on the development and productivity of individual crop species (Vandermeer, 1989). Researchers from all over the world have studied the effects of intercropping and found that it may be more productive than monocropping due to the facilitative linkages between the plants. If we can learn more about these relationships, we might be able to improve agroecosystems' productivity and efficiency in using their resources. Many studies have found that intercropping is more lucrative than monocropping since it increases total production and economic gains. Due to the pressures of sustainable agriculture and the environmental issues that arise from conventional farming methods, intercropping systems have been re-envisioned. Considering the various benefits of intercropping systems, research into the feasibility of producing many crops in a single space should continue (Lithourgidis et al., 2011). Unfortunately, the full potential of enhancing agricultural yield via optimal intercrop patterns has not yet been tapped. Better productivity may be achieved in intercropping systems by considering complementarities in models of resource utilization. Crop cultivars chosen for use in intercropping systems should enhance the synergistic benefits gained by growing different species next to one another. In addition, the species or cultivars, spatial geometry, and planting density of component plants utilized in intercropping must be determined to reap the benefits of this technique.

References

Abusuwar, A.O., Al-Solimani, S.J. Effect of Chemical Fertilizers on Yield and Nutritive Value of Intercropped Sorghum bicolor and Lablab purpureus Forages Grown under Saline Conditions. *The Journal of Animal and Plant Sciences*, 2013, 23, no. 1, (2013): 271-76.

Aghaalikhani, M., Zaefarian, F.Z., Mashhadi, R.H., Rezvani, M. Corn and Soybean Intercropping Canopy Structure as Affected by Competition from Redroot Pigweed (Amaranthus retrofelxus L.) and Jimson Weed (Datura srtramonium L.). *Iranian Journal of Weed Science*, 5, (2009): 39-53.

Ahmed, H, Herren, H, Khan, Z.R, Pickett, JA., Woodcock CM. Integrated Pest Management: The push-pull approach for controlling insect pests and Weeds of cereals and its potential for other agricultural systems including animal husbandry. *Philos Trans R Soc. Lond B Biol. Sci.*, 363, no. 1491, (2008): 611-21.

Akyeapong, E., Hitimana, L., Munyemana, P. C. Multistrata Agroforestry with beans, bananas and Grevillarubusta in the highlands of Burundi, *Expt. Agric.*, 35 (1999): 357-69.

Allen, J. R. and Obura, R. K. Yield of corn, cow pea, and Soya bean under different intercropping systems. *Agron. J.*, 75 (1983): 1005-9.

Altier, M. A., Liebman. 1994. Insect, Weed and Plant Disease Management in multiple cropping systems. p. 383. New York: Macmillan Co.

Altieri, M.A. 1995. Agroecology: The Science of Sustainable Agriculture, (2nd edition). Westview Press.

Aminifar, J., Ghanbari, A. Biological Facilitative Interactions and Their Roles on Maximize Growth and Productivity of Crops in Intercropping Systems. Scientia Agriculturae, 2, no. 2, (2014): 90-5.

Anil, L., Park, J., Phipps, R.H., Miller, F.A. Temperate Intercropping of Cereals for Forage: Review of Potential for Growth and Utilization with Particular Reference to the UK. *Grass and Forage Science,* 53, (1998): 301-17.

Barley uptake of N Deposited in the Rhizosphere of Associated Field Pea. *Soil Biol Biochem*, 28, (1996): 159-68.

Baumann, D.T., Bastiaans, L., Kropff, M.J. Competition and Crop Performance in a Leek-Celery Intercropping System. *Crop Science,* 41, (2001): 764-74.

Bekunda, M. 1999. Farmers Responses to Soil Fertility Decline in Banana-based cropping Systems in Uganda. Managing Africa's Soils No. 4 Russel Publishers No Hingham 2 Pp.

Bekunda, M. and Woomer, P. L. Organic Resources Management in Banana-based cropping systems of the Lake Victoria Basin, Uganda. *Agriculture, Ecosystems and Environment,* 59, no. 3, (1996): 171-80.

Betencourt, E., Colomb, B., Justes, E., Souche, G. and Hinsinger, P. P for Two- Intercropping as a Means to Better Exploit Soil P Resources under Low Imput Conditions. 19th World Congress of Soil Science, Soil Solutions for a Changing World, 1-6 August, 2010, Brisbane, Australia.

Buresh, R. J. Sanchez P. A., Calhoon, F. 1997. Replenishing soil fertility in Africa. Soil Science Society of America Spec. Public No. 51, Madison, U. S. A. Pp. 251.

Clark, E., Francis, C. A. Transgressive yielding bean; maize intercrops interference in time and space field crops Res., 11 (1985): 37-53.

Collard-Cowpea Intercrop Response to Nitrogen Fertilization, Redroot Pigweed Density, and Collard Harvest Frequency. *Hort Science*, , 32, no. 5, (1997): 850-53.

Coolman, R.M., Hoyt, G.D. 1993. Increasing Sustainability by Intercropping. *Hort Technology,* 3, (1993): 309-11.

Emuh, F.N., Ofuoku, A.E., Oyefia, E. Effect of Intercropping Okra (Hibiscus esculentus) with Pumpkin (Cucurbita maxima) on some Growth Parameters and Economic Yield of Maize (Zea mays) and Maximization Land use in a Fadama Soil. *Research Journal of Biological Sciences*, 1, no. 1-4, (2006): 50-4.

Eskandari, H. Intercropping of Wheat (Triticum aestivum) and Bean (Vicia faba): Effects of Complementarity and Competition of Intercrop Components in Resource Consumption on Dry Matter Production and Weed Growth. *African Journal of Biotechnology*, 2011, 10, no. 77, (2011): 17755-62.

Fukai, S. and Midmore D.J. 1993. Adaptive Research for Intercropping: Steps Towards the Transfer of Intercrop Research Findings to Farmers' Fields. *Field Crops Research*, 1993, 34: 459-467.

Fukai, S. and Trenbath, B.R. 1993. Processes Determining Intercrop Productivity and Yields of Component Crops. *Field Crops Research*, 34, (1993): 247-71.

Ghanbari, G., Dahmardeh, M., Siahsar, B.A., Ramroudi, M. Effect of Maize (Zea mays L.) - Cowpea (Vigna unguiculata L.) Intercropping on Light Distribution, Soil Temperature and Soil Moisture in Arid Environment. *Journal of Food, Agriculture and Environment,* 8, no. 1, (2010): 102-108.

Gardiner, T. R., Craker, L. E. Bean growth and light interception in maize-bean intercrop. *Field Crop Res.*, 4, (1981): 313-20.

Gold, C. S., Karamura, E. B., Kiggunda, A., Baganda, F. A. Geographic shifts in highland banana (Musa group A A A-E). Production in Uganda. *Afric. Crop. Sc. Jnl.* 10, no. 2, (1995): 20-5.

Grossman J., Quales, W. 1993. Strip intercropping for biological control. I PM Practitioner 1-11.

Guvenc, I., Yildirim, E. 2006. Increasing Productivity with Intercropping Systems in Cabbage Production. *Journal of Sustainable Agriculture*, 28, no. 4, (2006): 29-44.

Hauggaard-Nielsen, H., Jensen, E.S. Facilitative Root Interactions in Intercrops. Plant and Soil, 274(2005): 237-50.

Hauggaard-Nielsen, H., Jørnsgaard, B., Kinane, J., Jensen, E.S. Grain Legume-Cereal Intercropping: The Practical Application of Diversity, Competition and Facilitation in Arable and Organic Cropping Systems. *Renewable Agriculture and Food Systems*, 23, (2008): 3-12.

He, X.H., Critchley, C., Bledsoe, C. Nitrogen Transfer Within and Between Plants Through Common Mycorrhizal Networks (CMNs). *Critical Reviews in Plant Sciences*, 22, (2003): 531-67.

Henzel, E. F., Vallis, T. 1977. Transfer of nitrogen between legumes and other crops. In Ayananaba and P. J. Dort. (eds). Pp. 73-88. Biological Nitrogen Fixation in Farming Systems of the Tropics. New York: Wiley.

Hiebsch, C. K. Principles of intercropping: Effects of N. Fertilization, Plant population and crop duration on equivalent ratios in intercrop versus Monoculture PhD dissert. North Carolina State University. *Raleigh Abstr,* 413, (1980): 4337.

Horwith, B. A Role for Intercropping in Modern Agriculture. *BioScience*, 1985, 35, no. 5, (1985): 286-90. Itulya, F.M., Mwaja, V.N. and Masiunas, J.B. 1997.

Jensen, E.S. (1996a). Grain Yield, Symbiotic N2-Fixation and Interspecific Competition for Inorganic N in PeaBarley Intercrops. *Plant Soil,* 182, (1196): 25-38.

Jensen, E.S. (1996b). Barley uptake of N Deposited in the Rhizosphere of Associated Field Pea. Soil Biol Biochem, 28, (1996): 159-168.

Jodha, N. S. Intercropping in traditional farming systems. Proc. Intl. Workshop on intercropping 10-13 Jan 1979 Hyderabad, India.

Kanton, R.A.L., Dennett, M.D. Water uptake and use by Morphologically Contrasting Maize/ Pea Cultivars in Sole and Intercrops in Temperate Conditions. *Experimental Agriculture*, 40, (2004): 201-214.

Karlidag, H., Yildirim, E. (2007). The Effects of Nitrogen Fertilization on Intercropped Strawberry and Broad Bean. *Journal of Sustainable Agriculture*, 29, no. 4, (2007): 61-74.

Karlidag, H., Yildirim, E. (2009a). The Effect of Vegetable Intercropping on Plant Growth, Yield, Land Equivalent Ratio and Economic Income in Sapling Growing. YuzuncuYıl University, *Journal of Agricultural Science*, 19, no. 2, (2009): 71-7.

Karlidag, H. and Yildirim, E. (2009b). Strawberry Intercropping with Vegetables for Proper Utilization of Space and resources. *Journal of Sustainable Agriculture,* 29, (2009): 61- 74.

Lambers, H., Chapin, F.S. and Pons, T.L. 1998. Plant Physiological Ecology. New York: Springer-Verlag.

Li, L., Tilman, D., Lambers, H., Zhang, F.S. Plant Diversity and Over Yielding: Insights from Belowground Facilitation of Intercropping in Agriculture. *New Phytologist*, 203, (2014): 63-9.

Liebman, M. and Dyck, E. Crop Rotation and Intercropping Strategies for Weed Management. *Ecological Applications*, 3, no. 1, (1993): 92-122.

Liebman, M., Elizabeth, D. Crop rotation and intercropping. Strategies for weed management. *Ecol. Appl.,* 3, no. 1, (1993): 92-112.

Lima, A. E., Lopez, L. H. 1979. Plant population and spatial arrangement study on intercropping maize and beans (P. vulgaris) in North East Brazil. Proc. Intl. Workshop on intercropping Pp. 400-12.

Lithourgidis, A.S., Dordas, C.A., Damalas, C.A, Vlachostergios, D.N. Annual Intercrops: An Alternative Pathway for Sustainable Agriculture. *Australian Journal of Crop Science,* 5, no. 4, (2011): 396-410.

Magaguda, G. T., Haque, I., Godfrey, W., Fendu, I., Masina, G. T. 1979. Intercropping studies in Swaziland: Present status and future projections. Proc. Intl. Workshop on intercropping 10-13 Jan., 1979, Hyderabad, India, Pp. 98-104.

Mahant, H.D., Patil, S.J., Bhalerao, P.P., Gaikwad, S.S., Kotadia, H.R. Economics and Land Equivalent Ratio of Different Intercrops in Banana (Musa paradisiacal L.) cv. Grand Naine under Drip Irrigation. *The Asian Journal of Horticulture,* 7, no. 2, (2012): 330-332.

Mandal, B. K., Roy, P. K. Yield of wheat mustard and chick pea grown as sole and intercrops with four moisture regimes. *India J. Agron Sci.* 56, (1986): 577-83.

Mashingaidze, A.B., Nyakanda, C., Chivinge, O.A., Mwashairen, A., Dube, K.Y. (2000). Influence of Pumpkin live mulch on weed dynamics and Maize yield. *African Crop Prot.*, 6, no 1, (2000): 57-63.

Mobasser, H.R., Vazirimehr, M.R., Rigi, K. Effect of Intercropping on Resources use, Weed Management and Forage Quality. *International Journal of Plant, Animal and Environmental Sciences*, 4, no. 2, (2014,): 706-13.

Moradi, M., Noori, M., Sobhkhizi, A., Fahramand, M., Rigi, K. Effect of Intercropping in Agronomy. *Journal of Novel Applied Sciences*, 3, no. 3, (2014): 315-20.

Morris, R.A., Garrity, D.P. Resource Capture and Utilization in Intercropping: Non-Nitrogen Nutrients. *Field Crops Research*, 34, (1993): 319-334.

Natarajaa, M. and Nairk, D. M. Competitive effects of short duration, bush type cowpea when intercropped with cotton in Zimbabwe Expt. Agric., 28, (1992): 41-8.

Ndungu, J. C. M. 1987. Regional Needs for banana and plantain improvement in East Africa in Banana and Plantain Strategies. In: Parsley, E. J. and De Longhoe, (eds). Pp. 36-38.

Ndungu, K. W., Kifuko, M. N. and Okatebo, Jr. 2003. Producing fortified compost from crop residues. In: Scvala, E. N., Omare, M. N. and Woomer, P. L (eds). Organic Resource Management in Kenya. Perspectives and Guidelines: Forum for Organic Resource Management and Agricultural Technologies, Nairobi, Pp. 71-79.

Nornecke, I.L. 1989. vegetable Production. 657pp, New York: Van Nostrand Rainhold.

Olaniyan, A. A., Fagbayidd, J. A., Oladapo, M. O. and Amin, C. A. (2006). Productivity of Cleopatra mandarin rootstock seedlings intercropped with cucumber. Asian Jnl. of Plant Sciences 5 (3): 534-536.

Ouma, G. and Jeruto, P. Sustainable Horticultural Crop Production through Intercropping: The Case of Fruits and Vegetable Crops: A review. Agriculture and Biology *Journal of North America,* 1, no. 5, (2010): 1098-105.

Owuor, C. Tenywa, J. S., Muwanga S., Woomer P. L., Esele, P. 2002. Performance of a sorghum-legume intercrops in response to row levels of nitrogen. Program and extended abstracts. FORUM Working Document. No. 5. The Forum on Agriculture Resource Husbandry, Nairobi, Pp. 542-44.

Prabhakar, B. S., Shulka, V. and Srinwa, K. Nutritional potential of vegetable intercropping system. *Indian J. Hort.,* 2, no. 3, (1983): 258-68.

Reddy, K. Visser, C. P., Buckner, P. Pearl/Millet and cowpea yields in sole and intercrop systems and their after effects on soil and crop productivity. *Field Crop Res.* 4, (1992): 13-124.

Reddy, M. S., Wiley, R. W.. Growth and Resource use studies in an intercrop of Pearl millet / groundnut. *Field crop Res.* 4, (1981): 13-24.

Rehman, H., Ali, A., Waseem, M., Tanveer, A., Tahir, M., Nadeem, M.A., Zamir, M.S.I. (2010). Impact of Nitrogen Application on Growth and Yield of Maize (Zea mays L.,) Grown Alone and in Combination with Cowpea (Vigna unguiculata L.,). *American-Eurasian Journal of Agriculture and Environment Science*, 7, no. 1, (2010): 43-47.

Roger, F. Dennis R. D. (1993). Developing an effective Southern Pea and sweet corn crop system. *Hort Technology,* 3, no. 2 (1993): 178-84.

Sanchez, P. A. Shepherd, K. Soule, M. J., Ndiritu G., Woomer, P. L. (1996). Soil fertility in Africa. In Buresh R. T., Sanchez, P. A. and Calhan, F. (eds) Pp. 1-46. Madison, Soil Sci. Soc. Amer.

Santos, R.H.S., Gliessman, S.R., Cecon, P.R. Crop Interactions in Broccoli Intercropping. *Biological Agriculture and Horticulture,* 20, (2002): 51-75.

Sharaiha, R.K., Hattar, B., (1993). Intercropping and Poultry Manure Effects on Yields of Corn, Watermelon and Soybean Grown in a Calcareous Soil in the Jordan Valley. *Journal of Agronomy and Crop Science*, 171, (1993): 260-67.

Sileshi, G., Mafongoya, P. L. Effect of rotational fallows on abundance of soil insects and weeds in maize crop in Eastern Zambia. *Appl. Soil Eco.*, 23, (2003): 211- 22.

Snapp, S.S., Silim, S.N. (2002). Farmer Preferences and Legume Intensification for Low Nutrient Environments. *Plant and Soil*, 245, (2002): 181-192.

Sobkowicz, P., Sniady, R. Nitrogen Uptake and Its Efficiency in Triticale (Triticosecale Witt.) Field Beans (Vicia faba var. minor L.) Intercrop. *Plant Soil Environment*, 50, no. 11, (2004): 500-6.

Sogbedji, J. M., Van, H. M. Melkonian, J. R. R., Schindelbeck R. R. Evaluation of the P. N. M. model for simulating drain flow nitrate-N concentration under manure fertilized maize plant and soil 282 (2006): 343- 60.

Spiejer, P. D., Kajumba, C., Ssango, F. East African Highland banana as influenced by nematodes and crop management in Uganda. *International Journal of Pest Management,* 45, no. 1, (1999): 1-49.

Trenbath, B. R. 1976. Plant interactions in mixed crop communities. In R. I. Papendick, P. A. Sanchez and G. B. Triplett (eds). Pp. 129-169. *Multiple cropping Amer. Soc. Agron. Spec.,* Public No. 22.

Trenbath, B. R. 1979. Light use efficiency of crops and potential for improvement through intercropping. Proc. Intl. Workshop on intercropping 10-13 Hyderabad, India Pp. 200-215.

Vandermeer, J. 1989. The Ecology of Intercropping. p 237, Cambridge, UK: Cambridge University Press.

Varghese, L. Indicators of Production Sustainability in Intercropped Vegetable Farming on Montmorillonitic Soils in India. *Journal of Sustainable Agriculture*, 16, no. 4, (2000): 5-17.

Whitmore, A.P., Schröder, J.J. Intercropping Reduces Nitrate Leaching from under Field Crops without Loss of Yield: A Modelling Study. *European Journal of Agronomy*, 27, no. 1, (2007): 81-88.

Woomer, P. L., Longat, M. and Tungami, J. O. Innovative maize-legume intercropping results in above and below ground competitive advantages for under storey legumes. *West Africans Journal of Appl. Ecol.,* 6, (2004): 85-94.

Yildirim, E., Guvenc, I. Intercropping in Cucumber (*Cucumis sativus*) under Greenhouse Conditions. *The Indian Journal of Agricultural Sciences*, 74 (2004): 663-64.

Yildirim, E., Guvenc, I. Intercropping Based on Cauliflower: More Productive, Profitable and Highly Sustainable. *European Journal of Agronomy*, 22 (2005): 11-18.

Yildirim, E., Turan, M. Growth, Yield and Mineral Content of Broccoli Intercropped with Lettuce. *The Journal of Animal and Plant Sciences*, 23, no. 3, (2013): 919-22.

Zhang, F., Li, L. Using Competitive and Facilitative Interactions in Intercropping Systems Enhance Crop Productivity and Nutrient-use Efficiency. *Plant Soil*, 248, (2003): 305-12.

Zimmermann, M.J.O. Breeding for Yield, in Mixtures of Common Beans (*Phaseolus vulgaris* L.) and Maize (*Zea mays* L.). *Euphytica,* 92 (1996): 129-34.

6

Horticultural Wastes: Processing and Applications for Future Sustainability

***Navneet Singh*[1]*, Lakshay Kaushal*[2]*, and Chongtham Tania*[3]**

[1,2]*School of Agriculture, Lovely Professional University, Phagwara, Punjab, India*

[3]*Scientist (Horticulture), ICAR-RC for NEH Region, Mawbri, Meghalaya, India*

Introduction

Nowadays, horticultural fields are expanding due to the increasing demand for various utilizations. Increased production is leading to an increase in exports. The global horticulture market value of about 20.77 billion USD was estimated in 2021 and targeted to attain 40.24 billion USD by 2026 (Global Market Estimates, 2021). From the production of horticultural foods, various factors are involved in the hazardous environmental appearance. Uncontrolled use of pesticides and residual effects of various chemicals initiate the primary threats to nature by creating a negative impact on soil, water, humans, animals, and wildlife species too (Sanchez-Bayo, 2011). Alone agricultural practices are not responsible for the production of horticultural wastes some other aspects are also included such as horticultural food processing industries as well as urban and peri-urban wastes. Food processing is an important sector of agriculture. Being highly perishable horticultural food processing became more necessary to increase the self-life as well as the value addition of these crops. But at the same time, these food processing industries where several processes such as production, processing, and preparation of food take place again result in the generation of huge quantities of waste material causing health hazards due to environmental pollution. The huge waste of food is due to a lack of control and ignorance of higher officials in this agro-economic sector. Another important source where waste management is equally important is Urban and peri-urban areas. Being a sink for horticultural produce (raw or processed) creates huge waste at the same time. The urban population is increasing between 3% and 3.5% per annum and the yearly increase in waste generation is around 5% annually (Agarwal et al., 2005).

The following is a list of the several kinds of solid organic wastes (Table 1) that are produced by agriculture in urban and peri-urban areas:

1. Agricultural wastes
2. Fruits and Vegetables market wastes
3. Food processing industry
4. Kitchen, restaurant, and canteen food wastes
5. Yard waste from community households and shops
6. Old garbage dumps
7. Cow dung
8. Park and verge trimmings, swept leaves, etc.

Table 1: Annual distribution of wastes according to their function in Almeria

Function	Weight		Volume	
	tons	%	m^3	%
Greenhouses	39,215	43	49,798	27
Substrates	1219	1	1598	1
Water storage	576	1	730	0
Disinfection	21,061	23	24,066	13
Shading	10	0	10	0
Transplanting	698	1	40,714	22
Tunnels	2259	2	2429	1
Padding	4900	5	5065	3
Supporting system	6448	7	4891	3
Irrigation	4967	5	20,760	11
Plant protection	4034	4	17,333	9
Pollination	2469	3	26	0
Harvesting	2883	3	19,630	13
Total	90,738	100	187,050	100

Source: Gmada et al., 2019

So, the disposal of waste material has become a challenge to the processors, as different agencies are pressurizing for an environment-friendly treatment of waste material. This waste disposal is our major concern in the case of sustainable waste management.

Nature and Types of Horticultural Wastes from Different Sources

Wastes Produced during Horticultural Crop Production

During horticultural crop production, several types of waste are produced (Table 2), including:

(i) **Agricultural wastes**: This includes crop residues, leaves, stems, and other plant parts that are not utilized.

(ii) **Packaging wastes**: This includes containers, trays, and other materials used to transport and store crops.

Table 2: Types of wastes generated (in %) from various fruit and vegetable processing industries

Commodity	Percent waste (weight basis)	Nature of wastes
Apple	20–30	Pomace
Apricot	8–25	Stones
Banana	-	Peel
Bean, green	5–20	Strings, stem
Beet	7–4	Peel
Cabbage	5–25	Outer leaves
Carrot	18–52	Peel, top portion, pomace
Grapefruit /citrus	55–60	Peel, rag, and seed
Grapes	-	Stem, skin, and seeds
Guava	-	Peel and core and seeds
Lime	60	Peels, seeds, rag and pulp
Mango	40–60	Peel (12-15), pulp (5-10), stones (15-20)
Orange	50	Peel, rag, and seed
Peach	4–11	Stones
Pear	2–46	Peel, pomace
Peas	6–79	Shell
Pineapple	30–60	Peel, core, and coarse solids
Potatoes	10–15	Peel, starch, fiber.
Sweet potato	15	Peel
Tomato	20–30	Skin, core, and seeds

Source: Raj et al., (2016)

(iii) **Pesticide and fertilizer wastes:** These materials can potentially harm the environment if not properly disposed of.

(iv) **Water wastes**: Excessive irrigation and runoff can lead to water waste in horticultural crop production.

(v) **Energy wastes:** The use of energy-intensive equipment and practices in horticultural production can result in energy waste.

It is important for horticultural producers to properly manage these wastes to minimize their impact on the environment and ensure sustainable crop production.

Horticultural Post-Harvest Wastes

Post-harvest waste in horticulture refers to the loss of fruits, vegetables, and other horticultural products that occur after harvesting and before they reach the consumer. This waste can be caused by various factors, including improper handling and storage, inadequate transportation, and lack of market demand. The amount of post-harvest waste in horticulture can be significant, affecting both the food security and economic viability of farming operations. To reduce post-harvest waste, it is important to adopt good agricultural practices, improve supply chain management, and develop strategies for reducing food loss and waste along the entire food chain.(Aulakh et al., 2013)

Wastes in Horticulture Processing Industry

Waste in the horticulture processing industry refers to any material generated during the processing of horticultural products that is not used as a final product. This waste can come from various sources, including trimming and peeling of fruits and vegetables, spoilage due to poor storage and handling, and product defects. The type and amount of waste generated in the horticulture processing industry can vary depending on the type of products being processed and the processing methods used.

To reduce waste in the horticulture processing industry, it is important to implement efficient and sustainable processing methods, optimize production processes, and develop strategies for minimizing food loss and waste throughout the entire supply chain. Additionally, various technologies and solutions can be used to reduce waste in the horticulture processing industry, such as waste management systems, composting, and anaerobic digestion (Raj et al., 2016)

Fruit and Vegetable Market Wastes

Fruit and vegetable market waste is a growing issue that affects not only the environment but also communities and economies worldwide. The waste of unsold or unharvested produce in markets results from various reasons, including overproduction, cosmetic imperfections, and reaching the expiration date. This leads to a significant amount of food waste, which is both a financial loss for farmers and distributors and a waste of resources such as land, water, and energy used in the production process. The environmental impact of fruit and vegetable market wastes is substantial. When food is thrown away, it decomposes in landfills, producing methane, a potent greenhouse gas. Landfills are the largest source of methane emissions in the United States, and food waste is a significant contributor. In addition, the production of food also consumes large amounts of resources and contributes to deforestation, habitat loss, and water scarcity (Yazdanpanah *et al.,* 2014).

Different Strategies and Approaches of Waste Management

Wastes Utilization and Treatment

Waste treatment is a crucial aspect of managing horticultural waste, as it helps to reduce its impact on the environment and prevent the spread of diseases. There are several methods for treating horticultural wastes, including:

Composting

This is the process of breaking down organic waste into a nutrient-rich soil amendment. Horticultural waste, such as leaves, stems, and fruit and vegetable scraps, can be composted along with other organic materials, such as food waste and yard trimmings. The composting process can be done on a small scale in a backyard compost bin or on a larger scale at a commercial composting facility. The transformation of organic material, such as food scraps and yard trash, into a nutrient-rich soil amendment, is referred to as composting. Composting is a natural process. Composting is a common practice for the management of horticulture waste, and it can be done on a modest size in a compost bin in one's backyard or on a larger scale at a composting facility that is used for commercial purposes. The following are the stages that are included in the composting process:

(i) **Collection and preparation**: Horticultural waste, such as fruit and vegetable scraps, leaves, and stems, is gathered and chopped or shredded to increase its surface area. This makes it easier for composting organisms to break the trash down into its constituent parts.

(ii) **Mixing**: The horticulture waste is mixed with other biodegradable materials, such as yard trash, food scraps, and manure, to generate a balanced mix of carbon-rich "brown" material and nitrogen-rich "green" material. This process is referred to as the "mixing" step.

(iii) **Aeration:** The decomposition of organic waste in compost requires oxygen so that the aerobic bacteria can thrive and continue their work. This is accomplished by stirring the compost pile regularly, which allows air to circulate and keeps the composting process active.

(iv) **Moisture management**: Composting also requires the appropriate amount of moisture in order to maintain an active decomposition process within the compost pile. When there is insufficient moisture in the compost, aerobic decomposition can become anaerobic, which in turn slows down the composting process. When there is an excess of moisture, the process can come to a complete halt.

(v) **Maturation**: As the process of composting continues, the compost pile will gradually become hotter over the course of time. As the organisms responsible for composting go through their life cycles, the temperature of the compost pile will first increase and then decrease. After the compost pile has cooled down and the organic matter has decomposed into an amendment for the soil that is rich in nutrients, the compost is ready to be used.

Composting is an efficient method that can be used to handle waste from horticulture and this method is also kind to the environment. Composting turns organic waste into a beneficial soil amendment that can be used to increase soil health and fertility, reduce the need for artificial fertilizers, and preserve water. Composting also helps to minimize the amount of water needed for irrigation. By minimizing the breakdown of organic waste caused by a lack of oxygen, composting can additionally cut down on the amount of garbage that is transferred to landfills and lower the amount of greenhouse gas emissions. (Barker, 2019)

Incineration

This is the process of burning waste at high temperatures to reduce its volume and eliminate pathogens. Incineration is commonly used to treat medical and hazardous waste but can also be used for horticultural waste in certain circumstances. Incineration is a method of waste management in which horticultural waste, along with other types of waste, is burned at high temperatures to reduce its volume and convert it into ash and gases. The process of incineration involves the following steps:

(i) **Collection and transportation**: Horticultural waste, along with other types of waste, is collected and transported to an incineration plant.

(ii) **Waste pre-treatment**: The waste is pre-treated to remove any recyclable materials, such as metals and plastics, and to reduce the moisture content of the waste to improve combustion efficiency.

(iii) **Combustion**: The waste is burned in a furnace at high temperatures, typically above 800°C. The heat from the combustion process converts the waste into ash and gases.

(iv) **Ash treatment**: The ashes generated from the incineration process are typically treated to reduce their toxicity and stabilize them for disposal. The ash may be used as a construction material or sent to a landfill.

(v) **Gas treatment**: The gases generated from the incineration process are typically treated to remove any harmful pollutants, such as dioxins and furans before they are released into the atmosphere.

Incineration is a convenient method for managing horticultural waste, as it reduces the volume of waste and eliminates the need for landfills. However, incineration is not a sustainable solution in the long term, as it generates ash and gases that can have negative environmental impacts. In addition, incineration is energy-intensive and can contribute to air pollution and greenhouse gas emissions (Ji et al., 2016). Overall, while incineration is a useful method for managing horticultural waste in certain situations, it is not a long-term solution for waste management. Alternative methods, such as composting and recycling, should be considered to reduce the amount of waste sent to incineration and to minimize the environmental impact of waste disposal.

Landfilling

Landfilling is the practice of disposing of waste in a designated area, typically a trench or pit, and covering it with soil or other materials. This method of waste management is widely used for the disposal of horticultural waste and other types of solid waste (Agarwal et al., 2005; Bhat et al., 2014). The process of landfilling involves the following steps:

(i) **Collection and transportation**: Horticultural waste, along with other types of waste, is collected and transported to a landfill site.

(ii) **Site preparation**: The landfill site is prepared by excavating a trench or pit and lining it with a layer of impermeable material, such as plastic or clay, to prevent the escape of liquids and gases.

(iii) **Waste disposal**: The waste is placed in the landfill site and covered with soil or other materials. The waste is typically compacted to reduce its volume and increase its density.

(iv) **Monitoring and maintenance**: Landfills must be monitored and maintained to prevent the escape of gases, such as methane, and to control the generation of leachate, a liquid that can contain toxic substances. Landfills are also monitored to ensure that the waste is being properly contained and that the site is safe for the surrounding environment and community.

Landfilling is a common method for managing horticultural waste, but it is not without its drawbacks. Landfills can generate greenhouse gases, such as methane, as the waste decomposes. They also take up valuable land and can pose environmental risks if not properly managed. In addition, landfilling is a finite solution, as the available land for landfill sites is limited. Overall, while landfilling is a convenient method for disposing of horticultural waste and other types of solid waste, it is not a sustainable solution in the long term. Alternative waste management methods, such as composting and recycling, should be considered to reduce the amount of waste sent to landfills and to minimize the environmental impact of waste disposal.

Biogas production

Biogas production (Table 3) from horticultural waste involves the process of anaerobic digestion, where microorganisms break down organic waste in the absence of oxygen to produce biogas and a nutrient-rich fertilizer. Horticultural waste, such as fruit and vegetable scraps, leaves, and stems, can be an excellent feedstock for biogas production due to its high organic content and low levels of contaminants. The process of biogas production from horticultural waste typically involves the following steps:

(i) **Collection and pre-treatment**: Horticultural waste is collected and sorted to remove any contaminants, such as plastic or metal that could interfere with the anaerobic digestion process.

(ii) **Digestion**: The pre-treated horticultural waste is mixed with other organic materials, such as food waste, and fed into a closed anaerobic digestion vessel, where it is broken down by microorganisms. The biogas produced during this process is composed primarily of methane and carbon dioxide.

(iii) **Biogas purification**: The biogas produced in the anaerobic digestion vessel is then purified to remove impurities, such as hydrogen sulfide, that could negatively impact the performance of a biogas engine or generator.

(iv) **Utilization**: The purified biogas can then be used for a variety of purposes, including electricity generation, heating, or as a transportation fuel. The nutrient-rich fertilizer produced during the anaerobic digestion process can also be used to enrich soil and improve plant growth.

Table 3: Rate of biogas production from different horticultural waste

Substrates	**Organic dry matter in %**	**Methane yield in Nm^3/t ODM**
Banana peel	87–94	243,322
Citrus waste	89–97	433,732
Coriander waste	80–86	283,325
Mango peel	89–98	370–523
Oil palm fiber	94	183
EFB	79–84	200–400
Onion peels	88	400
Pine apple waste	93–95	355,357
Pomegranate	87–97	312–430
Sapota peels	96	244
Tomato waste	93–98	211–384
Water hyacinth	81	211–310
Coffee waste (pulp)		380 (biogas yield)

Source: Barz et al., 2019

Biogas production from horticultural waste can provide several benefits, including reducing the amount of waste sent to landfills, reducing greenhouse gas emissions, and providing a source of renewable energy. However, the cost and complexity of setting up an anaerobic digestion facility, as well as the need for a reliable and continuous supply of horticultural waste, can become a challenge to the widespread adoption of this technology. Nevertheless, biogas production from horticultural waste has the potential to be a significant contributor to the production of renewable energy and a more sustainable future.

It is important to note that different methods of waste treatment are better suited to different types and quantities of horticultural waste, and local regulations may dictate which methods are allowed. Additionally, some methods, such as composting and anaerobic digestion, can also help to reduce greenhouse gas emissions and provide a source of renewable energy. By properly treating horticultural waste, it can be managed in an environmentally responsible manner and prevented from becoming a problem for the environment and public health.

Vermicomposting

It is a type of composting that uses worms to break down organic waste into a nutrient-rich soil amendment. This process is well-suited for managing horticultural waste, such as fruit and vegetable scraps, because it can break down the waste quickly and effectively. The process of vermicomposting involves the following steps:

(i) **Collection and preparation**: Horticultural waste is collected and chopped or shredded to increase its surface area, making it easier for the worms to break it down.

(ii) **Bin preparation**: Vermicomposting is typically done in a bin or container, which can be made of wood, plastic, or other materials. The bin is filled with a mixture of composting material, such as shredded newspaper, leaves, and straw, and a bedding material, such as coconut coir or peat moss, to provide a suitable environment for the worms.

(iii) **Adding the worms**: Red wigglers, a type of composting worm, are added to the bin. These worms feed on organic waste and bedding material and produce nutrient-rich worm castings, or compost, as they move through the bin.

(iv) **Feeding**: The horticultural waste is added to the bin on top of the bedding material and is gradually consumed by the worms. It is important to monitor the moisture levels of the composting material and add water if necessary to keep the environment moist but not waterlogged.

(v) **Harvesting**: As the worms break down the organic waste, the compost material will become darker and crumblier, indicating that the compost is ready to be harvested. The compost is harvested by removing the top layer of the composting material and separating the worm castings from the remaining compost material. The worm castings can then be used as a soil amendment, while the remaining compost material can be returned to the bin to continue the composting process.

Vermicomposting is an effective and sustainable way to manage horticultural waste. It provides several benefits, including reducing the amount of waste sent to landfills, producing a valuable soil amendment, and reducing greenhouse gas emissions by preventing the decomposition of organic waste in the absence of oxygen. Vermicomposting can be done on a small scale in a home garden or on a larger scale at a commercial composting facility. (Bezboruah et al., 2003)

Uses of By-Products or Value-Added Products Bottom of Form

Waste Used as Animal Feed

The use of horticultural waste as animal feed (Table 4) is a sustainable and environmentally friendly alternative to traditional methods of waste disposal. Horticultural waste, such as fruit and vegetable trimmings, leaves, and stems, can be used as a source of nutrition for livestock, including poultry, pigs, cattle, and rabbits. The benefits of using horticultural waste as animal feed include:

(i) **Reducing waste**: By using horticultural waste as animal feed, the amount of waste that ends up in landfills or incineration facilities is reduced.

(ii) **Providing a source of nutrition**: Horticultural waste can provide a valuable source of nutrition for livestock, helping to reduce the cost of feed and improve the overall health and productivity of the animals.

(iii) **Improving soil fertility**: By using horticultural waste as animal feed, the nutrients in the waste are recycled back into the soil, helping to improve soil fertility and support sustainable agriculture.

(iv) **Lowering greenhouse gas emissions**: The use of horticultural waste as animal feed can help lower greenhouse gas emissions by reducing the amount of waste sent to landfills, which are significant sources of methane emissions.

Table 4: Different animal feed from horticultural waste

Plant origin	**Fermentation industry**
Bran	Grain
Waste flour	Sugarcane industry (molasses, bagasse)
Wastes from grain-cleaning process	Potato distillers soluble
Wheat	Brewery waste
Corn	Bacteria and fungi biomass
Rye germs	Winemaking industry (grape pomace)
By-products of oil industry	Citrus by-products (molasses, citrus-activated sludge)
By-products of sugar and starch industry	Anthocyanins
By-products of fruit and vegetable industry	Effluents from biogas production
Plant by-products (husk and pods)	Dairy industry

Source: (Lobo and Dorta, 2019)

(v) **Promoting local agriculture**: The use of horticultural waste as animal feed can support local agriculture by reducing the need for imported feed and helping to create a closed-loop system for food production.

Several different types of animal feed can be made from horticultural waste, including:

Silage: Silage is a fermented feed made from horticultural waste, such as corn stalks and leaves that have been chopped and stored in an airtight container. Silage provides a valuable source of nutrition for ruminants, such as cattle and sheep, and can help to reduce the amount of waste sent to landfills or incineration facilities.

Poultry feed: Horticultural waste, such as fruit and vegetable trimmings and leaves, can be used as a source of nutrition for poultry. Poultry feed made from horticultural waste can help to reduce the cost of feed and improve the overall health and productivity of the birds.

Pig feed: Horticultural waste, such as apple pomace and vegetable trimmings, can be used as a source of nutrition for pigs. Pig feed made from horticultural waste can help to reduce the cost of feed and improve the overall health and productivity of the pigs.

Cattle feed: Horticultural waste, such as corn stalks and leaves, can be used as a source of nutrition for cattle. Cattle feed made from horticultural waste can help to reduce the cost of feed and improve the overall health and productivity of the cattle.

Rabbit feed: Horticultural waste, such as carrot tops and lettuce leaves, can be used as a source of nutrition for rabbits. Rabbit feed made from horticultural waste can help to reduce the cost of feed and improve the overall health and productivity of the rabbits.

Overall, the use of horticultural waste as animal feed can provide a valuable source of nutrition for livestock and help to reduce the amount of waste sent to landfills or incineration facilities.

However, it is important to note that horticultural waste can also contain harmful substances, such as pesticides and herbicides, which can pose a risk to animal health. Therefore, it is important to ensure that horticultural waste is thoroughly washed and processed before it is used as animal feed.

Waste as a Source of Functional Foods

Waste from horticultural processing can be a source of functional foods, which are foods that provide health benefits beyond basic nutrition. By-products such as pomace, seed, and stalks are known sources of antioxidants (such as phenolic acids, quercetin, flavonoids, phytoalexins, and pterostilbene, etc.), counteracting the effects of saturated fat and reducing the incidence of coronary heart disease mortality (Joshi et al., 2008). Some examples include:

(i) **Fruit powders:** Dehydrated fruit waste can be ground into a fine powder that can be used as a natural sweetener or added to smoothies and baked goods (Table 5).

(ii) **Vegetable chips:** Vegetable waste, such as peels and stems, can be dehydrated and seasoned to make vegetable chips.

(iii) **Fermented products:** Horticultural waste, such as vegetable scraps and fruit peels, can be fermented to make functional foods such as kimchi, sauerkraut, and pickles.

(iv) **Juice pulp crackers:** Juice pulp, a by-product of juice production, can be combined with other ingredients to make crackers.

(v) **Herbal teas:** Horticultural waste, such as stems and leaves, can be dried and processed into herbal teas.

By converting waste into functional foods, horticultural processors can reduce waste, increase sustainability, and offer consumers healthy and unique food options.(Ambulkar et al., 2004)

Table 5: Possible by-products from wastes of fruit processing units

Fruit/vegetable	Nature of wastes	By-products
Apple	Pomace	Winery, waste, Juice, wine, vinegar, pigment, pectin/fibers, For use as an animal feed. ethanol, butanol, pectinase, citric acid, biogas, single-cell proteins, baker's yeast or industrial yeast,
Apricot, peach, plum	Stone and kernels	Oil, fuel, hawan samagri, essential oil, face cream
Banana	Peels	Animal feed, fiber
Cabbage	Sauerkraut wastes	Animal feed
Capsicum	Peels, seed purpose	Chutney, color, pickle

Grapes	Seed, sediment	Oil, animal feed, color, tartrate, wine fermentation, tartaric acid, and malic acid for use in confectionery
Mango	Peels and pulp	Pectin, cattle feed, alcohol
	Mango kernels (Pulper waste) —Hull and kernels	Fat, starch, Animal feed
	Mango peels—fibers	Wine, vinegar, juice
Oil palm	Fibers	Furniture making
Orange/lime	Peels, seeds, and pulp	Essential oil, pectin, cattle feed, candy, facial cream, cosmetics, pharmaceuticals, perfumes, etc.
Pineapple	Peel, core, trimming, shreds	Juice, wine, syrup, bromelain, cattle feed, biogas, candy
Potatoes	Peel, starch, fiber	Animal feed, custard powder, soup thickener
Tomato	Peels and seeds	Animal feed, seed oil, and colorant

Source: (Raj et al., 2016)

Production of Natural Colorants

Natural colorants can be produced from horticultural waste, providing a sustainable alternative to synthetic colorants. Some examples include:

(a) **Fruit and vegetable juices**: The pigments in fruits and vegetables can be extracted and concentrated to produce natural colorants.

(b) **Plant extracts:** The pigments in plant waste, such as leaves and stems, can be extracted and used as natural colorants.

(c) **Flowers:** Pigments from flowers, such as petals, can be extracted to produce natural colorants for food, cosmetics, and textiles.

(d) **Spices:** Spices, such as turmeric and paprika, can be used as natural colorants due to their high pigment content.

By using horticultural waste as a source of natural colorants, companies can reduce waste and support sustainable production while providing consumers with healthier and more environmentally friendly products.

Importance of Horticultural Wastes Management for Future Sustainability

Horticultural waste management is necessary for future sustainability for several reasons:

(i) **Environmental Protection:** Horticultural waste contains a significant amount of organic matter that can release methane and other greenhouse gases if left to decompose in landfills. Proper management of horticultural waste, through methods such as composting and anaerobic digestion, can reduce the release of these harmful gases and help protect the environment. (Gupta et al., 2005).

(ii) **Conservation of Natural Resources:** By recycling horticultural waste materials, such as plastic pots and trays, resources can be conserved. This not only helps to reduce waste but also conserves the resources needed to produce new materials.

(iii) **Soil Health:** Composting horticultural waste can improve soil health by adding nutrients and improving soil structure. This can support the growth of healthy plants and improve the overall sustainability of horticultural operations (Breś et al., 2016)

(iv) **Wastes Reduction:** Incorporating waste reduction practices into horticultural operations can help to minimize the amount of waste generated in the first place. This not only helps to reduce the environmental impact of waste but also reduces costs associated with waste disposal.

(v) **Renewable Energy:** Anaerobic digestion can be used to convert horticultural waste into biogas, a renewable source of energy. This can help to reduce reliance on non-renewable sources of energy and support a more sustainable future.

In conclusion, horticultural waste management is essential for ensuring a sustainable future by protecting the environment, conserving natural resources, improving soil health, reducing waste, and producing renewable energy. Individuals, businesses, and governments must work together to develop and implement effective horticultural waste management programs.

Advantages of Horticultural Waste Management for Future Sustainability

There are several advantages of horticultural waste management for a sustainable future, including:

(i) **Reduced waste:** Proper horticultural waste management reduces the amount of waste generated, thereby reducing the burden on landfills and reducing environmental pollution.

(ii) **Soil improvement**: Composting horticultural waste can improve soil structure and fertility, reducing the need for synthetic fertilizers and promoting sustainable agriculture.

(iii) **Climate change mitigation:** Composting organic waste sequesters carbon, helping to mitigate the effects of climate change.

(iv) **Conservation of natural resources:** By recycling horticultural waste, we conserve natural resources like water, energy, and raw materials used in the production of synthetic fertilizers and soil enhancers.

(v) **Lower greenhouse gas emissions:** By reducing the amount of waste sent to landfills, horticultural waste management reduces the production of methane and other greenhouse gases.

(vi) **Creation of new products:** Horticultural waste can be processed into new products like compost, mulch, and soil conditioners, creating new job opportunities and economic benefits.

Conclusion

Horticultural waste management is essential for ensuring a sustainable future. The proper disposal and utilization of horticultural waste can bring numerous environmental, economic, and social benefits. Firstly, it helps reduce the amount of waste sent to landfills, which minimizes the negative impact on the environment. Additionally, composting organic waste improves soil fertility and structure, reducing the need for synthetic fertilizers and promoting sustainable agriculture. This, in turn, helps mitigate the effects of climate change by sequestering carbon and reducing greenhouse gas emissions. Furthermore, horticultural waste management conserves natural resources such as water, energy, and raw materials. It also creates new job opportunities and economic benefits by processing waste into useful products like compost, mulch, and soil conditioners. By investing in and implementing effective horticultural waste management practices, we can create a cleaner, healthier, and more sustainable future for all. In conclusion, horticultural waste management is a critical aspect of ensuring a sustainable future. By taking the necessary steps to manage horticultural waste properly, we can promote environmental protection, economic growth, and social well-being. This, in turn, will help us secure a more sustainable future for generations to come.

References

Ambulkar, A.R., Shekdar, A.V. (2004). Prospects of biomethanation technology in Indian context: A pragmatic approach. *Journal of Resources, Conservation and Recycling.*40, (no. 2), (2004): 111-128.

Agarwal, A., Singhmar, A., Kulshrestha, M., Mittal, A.K. (2005). Municipal solid waste recycling and associated markets in Delhi, India. *Journal of Resources, Conservation and Recycling.* 44, no. (1), (2005): 73– 90.

Agrawal, S., Trivedi, C., Sengupta, R. (2006). Air pollution due to burning of residues. In proceeding of the workshop on air pollution problems due to burning of agricultural residues, held at PAU, Ludhiana Organized by the Indian Association for Air Pollution Control in collaboration with the Punjab State Pollution Control Board. New Delhi: Patiala and the Central Pollution Control.

Aulakh J, Regmi A. Post-harvest Food Losses Estimation - Development of Consistent Methodology. In: Paper presented at the Agricultural & Applied Economics Associations 2013. Washington DC, USA: AAEA & CAES Joint Annual Meeting, Purdue University, 4-6 August 2013. 2013

Barker, J. *Formulating an Organic Compost Recipe. Water Quality and Waste Management.* North Carolina Cooperative Extension Service. North Carolina State University.

Barz M, Delivand MK, Dinkler K. Agricultural wastes-a promising source for biogas production in developing countries of the tropical and subtropical regions. *Revista Forestal Mesoamericana Kurú.*, 2019;16, no. (38(2019): 2-12.

Bezbaruah, D., Kotoky, P., Baruah, J. and Sarma, J. N. Geomorphological Explanation of Swamps along the Brahmaputra River Channel, Assam. *Journal Geological Society of India,* 62 no. (2003): 605-613.

Bhat R. A., Nazir, R., Ashraf, S., Ali, M., Bandh, S. A., Kamili, A. N. (2014). Municipal solid waste generation rates and its management at Yusmarg forest ecosystem, a tourist resort in Kashmir. *Waste Management & Research.* 32, (no. 2,) (2014): 165-169.

Breś W, Politycka, B. (2016). Contamination of soils and substrates in horticulture. In: Soil Contamination-Current Consequences and Further Solutions. Rijeka: IntechOpen.;

Central Pollution Control Board (CPCB), 2004. Management of Municipal Solid Waste. Ministry of Environment and Forests, New Delhi, India.

Devrajan A., Joshi V. K., Gupta K., Chander Shakhar C, and Lal B. B. (2004). Evaluation of apple based reconstituted feed in rats after solid state fermentation and ethanol recovery. *Brazilian Arch. Bio. Tech.*, 47, no. (1,) (2004): 93-106.

Global Market Estimates. (2021). Global Horticulture Market Analysis | Size & Forecasts. Global Market Estimates Research & Consultants. Retrieved March15March 15, 2023, from: https://www.globalmarketestimates.com/market-report/horticulture-market-3414

Gupta PK, Sahai S. (2005). Residues Open Burning in Rice-Wheat Cropping System in India: An Agenda for Conservation of Environment and Agricultural Conservation. by IP Abrol, RK Gupta and RK Malik. Centre for Advancement of Sustainable Agriculture, National Agriculture Science Centre: New Delhi, 50-54.

Gupta, P. K. and, Sahai, S. (2005). Residues Open Burning in Rice-Wheat Cropping System in India: An Agenda for Conservation of Environment and Agricultural Conservation. by IP Abrol, R. K., Gupta and R. K. Malik. Centre for Advancement of Sustainable Agriculture, National Agriculture Science Centre: New Delhi, 50-54.

Ji, L., Lu, S., Yang, J., Du, C., Chen, Z, Buekens A. and, Yan, J. (2016). Municipal solid waste incineration in China and the issue of acidification: A review. Waste Management & Research. 34, no. (4,) (2016): 280-297. 10.1177/0734242X16633776

Joshi, V. K. and, Devrajan, A. (2008). Ethanol recovery from solid state fermented apple pomace and evaluation of physico-chemical characteristics of residue. *Natural Product Radiance.* 7, no. (2)(2008): 127-132.

Lobo MG, Dorta E. Utilization and management of horticultural waste. In: Postharvest Technology of Perishable Horticultural Commodities. Woodhead Publishing; Elsevier 2019. pp. 639-666

Raj, D., Senapati, A.K.,and Patel, N.L. (2016). Waste Management in Horticulture Processing Industry, Navsari Agricultural University, Navsari (Gujarat). *NIPA*,. 391-401.

Raj, D., Sharma R. and, Joshi V. K. (2011). Quality control for value addition in food processing.

Reduce Wasted Food by Feeding Animals. (2021). Retrieved March20, 2023, US EPA. Available from: https://www.epa.gov/sustainable-management-food/reduce-wasted-food-feeding-animals.

Sánchez-Bayo F. Impacts of agricultural pesticides on terrestrial ecosystems. Ecological Impacts of Toxic Chemicals. 2011. pp. 63-87.

Sayadi-Gmada, S., Rodríguez-Pleguezuelo, C.R., Rojas-Serrano, F., Parra-López, C., Parra-Gómez, S., García-García, M.d.C., García-Collado, R., Lorbach-Kelle, M.B., Manrique-Gordillo, Tet al. (2019). Inorganic Waste Management in Greenhouse Agriculture in Almeria (SE Spain): Towards a Circular System in Intensive Horticultural Production. *Sustainability*. 11, (2019): 3782. Available from: https://doi.org/10.3390/su11143782/.

Yazdanpanah, M., Hayati, D., Thompson, M., Zamani, G.H. and Monfare, N., et al., Policy and plural responsiveness: Taking constructive account of the ways in which Iranian farmers think about and behave in relation to water., *J. Hydrol.* 514, ((2014): 347-357. Available from: https://doi.org/10.1016/j.jhydrol.2014.04.015)

7

Organic Farming in Regenerative Agriculture

Prabhjot Kaur and Y. S. Bagal

[1]*School of Agricultural Sciences and Technology, RIMT University, Mandi Gobindgarh, Punjab, India*

[2]*School of Agriculture, Lovely Professional University, Phagwara, Punjab India*

Introduction

The goals and ideas of organic farming and regenerative agriculture are quite similar. Organic farming is a method of cultivating crops that prioritizes ecological sustainability by eschewing the use of artificial chemicals and genetically modified organisms in favor of more traditional practices. Producing residue-free food while also improving soil health, biodiversity, and ecological balance is the primary goal of this method. To restore and improve the health of the soil, ecosystems, and communities involved in agricultural systems, regenerative agriculture goes above and beyond organic farming. The ultimate goal is the development of resilient and sustainable agricultural systems that can deal with the effects of climate change.

Regenerative agriculture and organic farming share a common set of values, including an emphasis on the importance of good soils to successful and sustainable farming. Crop rotation, cover crops, and composting are some of the methods advocated by these people to improve soil fertility, structure, and biological activity. Preservation and promotion of a wide variety of plant and animal life on and around farms is an aim of both organic and regenerative farming practices. Pest and disease management, pollination, and ecosystem stability are all bolstered by biodiversity. Both methods encourage using less or no man-made chemicals, such as those found in pesticides, herbicides, and fertilizers. Organic matter, compost, and biological pest management are just some of the natural inputs that these farmers rely on. Water, energy, and material resource conservation are fundamental to both organic farming and regenerative agriculture. To lessen agriculture's impact on the environment, they advocate for methods including water management, efficient irrigation systems, and the use of renewable energy. Strong relationships and collaborations between farmers, customers, and local communities are prioritized by these methods. All along the agricultural supply chain, they are committed to fair trade, social equity, and transparency. Farmers may grow healthy, resilient, and sustainable farming systems by applying the holistic concepts of regenerative agriculture to conventional organic farming practices. In addition to strengthening the bond between agriculture and its natural surroundings, these methods also improve soil quality, lessen negative effects on the environment, increase biodiversity, and result in more nutritionally dense food.

Increasing pollution levels in every aspect of contemporary life pose the greatest threat to our planet's long-term viability. Humanity's ever-increasing needs and wasteful habits are wreaking havoc on the planet. The agriculture sector is similarly vulnerable to environmental degradation in the United States. The use of synthetic fertilizers, insecticides, and other agricultural chemicals to increase crop yields has a negative impact on the environment. Overall, these methods lead to lower soil fertility because they disrupt the natural equilibrium of soil nutrients. To address the current issue, organic farmers use locally sourced organic resources from animals and plants that are safe for the environment and rich in the nutrients that agricultural plants need to thrive. Soil health is improved and microbial activity is stimulated. In terms of environmental sustainability, organic farming is an effective and promising agricultural technique since it ensures stable yields, enhances soil health, causes no harm to the environment, produces organic food, and reduces the need for synthetic fertilizers. Organic farming is, without question, the most environmentally benign kind of agriculture now available for preserving the natural harmony of our agricultural and ecological systems. Organic farming relies on several different foundations to function. Key components include organic manure, crop rotation, vermin-composting, nitrogen-fixing microorganisms, organic residue, crop residue, bio-fertilizers, biopesticides, kitchen waste, sludge, and biogas. These are quite helpful in preserving the quality and vitality of the soil. They are used in a way that is good for the environment and contributes to the growth of environmentally responsible farming. Beyond these, the organic farming method is guided by the values of health, ecology, fairness, and care. Sustainable development relies on these tenets, which form the basis of organic agriculture. Organic farmers are more concerned with ecological sustainability. Its eco-friendly methods contribute to cleaner air, water, and soil. Hence it functions as an organic resource for green growth and preservation. Farmers keep nitrogen and other key crop nutrients abundant through practices including crop rotation and the use of bio-fertilizers. Chemicals are not allowed out in the field, and only environmentally friendly options are considered. Veggies are also field-grown utilizing the organic farming method, guaranteeing the supply of organic vegetables in the food markets. The growth of organic farming has been the subject of many global laws and policies, yet there are still places where they have not been properly implemented on a local or national scale. Over time, more and more individuals from all over the world will become interested in organic farming. Effective though it may be, the organic industry nonetheless faces persistent problems. Because of this, the organic agricultural sector's policies must be formulated and implemented with the help of society's stakeholders. It is determined that the key to the success of any policy is good follow-up, implementation on the ground, impact analysis, and public awareness. When it comes to environmental sustainability, better policies lead to better outcomes.

Both minimizing the effects of climate change and feeding the world's expanding population rely heavily on agricultural practices that are more environmentally friendly. The World Bank estimates that now, 30% of all global greenhouse gas emissions come from agriculture. Transportation, planting, harvesting, processing, and the raising of animals all contribute to greenhouse gas emissions from the agriculture infrastructure. Add to that the contamination of water supplies by synthetic fertilizers, pesticides, and herbicides, and the problem becomes much more severe. A farm manager examines crop information on a tablet computer. It's obvious that there's a pressing need to both lessen agriculture's ecological footprint and boost output so that a growing global population can be fed. Those who are up for the challenge of learning more about the significance of sustainable agriculture can think about furthering their education in an area like sustainability. Agricultural land is being lost to rising metropolitan areas and climate change, while the global population is projected to rise from 7.7 billion now to 9 billion by 2050. According to the World Bank, food output would need to rise by 70% by 2050. That's why it's crucial to use sustainable farming methods. Sustainable agriculture is defined by the United States Department of Agriculture as "activities that conserve and restore natural resources, increase agricultural productivity, and benefit the environment." The desired outcomes are to achieve the following demands:

- Satisfy human food and fiber needs
- Enhance environmental quality and the natural resource base upon which the agriculture economy depends
- Make the most efficient use of nonrenewable resources and on-farm resources and integrate, where appropriate, natural biological cycles and controls
- Sustain the economic viability of farm operations
- Enhance the quality of life for farmers and society as a whole

Sustainable Agriculture Benefits

Some elemental sustainable agriculture methods can help reduce the environmental impact of farming. Benefits include the following:

(a) **Maintaining soil carbon levels**: The no-till farming practice keeps carbon where it belongs, in the soil, rather than releasing it into the atmosphere. In no-till farming, the farmer does not till the soil after harvest but instead leaves the crop residues in the soil. This can even extend to the planting process, when the farmer sprinkles seed over the earth instead of burying it. It also lessens the number of times machines have to go across a field.

(b) **Using fewer chemical fertilizers, insecticides, and herbicides:** Crop rotation is the process of switching crops that are planted in a given field from year to year to maintain a balanced and fertile soil. Weed and insecticides, as well as fertilizers, may be used less often as a result of this.

(c) **Repairing and keeping up pastures**: By moving cattle from one field to another, we may increase the soil's fertility with the manure they produce and protect the health of many pastures at once. Because of this, the soil can hold more carbon.

(d) **Cutting down on gas usage**: Perennial crops cut down on the number of times farmers need to use heavy machinery in the field for planting and chemical applications.

Trends in Shaping the Future of Sustainable Agriculture

If not the earliest adopters, farmers have always been reliable in introducing technology to the farm. These days, farmers use satellites to guide their tractors, sensors to learn more about their soil, and complex computer programs to oversee their operations. Improved field and weather data, less need for fertilizers and pesticides, more efficiency, and less fuel consumption are just a few ways in which agricultural technology has helped boost crop yields. The following are examples of technologies which may help in achieving agricultural sustainability in the future:

Artificial Intelligence (AI)

By analyzing this information, AI systems advise farmers on the optimal times to sow and harvest their crops, as well as the ideal times to feed their livestock and maximize their profit margins when selling their goods. Farmers may use data to improve the timeliness and precision of fertilizer applications. Farmers may save money on fertilizers and energy if they have a better idea of what they'll need ahead of time.

Biotechnology

When it comes to crop improvement, biotechnology is as ancient as farming itself. Plants and cattle have both benefited greatly from crossbreeding through the ages, with the former becoming more drought-resistant and the latter becoming more robust. The amount of time it takes to crossbreed plants, adding or removing features to adapt to circumstances, has decreased thanks to contemporary, sophisticated procedures created in the lab. Proponents of these methods argue that crops can be made more productive, resistant to insects and disease, and better able to respond to local conditions, such as more severe droughts or higher amounts of moisture, even though opponents fear that modified crops could introduce unforeseen and potentially devastating effects. CRISPR, a revolutionary gene editing tool, may be used to modify certain features, including those that contribute to productivity and disease resistance.

Drones

Growers are finding more and more ways to put drones to work for them in crop management. In the beginning, drones were used to apply pesticides to fields. Taking aerial photographs for crop evaluation and collecting data from sensors that can be mined to ascertain crop and weed health are further potential uses. Sometimes unmanned aircraft will scatter tree seeds for reforestation efforts.

Blockchain Technology

Cryptocurrency trading is where most people first hear of blockchain technology, but the agricultural sector is also a potential user. Using blockchain technology, financial dealings may be tracked and executed reliably and safely.

Blockchain technology may be used to trace agricultural goods from the farm to the customer. It is already being used by agricultural supply chains to trace the origins of salmonella and other food poisoning outbreaks.

Challenges in Building Sustainable Agriculture Strategies

It is great if a single farmer adopts sustainable farming methods, but it's even better if their neighbors do the same, and it's best if sustainable farming is adopted on a regional or national scale. Yet there are challenges to implementing sustainable agriculture on a large scale. When farmers face challenging growing circumstances, they often respond by increasing their use of chemical fertilizers and pesticides. Some farmers, even in fertile regions, may believe that additional fertilizer is the only way to boost yields. The fact that sustainable agriculture challenges habits that have been established over decades, if not centuries, may be its greatest difficulty. Three aspects of modern agriculture have garnered international attention: (a) rising food needs to support a populous world; (b) the increasing dangers posed by agriculture-driven changes to the global environment; and (c) the increasing dangers posed by the contamination of the food supply and associated health risks. Global food shortages were considerably alleviated as a result of the green revolution; however, this was accomplished at a steep price to the ecosystem. Agriculture has multiple impacts on the environment and human health due to the usage of fertilizers and pesticides. Most soil fertility is depleted when crops are grown continuously and intensively without the corresponding supply of fertilizers. Because of this, the amount of nutrients needed by arable land to produce the same amount of crop is growing. There has been a significant increase in global food production thanks to the widespread use of chemical fertilizers and pesticides. Despite this, these practices have led to the degradation of agricultural lands, a decline in the quality and flavor of agricultural produce, and an increase in environmental risks. Agriculture that does not deplete natural resources or cause environmental damage is not considered sustainable. The phrase "sustainable agriculture" refers to the methods of agricultural and animal production management that provide continuous ecological productivity without depleting natural resources or compromising human well-being. Conservation tillage and residue management are two examples of sustainable agricultural management practices. Intercropping and agroforestry are two other examples. Windbreaks and terracing are two examples of practices that help prevent soil erosion. Crop rotation and efficient irrigation and drainage are two examples of practices that help strengthen biogeochemical cycles.

The concept of sustainable development of agriculture stems from the Rio Declaration on Environment and Development from 1992, Chapter 14 of Agenda 21 from the United Nations General Assembly, and the Rome Declaration on World Food Security from 1996. According to the Commission on Sustainable Development's resolution, "the purpose of sustainable development of agriculture is to grow food and strengthen food security in an environmentally sound way so as to contribute to sustainable natural resource management. " People's preferences have shifted dramatically toward safe, high-quality food as they become more aware of the health dangers linked with agrochemicals and sustainable agriculture. Organic farming and other nontraditional methods of crop production have so gained popularity.

The Concept of Organic Farming

Organic farming is a sustainable method that doesn't harm the environment and yields nutritious food. Instead of depending on synthetic chemical fertilizers and pesticides, organic farming focuses on creating healthy, rich soil and cultivating a diversity of crops. The biodiversity of predatory insects and other animals keeps crop pests in check, while soil microbes and earthworms keep the farm healthy. Organic farmers reduce risks to human health and the environment by not using synthetic chemicals. Two of organic farming's numerous stated aims are (i) producing food that isn't harmful to humans in the long run through the elimination of toxic chemicals and (ii) closing the nutrient loop to restore depleted soil nutrient levels. Demand for organic food, which lacks harmful chemicals, has been on the rise in both developed and developing nations, increasing by an average of 20% to 25% per year. To accomplish sustainable agriculture, a closed nutrient cycle is necessary, and organic farming relies on "biologically driven nutrients."

Need for Organic Farming

The environmental impact of industrialized farming is substantial. The widespread application of pesticides has resulted in a toxic accumulation of chemicals in every aspect of our ecosystem, including plants, insects, animals, and humans. There has been an uptick in the prevalence of several diseases, including cancer and lowered resistance to infection, attributable in large part to pesticides and other chemical residues in our food and a general decline in food quality. The practice of organic farming addresses this issue. All these issues may be solved by switching to organic farming. Organic farming not only has direct, good consequences on environmental sustainability and food quality, but it also considerably aids in a farmer being independent of outside agro inputs and decreasing production costs.

Principles of Organic Farming

Organic farming is successful because of its adherence to three fundamental principles. To begin, there is the idea of mutual dependence. To the organic farmer, the farm is like an ecosystem, and any changes to one component might have unintended consequences elsewhere. Organic farmers, for instance, will deal with the issue of excessive nitrogen in the soil, which can promote weed development, in a certain way. To remedy this, farmers will introduce a crop that requires more nitrogen into the soil. Organic farmers value diversity as their second guiding principle. Crop rotation is a key component of organic farming that helps keep the environment as unaltered as possible. Organic farmers adhere to a third guideline called "recycling." Organic farmers use byproducts from plants and animals to become more self-sufficient in terms of nutrient production.

Objective of Organic Farming

(i) To produce healthy, nutritious, and quality food.

(ii) To encourage and enhance biological cycles involving microorganisms, soil flora, fauna, plants, and animals.

(iii) To maintain and enhance the long-term fertility of soils

(iv) To help in soil and water conservation

(v) To minimize all forms of pollution that may result from agricultural practices

(vi) To use farm resources as far as possible

(vii) To preserve and enhance traditional and indigenous knowledge in farming, seeds, and varieties

(viii) To consider the social and ecological impact of the farming system

Favorable Effects of Organic Farming on Environment and Human Health

When compared to conventional farming, organic agriculture has significant positive environmental impacts. Conventional agricultural practices are a major contributor to pollution and climate change, while organic farming practices significantly reduce these impacts. In reality, the organic agricultural method is around 7% more efficient in terms of energy use. Organic farming also benefits the environment since it uses less fertilizer and eliminates the use of synthetic fertilizers, which are bad for the soil, water, animals, and humans. Additionally, since soluble fertilizers are not used in organic farming, nitrate levels in organic fields are far lower than those of conventional farms. Algal blooms in water that suffocate other aquatic species can be caused by excessive amounts of nitrogen, which can also disrupt the soil ecology. Biodiversity, the rich variety of animal and plant species that is vital to the survival of all life on Earth, is also fostered in organic farming areas. The carbon dioxide emissions from organic farming are much lower than those from conventional farming. When it comes to the greenhouse effect, carbon dioxide is king. Salicylic acid levels in organic vegetables were 6 times higher than in conventionally grown vegetables. Salicylic acid's anti-clotting and anti-cancer properties make it a valuable tool in the fight against atherosclerosis and bowel cancer.

Requirements of Organic Farming

In an organic farming system, certain minimum requirements are to be met so that the farms are certified as organic. The following processes are required for any conventional farming system to be considered organic.

Conversion

The term "conversion" is used to describe the process through which a farmer moves from conventional to organic farming. The conversion period refers to the time frame between the introduction of organic management and official certification. Actions should be taken to preserve biodiversity during the conversion process, like swamps, mucky fields, dense woodlands, etc.

Mixed Farming

Animal husbandry, poultry, fisheries, etc., should be practiced in addition to agricultural farming. Shifting cultivation is not allowed.

Cropping Pattern

If planting annual crops, it's important to rotate them regularly. When planting perennials, it's important to alternate plantings every few years. Green manure and fodder crops should both be rotated regularly.

Planting

Cultivated species and cultivars ought to be able to thrive in the given soil and climate, as well as be immune to common pests and illnesses. It's important to use only organic seeds and other planting supplies. Seeds and other planting materials that have not been treated with chemicals can be used once if available. Transgenic plants, tissue culture plants, and pollen culture plants are prohibited, as are genetically altered seeds.

Manuring

Green manure crops, leguminous crops, etc., should be raised to maintain or increase soil fertility. As much of the plant material left behind after harvest as possible should be worked back into the ground. Manures must be applied from biodegradable materials of microbial, plant, or animal origin (such as manure, worm compost, compost, sheep pen waste, etc.). It is forbidden to use artificial or chemical fertilizers. In cases of dire necessity, small doses of mineral-based products such as rock phosphate, gypsum, lime, etc. may be used.

Organic Farming in India

As a long-term and eco-friendly production method, organic farming is gaining popularity across the country. In two rain-fed regions, tribal areas, the north-east, and the hilly parts of India, where agricultural techniques and production systems are essentially organic owing to nil or low fertilizer and pesticide use, this trend is quickly gaining traction among Indian farmers and businesspeople. The system's focus on boosting natural processes and cycles in tandem with the natural environment bodes well for its ability to keep soil fertile and to keep pests and illnesses at bay. Organic manures have been employed in Indian agriculture at least as far back as the Vedic period, and organic farming was pioneered in the UK by British agronomist Sir Albert Howard in the 1990s. Since then, it has been common practice among farmers in certain regions of India due to a lack of alternatives. Organic commercial farming as it is now practiced is still in its infancy. The area under organic management in India is quite small, accounting for just approximately 0.05% of the country's total agricultural land, according to a survey conducted in February 2005 by the International Federation of Organic Agriculture Movement (IFOAM) and Stiftung Oekologie and Landbau (SOEL). According to the results of this survey, there are about 5147 organic farms in India.

A recent estimate places the value of India's organic agricultural sector at $20 million, with nearly all of that money coming from exports. About 6792 tons (or 72 million rupees) of organic goods are exported from India annually, as reported by the Agriculture and Processed Food Products Export Development Authority (APEDA), a central organization supporting organic agriculture in India.

Areas of Concern

The production of organic food does not "cost the environment." Costing "the planet" to customers is unacceptable. While it is true that consumers are aware that organic food is superior to chemically-grown food in all respects, including taste, flavor, and for their own health, in addition to that of the earth, it is only fair that an organic farmer should receive at least a marginally higher price for his produce and his efforts. The absence of a convenient market and frequently unremunerative rates for their goods is another concern that organic farmers confront, at least in India. In many instances, organic food merchants and other middlemen fail to pay the producer in a timely manner. On the other side, those looking to purchase organic foods are unable to do so at affordable costs. Supplies are typically sporadic at best, and in other situations, shoppers may not even be confident that the food they are purchasing is indeed organic. Some Indian farms have never used chemical fertilizers or pesticides, while others have switched back to organic methods for ethical or financial reasons. Many thousands of farmers are growing crops on tens of thousands of acres of land, but they are not recognized as organic. Either their goods are sold at the same price as conventionally cultivated goods in the general market, or they are sold as organic goods through a small number of specialty stores and markets. Due to the high expenses and considerable paperwork required by certifiers, these farmers will never choose to pursue certification.

Conclusion

Soil, minerals, water, plants, microorganisms, insect-eating animals, and humans all work together in harmony on an organic farm. It makes for good landscape design and manages to strike a balance between agriculture and the preservation of the natural world. Ecological carrying capacities are respected while natural resource processes are bolstered through organic management's use of local human resources and expertise. Ecosystem resilience is improved, food security is bolstered, and incomes are raised as a result of less reliance on non-farm inputs and more balanced fluxes of nutrients and energy. Maintaining soil fertility, increasing crop yields, and bettering farmers' economic and social standings are just a few of the many ways in which organic farming contributes to sustainable agriculture and rural development.

References

Allen, P. 1996. Social Imperatives in Organic Agriculture. In: 11th IFOAM Scientific Conference. Copenhagen, Denmark. 11-15 August. Tholey-Theley, Germany: International Federation of Organic Agriculture Movements.

Allen, P. and M. Kovach. The capitalist composition of organic: The potential of markets in fulfilling the promise or organic agriculture. *Agriculture and Human Values*, 17, no. 3, (2000): 221-232.

Altieri, M. A. 2000. The ecological impacts of transgenic crops on agro ecosystem health. *Ecosystem Health*, 6, no. 1, (2000): 13-23.

Biao, X., Xiaorong, W., Zhuhong, D., and Yaping, Y. Critical impact assessment of organic agriculture. *Journal of Agricultural and Environmental Ethics*, 16, no. 3, (2003): 297-311.

Brown, L. 2004. Outgrowing the earth, the food security challenge in an age of falling water tables and rising temperatures. New York: W.W. Norton.

FAO. 1998. Evaluating the potential contribution of organic agriculture to sustainability goals. Food and Agriculture Organization of the United Nations. Rome. 24 p.

FAO. 1999. Guidelines for the production, processing, labelling and marketing of organically produced foods. Joint FAO/WHO Food Standards Program Codex Alimentarius Commission. Rome. CAC/GL 32-1999.

FAO. 2000. Food Safety and Quality as Affected by Organic Farming. Agenda Item 10.1. In: Twenty-second FAO Regional Conference for Europe. Porto, Portugal. 24-28 July. Food and Agriculture Organization of the United Nations. Goodman, D. 2000.

Haas, G., F. Wetterich, and U. Kopke. Comparing intensive, extensified and organic grassland farming in southern Germany by process life cycle assessment. *Agriculture, Ecosystems and Environment*, 83, (2001): 43-53.

Hamm, U. 1996. Organic agriculture in a market economy. In: 11th IFOAM Scientific Conference. Copenhagen, Denmark. 11-15 August. Tholey-Theley, Germany: International Federation of Organic Agriculture Movements.

Hamm, U. 1997. Organic trade: The potential for growth. In: The Future for Organic Trade: The 5th IFOAM International Conference on Trade in Organic Products. Oxford, England. 24-27 September. Tholey-Theley, Germany: International Federation of Organic Agriculture Movements,18-21.

Hamm, U. 1998. Organic perspectives in international markets. In: Organic agriculture the credible solution for the XXIst Century: Proceedings of the 12th international IFOAM scientific conference. Mar del Plata, Argentina. Tholey-Theley, Germany: International Federation of Organic Agriculture Movements. 266.

Hansen, B., H. Fjelsted, and E. S. Kristensen. Approaches to assess the environmental impact of organic farming with particular regard to Denmark. *Agriculture, Ecosystems and Environment,* 83, (2001): 11-26.

Hayo, M. G. and J. Petit. 2002. Evaluation of the environmental impact of agriculture at the farm level: A comparison and analysis of 12 indicator-based methods. Agriculture, Ecosystems and Environment, 1922: 1-15.

Holt-Gimenez, E. Hurricane Mitch reveals benefits of sustainable farming techniques. Global Pesticide Campaigner, 10, no. 3, (2000): 1-13.

Lohr, L. 1998. Implications of organic certification for market structure and trade. *American Journal of Agricultural Economics*, 80, no. 5, (1998): 1125-29.

Lotter, Donald W. 'Organic Agriculture'. *Journal of Sustainable Agriculture*, 21, no. 4, (2003): 59-128.

Organic and conventional agriculture: Materializing discourse and agroecological managerialism. Agriculture and Human Values, 17 no. 3, (2012): 215-219.

Pesticide Action Network. Liebhardt, W. C. Get the facts straight: Organic agriculture yields are good. Organic Farming Research Foundation Information Bulletin, Summer 10, (2001): 1-7.

Pretty, J. and R. Hine. 2001. Reducing food poverty with sustainable agriculture: A summary of new evidence. SAFE Research Project, University of Essex. Wivenhoe Park, UK. 136 p.

Rigby, D. and D. Caceres. Organic farming and the sustainability of agricultural systems. *Agricultural Systems*, 68, (2001): 21-40.

Tellarini, V. and F. Caporali. An input/output methodology to evaluate farms as sustainable agroecosystems: An application of indicators to farms in central Italy. *Agriculture, Ecosystems and Environment,* 77, no. 1-2, (2000): 111-23.

8

Prospects and Applications of ITK for Environment-Friendly Pest Management

I Yimjenjang Longkumer[1] *and Devina Seram*[1]

[1]*Department of Entomology, School of Agriculture, Lovely Professional University, Phagwara, Punjab, India*

Introduction

Farmers have long used biorationals such as Indigenous Technical Knowledge (ITK) in both crop production and crop protection. This understanding is the result of many generations of interaction within natural and physical environments (Rajasekaran et al., 1991 and Kolawole, 2001). ITK is the unique and local traditional knowledge that is specific to a particular culture or community and is passed down from one generation to the next and every individual has inherited this information from their ancestors. Except for a handful, ITKs are passed down from generation to generation verbally, with no validated written documents. Many definitions of ITK systems have been proposed, but all of them are incomplete because the notion is still relatively young and changing. The ITKs are environmentally friendly and compatible with pest management approaches (Deka et al., 2006). ITK was defined by Wang (1988) as "the whole body of knowledge and practices based on people's cumulative experiences in resolving situations and issues in a variety of spheres of life and such information and practices are unique to a specific culture." According to Haverkort, 1995, ITK is the actual knowledge of a given population that reflects the experiences based on tradition and includes more recent experiences with modern technologies. The phrase "indigenous technological knowledge" is frequently disguised with the idea that it is connected to upcoming events and the innovations developed by farmers to address certain issues. Among the associated terms are (i) Indigenous Knowledge (IK), which refers to the participant's understanding of their social and temporal context. The term "indigenous knowledge" as used here refers to any identifiable community's knowledge, not just which of indigenous peoples; (ii) The Indigenous Knowledge System (IKS) describes a conceptual framework for theories and perceptions of nature and culture. The physical, ecological, social, economic, and ideational surroundings are thus included, together with their definitions, categorizations, and concepts. The cognitive and empirical levels both have a role in IKS dynamics. IKS is evident in institutions, artifacts, and technology on an empirical level. ITK is expressed through stories, songs, folklore, proverbs, dances, myths, cultural values, beliefs, rituals, community regulations, local language and taxonomy, agricultural techniques, tools, materials, plant species, and animal breeds. ITK is stored in people's memories and activities, and is expressed in the form of stories; ITK is shared and communicated orally, by specific examples, and through culture. An African proverb says "When an old knowledgeable person dies, a whole library dies" indicating the importance of ITKs. It serves as the foundation for local decisions in rural communities on agriculture, health care, food preparation, education, and a variety of other activities. Indigenous knowledge serves as a society's information hub and helps in decision-making and communication. Indigenous information systems are dynamic and constantly affected by both internal innovation and experimentation and interactions with outside systems. The phrase "indigenous technological knowledge" is frequently disguised with the notion that it is connected to upcoming events and the innovations produced by farmers to address certain issues.

Sources of ITK

(i) Farmers, community members, especially elders are the best source of ITK

(ii) Community records some indigenous forms or records that include writings, paintings, and carvings

(iii) People working with communities and extension functionaries, and village panchayat sarpanch.

(iv) Secondary sources include published and unpublished documents, databases, videos, photos, museums, and exhibits.

Objectives of ITK

(i) To evoke a sound ambient

(ii) No pesticidal zone

(iii) Healthy environment

(iv) Less labor cost

(v) Use of local resources

Characteristics of ITK

(i) ITK is dynamic rather than static

(ii) ITK is transformed by endogenous creativity and exogenous knowledge.

(iii) ITK thinks intuitively.

(iv) The primary focus of ITK is on quality.

(v) ITK investigation requires a comprehensive strategy

(vi) If appropriately utilized, ITK can offer insightful information about resources and procedures

(vii) Oral tradition is used to record and transmit ITK.

(viii) ITK is taught by direct experience and observation

(ix) ITK creates a database of information for various

(x) ITK honors regional customs

Source: Prakash et al. (2012)

Classes of ITK in Agriculture

(i) Climatology

(ii) Local soil and taxonomy

(iii) Soil fertility

(iv) Primitive cultivar

(v) Intercropping

(vi) Agronomic practices

(vii) Irrigation and water management

(viii) Plant protection

(ix) Post-harvest technology and methods.

Source: Prakash *et al.*, (2012)

Constraints of Using Indigenous Technical Knowledge

(i) Education and exposure especially of the young generation to modern training have biased people's attitudes toward using ITK

(ii) Some farmers feel it is time-consuming, exhausting, and sometimes dangerous to hunt for herbs in the forests and bushes

(iii) ITK was reported as not effective in large-scale production

(iv) Some religious beliefs do not encourage traditional beliefs and technologies regarding them as demonic and superstitious

(v) Regressive government laws prohibiting some practices like growing cannabis, unauthorized tree cutting, and rudimentary castration of livestock

(vi) Depletion of most trees and herbs that are sources of local medicine

(vii) Lack of standardization and documentation of indigenous technologies and practices

(viii) Variation in prescription such as quantity to administer at a time and for how long

(ix) The educated people despise some indigenous methods referring to them as ineffective and dirty

(x) Selfishness that inhibits people from passing on knowledge to others

The fundamental traits of the ITKs support resource conservation and efficient resource use through environmental friendliness, reduced capital expenditure, economic efficiency, and successful by-product and waste recycling and use. The majority of ITKs are location-specific and made from materials that may be found nearby. ITKs have been proven to be resource-conserving and socially desirable, as well as economically accessible, sustainable, and posing little risk to users. ITK thus provides a foundation for community-based problem-solving techniques. Furthermore, ITK utilization ensures that the end user of particular development projects is involved in creating technologies that meet their demands. There are sources of ITK hidden in our villages, communities, and countryside. The main sources are folklore, song, poetry, and ancient records. NGO extension agencies published materials in different languages (Satapathy et al., 2002). According to the World Bank, the salient features of indigenous knowledge are as follows:

- Local, in that it is a collection of experiences created by individuals who live in those places and it is founded in a specific community and located within larger cultural traditions. It could be difficult to distinguish between the technical and non-technical, the reasonable and irrational. As a result, there is a chance that spreading indigenous knowledge could displace it.

- Tacit knowledge and, therefore, cannot be easily codified.
- Transmitted orally, or by emulating and demonstrating. It might lose some of its properties if it is codified.
- Experiential knowledge, as opposed to academic knowledge Indigenous wisdom, is constantly reinforced by experience, trial and error, and the demanding survival laboratory of local communities.
- Acquired by repetition, which distinguishes tradition even when new information is introduced. The retention and reinforcement of indigenous information are aided by repetition.
- Constantly evolving, produced and replicated, discovered and lost, while external onlookers frequently mistake it for being rather static.

Traditional Knowledge System (TKS)	**Scientific System (SS)**
The majority of knowledge is communicated orally.	The written word is the main medium for the transmission of knowledge.
TKS is built on observation and actual experience.	Knowledge gained in a situation that is unrelated to its application context
Knowledge is holistic, practical, intuitive, and qualitative.	Fundamentally, knowledge is a quantitative, reductionist, analytical, and theoretical concept.
Users of resources produce knowledge across a diachronic (long-term) time span.	Specialist researchers produce the majority of new knowledge in a synchronic (short-term) time period.
Spiritual beliefs have an impact on the character and status of some knowledge and are widely believed.	Peer review and specific experts have an impact on the nature and status of a given body of knowledge.
Many times, explanations for observed occurrences are based on subjective spiritual beliefs.	The explanations for observed occurrences are primarily rational and impartial.
Knowledge is used to make suitable decisions under variable conditions	Utilizing knowledge, one can test hypotheses and establish underlying rules and constants.

Source: Prakash et al., (2012).

Some aspects of indigenous knowledge that are important for conservation and sustainable development include the following:

1. Native knowledge is specifically tailored to the needs of the local environment since it represents a way of life that has developed alongside the local ecosystem.
2. Resource exploitation is limited to what is necessary for immediate survival; production is only done to meet these necessities.
3. Diversified production systems: Risk is frequently distributed by using a range of subsistence techniques; there is no overexploitation of a single resource.
4. Respect for the environment: This is known as a "conservation ethic." All animals are interconnected, the earth is revered, and mankind depends on nature to survive.
5. Flexible: Native knowledge can consider outside knowledge and adjust to changing circumstances.
6. Social responsibility: There are strong ties to the community and family, and these connections bring with them a sense of duty and responsibility to protect the land for future generations.

Case studies of effective ITKs in India

In a study conducted by Swangla et al., (2019) at Lahaul-Spiti districts of Himachal Pradesh, various ITKs were identified and validated by the Quantification of Indigenous Knowledge method. On a 5-point scale, the validation criteria were utility, availability, cost-effectiveness, ease of use, time consumption, and labor intensiveness (Table 8.1). It was concluded that based on the response from the informants/farmers ranked Kothar for food storage as the most useful ITK in the tribal areas of Kinnaur district followed by Graat for a hydro flour mill, Suttar for soil conservation, tying of branches of the apple tree at tender age, and chicha for drying and storing of fodder.

ITK also have a significant role in the Veterinary field (Table 8.2), according to a study by Amitendu et al., (2004) in the Paschim Midnapore area of West Bengal, ITKs were thought to be superior to Modern Veterinary Drugs (MVDs) in terms of availability, side effects, and cost when used to treat various cattle illnesses. However, in terms of the quantity of animals they were able to cure and the speed at which they recovered, these were thought to be less effective than MVD. Regarding accessibility, side effects, and price, Indigenous Practices were thought to be preferable to MVD. Another study by Nyando et al., (2013) examined the awareness of and use of indigenous technical knowledge as a supplement to the adoption and expansion of agroforestry in Maseno. This study's key conclusion was that although farmers in the Maseno region use indigenous technological knowledge, adoption is still limited. It has been crucial in the adoption of modern agroforestry techniques that aid in the control of weeds, soil, and fuel wood. This was clear from the farmers' improved agroforestry practices. Farmers share their knowledge of the new technology with one another despite the limited conventional agroforestry extension. Farmers have modified agricultural practices to maximize the benefits of these technologies. Farmers claim that these methods when used as advised, have one or more negative effects on the land resource and agricultural output. In addition, some technologies, like biomass transfer, require a lot of labor when used in accordance with extension advice.

Table 8.1: Ranking of ITKs in Lahaul-Spiti district of Himachal Pradesh (HP), India

ITKs	Utility	Availability	Cost effectiveness	Ease in use	Time consumption	Labor intensiveness	$\bar{x}$	Ranking
T1	4.4	4.6	4.2	4.2	3.6	3.5	4.08	III
T2	4.6	3.5	3.8	4.5	4.7	4.2	4.22	II
T3	4.6	4.4	3.8	4.4	4.1	4.3	4.27	I
T4	4.3	4.1	4.1	3.9	3.8	3.9	4.01	IV
T5	4.1	4.3	3.9	3.7	3.9	3.3	3.86	V
T6	4.7	4.5	4.3	4.3	3.5	3.7	4.16	II
T7	4.4	4.4	4.1	3.8	3.9	4.1	4.12	III
T8	4.2	3.8	4.1	4.3	4.0	4.0	4.06	IV
T9	4.4	4.2	3.9	3.6	3.5	4.0	3.93	V
T10	4.6	4.5	4.3	4.1	4.3	4	4.30	I

T1-Suttar: For soil conservation; T2-Graat: Hydro flour mill; T3-Kothar: Food grain storage structure; T4-Tying of branches of younger apple trees; T5-Chicha: For drying and storage of fodder; T6-Potato storage in pits; T7-Sumbhu: Storage of fodder; T8-Shaving off of Yak's hair in summer; T9-Sem: Storage of food grains; T10-Khudh: For keeping the livestock

According to Khatri et al., (2021), majority of respondents received the knowledge on indigenous technological knowledge through Parents/relatives (27.53%), followed by community leaders/elders and farmers. Such knowledge was reported to be used in various fields such as nutrient management, plant protection, plant storage, animal husbandry, and weather forecasting. Use of farmyard manure, Intercropping/mixed cropping, green manuring, burning

Table 8.2: Usage of traditional plants in the ITK applications for the treatment of different ailments in animals

Sl. no.	ITKs	Diseases treated	References
1	100 to 150 g of *Hemidesmus indicus* leaves (sarsaparilla) are ground and juice is extracted and mixed with honey	Dysentery	Amitendu et al., (2004)
2	Three pieces of *Glcyrrhiza glabra* (licorice) and *Piper nigrum* (black pepper), 2 teaspoons full ghee, and 50 g smashed licorice mixed with 250 ml cold water	Dysentery	
3	Latex of *Alstonia scholaris* (devil's tree) is mixed with *Piper nigrum* in the ratio of (3:2)	Dysentery	
4	Bark of *Butea monosperma* (palas tree) is boiled with 250 ml water and then drenched the cattle for 3–4 days	Dysentery	
5	Decoction of the root of *Acacia arabica* (gum Arabic) is mixed with mustard oil in a ratio of 1:3	Arthritis	
6	Mix of the roots of *Mimusops elengi* (Bakul tree), *Datura stramonium* (devil's trumpet), *Allium sativum* (garlic)—a paste is made and applied on the affected part	Arthritis	
7	Roots of *Hoebe hainesiana* (Bonsom tree) are mixed with 21 pieces of *Piper nigrum* (black pepper) and the paste is fed to the animal	Dog bite	
8	Two hundred g of *Saraca asoca* bark (Ashok tree) is to be boiled with 1 l milk, then, every day it is to be drenched with water for 15–20 days	Anoestrus	
9	Fifty ml sap of *Psidium guajava* leaves (gauva) is fed. It is effective, especially for goats	Diarrhea	
10	Paste of 'Geru (mixture of ferric oxide and varying amounts of clay and sand) + turmeric + mustard oil' is applied to the affected portion of the animals. Ash of cow dung cake is applied on the mouth of goats to control the mouth disease within a week	Foot and mouth disease	Tyagi et al., 2018
11	Raw ground kalabhatt grains (black soybeans) are fed to animals	Ringworm	

the residue of the crop, use of animal products (dung/urine), storing seeds in an earthen pot with ash, clearing and incorporating terrace riser, erecting scare-crow in standing crop to scare birds, and so on are the commonly practiced traditional practices in the study area. Also, rainfall patterns, wind direction and velocity, and animal movements were useful in predicting the weather conditions. Attributes of indigenous practices i.e., relative advantage, compatibility, complexity, triability, and observability were studied to find out the influence of attributes of ITK itself on the adoption of indigenous practices. The results showed that on average use of farmyard manure and erecting scarecrows to protect from birds were perceived as the most compatible, more triable, more observable, and least complex practice to farmers. However, several constraints limit farmer's adoption of ITKs such as the preference of farmers for sophistication with reliance on readymade inputs, followed by more time required to get the desired results from the adoption of ITKs, and sociological constraints.

Application of ITKs in Various Fields of Agriculture

(i) ITKs for nutrient management in plants

ITKs	Methods	Effects
	½ kg dry Bulb powder or 5 kg fresh Bulb dissolved in 20 l of water and add 1 kg jaggery	Increase in Potassium
Gaajar Ghaas Svaras	2 kg Gaajar Ghaas Svaras in 20 l of water then add 20 g powdered fitkari (Alum)	Increase in Nitrogen
Phusp Rasayan	Add 2 kg flowers in 2 l of cow urine and add 2 l of water	Increase Boron
Calcium Aark	Add 150 dm dry turmeric clots in 100 g calcium water, add 50 ml milk. After 4 days take turmeric clots and dry. Then add 1 g turmeric powder/liter water	Increase Calcium
Harad Rasayan	Add 2 kg Harad in 10 l of water then add 20 g fitkari. After 6 days, filter it	Increase Iron

Source: Sow and Ranjan (2021)

(ii) ITKs for disease management in plants

Sl. no.	ITKs	Plant diseases controlled	Reference
1	A mixture of ash (2–3 kg) and 1 l of castor oil is spread on a seed bed of a size of about 100 m^2	Soil-borne diseases in tobacco nurseries.	Sow and Ranjan (2021)
2	fresh cow dung near the collar region of chilli	damping off and dieback	Roy et al. (2015)
3	20 kg of *Casuarina equisetifolia* leaves are boiled in water for 20 min. After cooling, the solution should be filtered. Then the extract is diluted with water and applied	bacterial and fungal disease	Roy et al. (2015)
4	Prepare solution from 2 kg fresh leaves of papaya in 3–4 l of water and keep in overnight. After filtration, this is diluted with 50–60 l of water and 250 ml soap solution added	Brown spot disease of rice	Roy et al. (2015)
5	Blend 30 kg seed of chickpea with 0.5 mg Heeng + 200 g Salt in 1 l of Buttermilk.	chickpea wilt disease management	Sahu and Satapathy (2021)

(iii) ITKs for insect pests management in plants

Sl. no.	ITKs	Target insect pests	Reference
1	Leaves of Rambans (*Agave Americana*) crushed in water and their extracts are used as spray	Aphids in oilseeds	Tyagi et al. (2018)
2	Leaf decoction of nirgundi (about 30–40 leaves in 10 l of water) is prepared till it is left with 1 l. This mixture is filtered properly. About 10 g heeng is mixed in 1 l water and then the above ingredients are mixed in about 5 l of cow urine and sprayed over affected crops.	Insect pests of wheat crop.	Roy et al., 2015,

3	Leaf decoction (1 kg) of Gwarpatha and tobacco powder extract (200 g) is prepared in 5 l of water for 3–4 hours to make a 2 l solution. Neem leaf extract (200 ml) is added after evaporation process and decoction of 50 g aritha powder is added to the above solution and mixed thoroughly.	Insect pests of mustard crops	Roy et al., 2015,
4	Crush 1 kg *Ipomea* leaves, 500 g chilli, 500 g garlic, and 5 kg neem leaves in 10 l cow urine. Boil the suspension 5 times till it becomes half. Filter the extract and store in glass or plastic bottles, 2–3 l extract diluted to 100 l is used for one acre	leaf roller, stem/ fruit/pod borer	NCOF, Ghaziabad (2011-12)
5	2–3 sprays of 50 ml cow urine, 20 g cow dung, 40 g wood ash mixed in 15 l of water during bud formation	Aphids and Jassids infestation in cotton	Yadav et al., 2016
6	Fresh cow dung (20 kg) + cow urine (10 l) +jaggery (1 kg) + chickpea flour (1 kg) and 2 kg of soil from the root zone area of well-grown tree. The mixture is fermented for 8 days and is used as a 200 l solution to spray in 1 acre of soil	Various crop pest	Patel, 2016

Conclusion

The creation of sustainable adaptation policies to support rural-based communities could benefit from the ITK, as well as the accomplishment of food security at the household level. The framework of local knowledge systems, which combines traditional knowledge, culture, and artifacts with contemporary information, viewpoints, and technologies, has existed for a long time. Since the beginning of time, new crops have been brought from one region of the earth to another. As these crops, animals, trees, tools, etc. were adapted to their new environments, cultural and ecological knowledge systems developed. As a result, it is urgent to establish a national foundation that aids in the development of a national register of innovations, files patent applications, provides micro-venture capital support for businesses based on indigenous knowledge, as well as nonmaterial incentives like recognizing or honoring innovators and communities that hold indigenous knowledge. The documented ITKs would be used as a ready reference for agricultural scientists to further study indigenous technologies used for insect pest prevention, as well as to determine their scientific rationality and effectiveness. ITKs are organic in nature; however, the documented ITKs may be useful for extension personnel in planning and implementing various IPM modules. Farmers can easily adopt ITKs because they are environmentally safe. It contributes significantly to the efficient use of natural resources. As a result, it can be concluded that ITKs, which have been used by farmers since time immemorial, may aid in various fields of Agriculture if organized and used scientifically. ITK plays an essential role in sustainable grassroots innovations. Such grassroots innovation largely differs across different sectors with respect to the characteristics, sources, actors involved, etc. In many cases, the indigenous communities are not well aware of the value of their indigenous knowledge which has been passed from generation to generation. Extension agents such as scientific institutions and NGOs could play a crucial role in this regard for capacity building among the indigenous community and popularization of traditional methods and techniques. In today's context, there is an urgent need to evaluate and popularize indigenous innovation. Government schemes and Research and Development activities should reach indigenous innovators. As most of the traditional knowledge and technologies are undocumented, there is also a need for more research in this field. Otherwise, this valuable knowledge will be extinct in the near future. There should be a proper collaboration between indigenous knowledge and modern knowledge. Such a type of alliance is occurring in some of the sectors such as ethnomedicine. The same should happen to all other sectors so that the contributions of the indigenous innovators appropriately reach and serve society.

References

Amitendu De, Arya, H. P. S., Tudu, B. & Goswami, A. 2004. Indigenous technical knowledge in animal husbandry. Livestock Research for Rural Development. Vol. 16, Art.#59. Retrieved December 15, 2022, from http://www.lrrd.org/lrrd16/8/arun16059.htm

Deka, M.K., Bhuyan, M. & Hazarika, L.K. Traditional pest management practices of Assam. *Indian J. Traditional Knowledge*, 5, no. 1, (2006): 75-8.

Haverkort, B. 1995. Agricultural Development with a Focus on Local Resources: ILEIA S view on Indigenous Knowledge. In The Cultural Dimensions of Development: Indigenous Knowledge Systems (Eds. D. M. Warren, L. J. Slikkerveer and D. Brokensha), London: Intermediate Technology Publications Ltd, pp. 454-57.

Kolawole, O.D.. Local knowledge utilization and sustainable rural development in the 21st century. *Indigenous Knowledge Develop. Monitor*, 9, no. 3, (2001): 13-15.

Khatri, S, Saugat, K. & Santosh, K. Perceived attributes and adoption of Indigenous Technological Knowledge on agriculture - a case study from Bhirkot municipality of Syangja District, Nepal, *Cogent Food & Agriculture*, 7, (2021): 1, 1914384. DOI:10.1080/23311932.2021.1914384.

Nyando, V, Tsingalia, H.M. & Onyango, C.A. Awareness and adoption of indigenous technical knowledge in agroforestry practices in Kenya's Maseno Region. *Research Journal of Agriculture and Environmental Management*. 2, no. 11, (2013): 467-73, November, 2013.

Patil, R. B. 2016. Documentation of ITK practices and formulations used in organic farming as IPM in Nashik District of Maharashtra. KVK, YCMOU, Nashik. (eds.), pp 37-46.

Prakash, N., Roy, S.S. & Ngachan, S.V. 2012. Role of ITK in Conservation Agriculture: Blending Indigenous and Scientific Knowledge. Retrieved from chrome-extension://efaidnbmnnnibpcajpcglclefindmkaj/http://www.kiran.nic.in/pdf/publications/ITK.pdfs

Rajasekaran, B.D., Warren, D.M. & Babu, S.C. Indigenous natural resource management systems for sustainable agriculture development - A global perspective. J. *Internat. Develop*., 3 no. 4, (1991): 387-401.

Roy, S., Rathod, A, Sarkar, S. & Roy, K. Use of ITK in Plant Protection. *Popular kheti*., 3, no. 2, (2015): 75-78

Sahu, H. P. & Satapathy, R. R. Indigenous Technical Knowledge in Plant Disease Management Asian Journal of Agricultural Extension, *Economics & Sociology*, 39, no. 10, (2021): 377-80.

Satapathy, C., Veeraswami, S. & Satapathy, B. Indigenous Technical Knowledge; method of Documentation and Rationalization, Souvenir on International Seminar Traditional Knowledge, Health and Environment, OUAT. 23-24 February, 2002. 12-15

Sow, S. & Ranjan, R. Indigenous Technical Knowledge (ITK) for Sustainable Agriculture in India. Agriculture and Food: E-Newsletter, 3, no. 1, (2021): 31-35.

Swangla, S., Vellaichamy, S., Singh, P., Burman, R. R., Priya, S., Palanisamy, et al. A note on indigenous technical knowledge in Kinnaur and Lahaul-Spiti districts of Himachal Pradesh. *Indian Journal of Traditional Knowledge*. 20, no. 2, (2021): 520-31.

Tyagi, S., M.K. Singh, B.D. Singh & Sunil Kumar. Conservation and Management of Indigenous Technical Knowledge for Livelihood Upliftment of Small and Marginal Farmers in Rural Areas. *International Journal of Inclusive Development*. 4, no. 2, (2018): 53-58.

Yadav, P.K., Husain, M. Singh, R, & Padaria, R.N. 2016. Indigenous Technical knowledge in agricultural sciences, Indian Agricultural Research Institute, (eds.) 15-22.

9

Augmentation: An Inevitable Environment-Friendly Insect Management Technique

Devina Seram[1] and I Yimjenjang Longkumer[2]

[1,2]Department of Entomology, School of Agriculture, Lovely Professional University, Phagwara, Punjab, India

Introduction

For many years, it has been acknowledged that "augmentative" or "inundative" biological management can decrease arthropod pests (DeBach, 1964). The goal of increased biocontrol is to bring about behavioral and numerical changes in the natural enemies that live within agricultural environments. This tactic calls for the large production of natural enemies as well as their periodic release, to maximize their potential for reproduction during the growing season. For instance, the use of augmentative releases may be required if naturally occurring populations of parasitoids or pests are either nonexistent, do not successfully colonize fields or orchards, or colonize at a point in the season that is too late to provide effective pest control. The increase is being carried out to establish a population of natural enemies capable of keeping the number of pests at economically hazardous levels until harvesting time.

Biological Control

Controlling problems caused by things like mites, insects, weeds, and plant diseases with the help of other organisms is called biological control or biocontrol. It uses natural mechanisms like predation, parasitism, herbivory, and so on, but human intervention is almost always present.

Parasitoids

The term "parasitoid" refers to a group of insects, the larvae of which develop on or within the bodies of other insects, which the parasitoid kills eventually. The phenomenon is known as parasitism. The vast majority of parasitoids are classified as either wasps (Hymenoptera) or flies (Diptera). These insects are numerically quite abundant and play a significant role in almost every terrestrial environment.

Predators

A predator can be defined as "an organism which feeds upon other animals (prey) that are usually smaller and weaker than itself, The phenomenon is known as predation. Insect predators consume several-to-many prey throughout their development.

Augmentation

Pest management by the periodic release of natural enemies. When using natural enemies for augmentation, keep in mind that they won't settle in and you'll have to keep introducing them whenever the pest population rises above a certain threshold.

Inundative Release

An "inundative application" of a biological control or natural enemy of a pest refers to the widespread release of a mass-produced biological control agent with the goal of either swiftly lowering a pest population or boosting the long-term survival of the biocontrol agent.

Inoculative Release

The term "inoculative biological control" refers to the practice of occasionally or seasonally releasing natural enemies to reestablish a balance that has either not been naturally maintained or has been upset by other forms of pest management. The control is expected from the progenies of released natural enemies.

Integrated Pest Management

Integrated pest management, or IPM, is an ecosystem-based strategy that employs a wide range of tools and techniques to reduce the prevalence of pests and the damage they cause over time. These include things like biological control, habitat manipulation, changes to cultural practices, and the introduction of pest-resistant plant varieties. Applying pesticides requires close observation prior to treatment.

Augmentation in Biological Control

Debach and Hagen first introduced the concept of augmentation to the scientific community in 1964. Increasing the number of parasites or predators in the environment, or the availability of resources that these organisms feed on, is what is meant by the term "augmentation," as this practice ensures that there are sufficient numbers of parasites and predators in the environment to achieve the desired level of pest control. In other words, it's the process of releasing a large number of captive-bred natural enemies of pests into the wild in order to "raise" natural enemy populations or "inundate" pest populations with natural enemies.

Augmentation Approaches

The inundative technique and the inoculative approach are the two primary types of augmentation strategies utilized in biological control. These strategies can be implemented in a variety of different ways, based not only on the adaptability of the natural enemy but also on the environment in which it will be used.

Inundative Release

Unlike other biological control approaches, this one doesn't rely on creating a sustainable population of natural enemies that eventually outnumbers the population of its hosts or victims. Instead, biological agents, like a pesticide, are used and released in large enough numbers to control insect populations. Most insect pathogens that are sold on the market are used by flooding an area with them. This technique of release is typically used to monitor an epidemic primarily by looking at the mortality caused by the natural enemies that have been released, rather than the natural enemies' offspring. Because the mortality of the pests is basically instantaneous and there is no extended population contact, this method of control is comparable to that which is produced by the application of pesticides. When pest populations approach threshold levels, bio-control agents are released for short-term management (Collier and Steenwyk, 2004). These agents are referred regarded as "biotic insecticides" because their effects are comparable to those of chemical pesticides. This technology is largely utilized in greenhouses because of the relatively high costs associated with it. The bacterium, *Bacillus thuringiensis* is the most effective agent in this group. The most well-known biological insecticides are those based on this bacterium. The digestive systems of insects are paralyzed by *Bt* spray, making them susceptible to death (mosquitoes, fungus gnats, elm leaf beetle larvae, or caterpillars, depending on the strain). It is a protein that is produced by the bacterium that is the active component, and in many products, there are no viable bacterial spores present; instead, there is only a formulation of the active protein. This causes the stomach to get paralyzed. This prevents further disease transmission throughout the insect population. It is used to manage pests such as lepidopterans, dipterans, and coleopterans. *Trichogrammas* sp. is the egg parasitoid most usually utilized through this method (Dhaliwal and Arora, 2004).

Beneficial Nematodes are one example of a live natural enemy that can be released in an inundative manner. These nematodes can move either through the soil or on the top of the soil, and they prey aggressively on the insects that they infect. Once they are inside, they will release bacteria that are symbiotic with them, which will then multiply and eventually kill the host. Nematodes consume bacteria and the tissue of the insects, after which they mate and generate offspring. After a week or two, new immature nematodes emerge from the decomposing insect in search of new hosts. Desiccation, ultraviolet light, and temperature extremes are all things that pose a significant threat to the survival of nematodes. They are especially effective against insects that live on or in the soil, as well as in other situations that are somewhat protected (such as tunneling inside plants). It is essential to have sufficient moisture, as well as temperatures ranging from roughly 53°F to 86°F (Lacey and Georgis, 2012). Large-scale releases of natural enemies of insects and mites in the field can be costly due to the high costs associated with mass breeding, storage, and transportation of live creatures. The cost of production, though, is getting lower as scientists learn more about the effects of artificial meals on natural predators. The inundative release method is recommended for use in various situations which are as follows:

(a) The insect control is designed for use on those insect pests that infest high-value crops in relatively low numbers.

(b) The targeted pest species are univoltine, i.e. insect pests having only one generation in a year and able to caused damage to plants.

(c) The cost of this endeavor is prohibitive if many releases are required for anything other than insect pathogens.

Inoculative Release

This natural enemy enhancement strategy relies on the offspring of the colonized beneficial species to control pest populations for multiple generations. When compared to inundative releases, the impact of this method often lasts for longer. In cases where an introduced species cannot persist in its new environment without the help of inoculative augmentation, natural enemies are brought in. Such releases are carried out before the start of the season to gain seasonal control, i.e., to colonize the region throughout the season or entire crop and so avoid pest buildup. In this case, the pest population does not decline rapidly, but it can be stopped before it causes significant economic harm. The release of predatory beetles like *Cryptolaemus* in guava and citrus, predatory mites like *Phytosieulus persimilis* against two-spotted spider mites, and predatory wasps like *Encarsia formosa* against greenhouse whitefly, *Vaporariorum trialeurodes*, etc., are all examples of inoculative releases that are commonly carried out (Hoodle et al., 1998). The production and distribution technologies of natural enemies are different in these two augmentation strategies. Given that released natural enemies can breed and spread, inoculative releases often have a lower economic impact in the field. However, the cost of the control mechanism goes up when natural enemies need to be mass-cultured or mass-reared for inundative release. Before releasing natural enemies for the augmentation process, one must consider the following points:

(a) When comparing bio-control to other control strategies (such as pesticides), we must consider both the expected costs and returns.

(b) The entire crop production system must be taken into account; if crop production practices are counterproductive to the utilization of natural enemies, then release efforts may be futile.

(c) It is necessary to have access to data on the species or strain best suited for the job, as well as the most efficient application strategy (such as the manner of release, the quantity of applications per unit area or per pest, and the measurement of their effectiveness, among other things).

*A Case Study on Managing Papaya Mealy Bug (*Paracoccus marginatus*) through Inundative Augmentation Release in Tamil Nadu (India) (Table 9.1)*

In the Southern part of India, damage to papaya plants was estimated at 90% during an outbreak of papaya mealybug (*Paracoccus marginatus*) in 2008. The National Bureau of Agricultural Insect Resources (NBAIR, Bangalore, Karnataka) brought in three hymenopteran parasitoids, *viz. Acerophagus papayae, Anagyrus loecki,* and *Pseudleptomastix mexicana* from Puerto Rico, United States, for checking their effectiveness against the papaya mealy bug. The Encyrtid parasitoid wasp, *A. papayae*, was shown to be the most efficient in reducing populations of the pest. The parasitoids were released on campus after the nucleus culture was collected from NBAIR for multiplication. The parasitoid wasp was rapidly reproduced within two weeks as part of a training program for plant protection scientists at 7 universities, 36 research stations, and 14 KVKs (Tanwar et al., 2010).

Table 9.1: Status of Papaya Mealy Bug (*Paracoccus marginatus*) after the release of *Acerophagus papayae* in Tamil Nadu

Sl. No.	District	Taluk	Crops	Status
1.	Coimbatore	Periyanaickan Palayam	Papaya, Mulberry, Brinjal, Tomato, Flower crops	Moderate
		Annur	Mulberry	Moderate
2.	Erode	Gobi	Mulberry, Tapioca, Vegetables	Severe
		Sathyamangalam	Papaya, Tapioca	Severe
3.	Tirupur	Avinashi	Mulberry, Papaya	Moderate
		Mudakkurichi	Mulberry	Moderate
		Udumalpet	Mulberry, Tapioca, Papaya	Moderate
4.	Trichy	Manikandam	Papaya, Mulberry, Flower crops	Moderate
5.	Karur	Karur	Papaya	Moderate
6.	Salem	Salem	Papaya, Tapioca, Mulberry	Severe
7.	Cuddalore	Cuddalore	Papaya	Low
8.	Tirunelveli	Senkottai	Papaya	Low
9.	Theni	Theni	Papaya	Moderate
10.	Namakkal	Namakkal	Papaya, Mulberry, Tapioca	Low

Source: Tanwar et al., 2010; Myrick et al., 2014

During a 6-month time frame, 5,65,000 *A. papayae* parasitoids were replicated and distributed at a rate of 100 parasitoids/field/village or block in every district except Nilgiris, where there was no occurrence of mealy bug. Farmers from a variety of districts had positive reactions to the newly developed technique. For the management of each particular insect pest, farmers historically applied an average of 2 insecticidal sprays to 20–24 sprays for papaya and 1 mulberry crop, and 10 sprays for cassava. By using this method of release instead of using pesticides, rural farmers have been able to save a combined total of almost ₹122 crores throughout the production of papayas, mulberries, and cassava over the preceding 6 months. It is estimated that the introduction of the mealybug parasitoid known as *A. papayae* prevented losses in the above three crops totaling ₹435 crores that would have been caused by an invasive pest. Additionally, the Kerala Agricultural University (KAU) was provided with a gift of 2000 parasitoids to control mealy bugs by their release (Myrick et al., 2014).

Strategies to Be Considered before Augmentation

When compared to the use of insecticides, biological control (bio-control) is a very different method of dealing with insect pests. It is more difficult to execute bio-control methods, but they are less harmful to the environment and to humans who come into contact with them. Bio-control organisms, often known as beneficials, are living and active, unlike conventional insecticides. Both they and the pest insects have complex activities that need explanation and are vulnerable to the same unfavorable environmental variables. The following points need to be considered while adopting augmentation procedures for the bio-control agents, especially parasitoids and predators (Day et al., 2022):

- Select insecticides that are less toxic to these predators. Determine what elements affect the fitness of the released natural enemies and include augmentation as part of coordinated pest control.
- Find and settle the most advantageous biotypes.
- Genetic refinement to boost performance.
- Establish an *in-vitro* breeding center for predators and parasites.
- The advantages of choosing hosts with a short life cycle

Requirements for Augmentation

Because augmentation entails the gathering methods for the capture, mass-rearing, and release of NEs must be devised for the manufacturing, storage, distribution, and deployment of these agents to guarantee that they continue to be viable. A further factor contributing to the high cost of rearing is the need for tritrophic cropping systems, which involves the cultivation of plants to maintain populations of pests, which in turn maintain populations of natural enemies. It is also extremely important that organisms that have been mass-reared or discharged should not suffer any loss of fitness as a result of being cultivated in a laboratory (Ballal, 2022). Keeping a living creature alive is significantly more difficult than maintaining the viability of a pesticide for around 2 years during storage, which is why storage is typically more expensive. The fulfillment of the following prerequisites is of the utmost significance in advancing the cause of augmentation:

(i) **Selection of natural enemy** The first thing that has to be done for a release program is to choose a natural enemy strain or species that is suitable for the management of the target pest that is being eradicated. The natural enemies should be amenable for storage and breeding in large numbers, and it also needs to be compatible with the environmental conditions in which it will be deployed.

(ii) **Mass-rearing technique** For an effective augmentative biocontrol, the development of a mass-rearing technology that is both efficient and cost-effective is required.

(iii) **Quality of the mass-reared natural enemies** If the released natural enemies are of low quality or the release amounts are low, the program will be unpredictable and the pest population will not be adequately suppressed. The effectiveness of natural enemies must be reliable and high enough to justify their expense as compared to the alternatives (like insecticides). Controlling production, monitoring general rearing procedures (equipment maintenance and environmental conditions), controlling processes, monitoring the quality of the unfinished product (proportion of hatch, larval, pupal weights, proportion of pupa), and monitoring the quality of the finished product (quantity, scale, age, ration, fecundity, and other biological and behavioral properties) are the three components of quality controls for insect mass production.

Costs Involved in Commercial Production of Natural Enemies for Augmentation

When it comes to developing biological control agents and producing them commercially, particularly for use in agricultural field crop production, there are a lot of obstacles and restrictions to overcome. Many bio-control agents are only required during certain times of the year and have a very limited period in which they can be used or released effectively. However, the producer is responsible for covering the costs of manufacturing for the few months leading up to the limited market opportunity. If climatic circumstances cause insect problems to be delayed or eliminated, or if an alternate way of controlling pests becomes available, the producer will have wasted time and resources developing a solution to a specific problem that no longer exists. In many cases, the size of the market is somewhat limited, but labor costs are high and overhead expenditures remain constant regardless of product sales because supplies have to be acquired continuously throughout the year (Collier and Steenwyk, 2004). If we consider the fact that it usually takes between 3 and 5 years of hard effort, the cost of research and development can become rather high because it takes much longer for a new species of natural enemy to become a marketable commodity. The producer's most significant expenditure goes toward labor costs, which include both development and production. The maintenance of all three trophic levels (flora, fauna, and natural enemies of pest insects) is necessary for the successful commercial production of bio-control agents, which can be mass-reared on natural or factitious host, i.e., host plant, insect host, or artificial diet (grain or soil mix), Everything has to be manufactured on a commercial basis for successful dissemination.

From the point of view of the producers, several concerns must be resolved before augmentative biological control can be developed and customer demands met. The economics of production need to be such that natural pest control products' unit cost in the marketplace can be considered reasonable by the consumers while still being maintainable for the producer. It is imperative that both the scale and efficiency of mass manufacturing technologies continue to advance. In order to create adequate revenue, the size of the market must be substantial. Particularly for smaller producers, the restrictions placed on packaging, shipping, and regulatory compliance can be burdensome. Small firms may be forced to switch from a direct-sales approach to one where they supply a greater corporation because of the expense of freight, losses that occur during shipment, and any future registration fees. It is possible that the small business will go bankrupt as a result of these charges if it is unable to modify its business methods to absorb them. When cultivating field crops, producers have a responsibility to examine, situation by situation, the moral implications and potential consequences of introducing nonendemic species, the entomophages, If we consider the paradigm of classical biological regulation (Collier and Steenwyk, 2004). To improve both the field efficacy and efficiency of the control, new release mechanisms must be developed for use in field crop production, which will ensure that the release is distributed evenly throughout the crops. The success of augmentative biological control is highly dependent on the education of the consumers as well as their receptiveness. The natural enemies that are being utilized in any IPM systems need to be compatible with any additional pest control technologies that are implemented (Hale and Elliott, 2002).

Insectaries for commercial rearing of entomophages are labor-intensive, expensive to produce, poor in profit margin, prone to failure, insufficient funding, and no access or protection as it relates to copyright. Moreover, it is absolutely necessary to find a highly qualified specialist who is willing to work for a lower wage and who is knowledgeable about the intricate biological systems of many organisms, including plants, insects, and their natural enemies (Collier and Steenwyk, 2004). In most cases, a scientist who became interested in the field is in charge of the investigation, rather than out of a desire to make money (Weicker and Company, 2000). Therefore, there needs to be an effort to promote partnership in order for augmentative biological control to be implemented on a commercial scale in field crops, collaboration is required between researchers, producers, administrators, and end consumers. (Hoffmann et al., 1998). Case studies that were successful in general were projects that received funding from outside sources, researchers studied the biology of natural enemies, devised techniques for raising them in the lab, licensed the technology to a manufacturer for mass production, and convinced end users to volunteer their crop for large-scale testing (Hale and Elliott, 2002). It is necessary to get a perspective on regardless of whether or not extensive pilot programs are essential in order to assist growers in determining whether or not various natural enemies may be beneficial when utilized appropriately. These kinds of tests have the potential to expand the market for products with limited applications to the point where manufacturing them would be economically viable.

Developmental Stages for Release Of Natural Enemies during Augmentation

When deciding when to release bio-control agents in the field, it is vital to consider not only how easily they may be transported and manipulated, but also when they will have the greatest impact on eliminating the pest. Typically, a non-mobile stage such as an egg or a pupa is chosen for transport and release since it is the least vulnerable to mechanical handling (Collier and Steenwyk, 2004). Multiple generations of entomophagous insects can be released into the field or greenhouse. Some common examples (Table 9.2) in which entomophages (parasitoids and predators) are released at various times include the following:

- Egg stage—*Chrysoperla cranea* (Predatory green lace wing insect, Neuroptera)
- Larval or nymph stage—*Chrysoperla carnea,* Predatory mites such as *Phytosieulus, Amblyseius, Orius*sp. (Acarina), etc.
- Pupal stage—*Aphidius* sp. (Braconid wasp, Hymenoptera), *Trichogramma* (Trichogrammatidae, Hymenoptera), *Encarsia*sp. (Aphelinid wasp, Hymenoptera), etc.
- Adults—*Phytosieulus, Orius* (Predatory spider, Acarina), *Cryptolaemus montrouzieri* (Australian lady bird beetle), *Rodolia cardinalis* (Australian vedalia beetle), etc.
- All stages together—*Phytosieulus* sp. (Predatory spider, Acarina)

Table 9.2: Examples of Successful Augmentative Biological Control in India

Pests (crop)	Bio-Agent (s)	Dose/ Amount Per ha area	Application Frequency	Application Methods
Chilo infuscatellus and other *Chilo* spp. (Sugarcane borers)	*Trichogramma chilonis*	50,000	@ 10 days interval, starting from 30 days old crop	Staple parasitized egg card bits on lower leaf (Tricho cards)
Pyrilla perpusilla (Sugarcane pyrilla)	*Epiricania melanoleuca*	1500 adult beetles/ grubs	Initiated before onset of monsoon	Egg masses/ cocoon spread in selected spots
Scirpophaga incertulas (Rice yellow stem borer)	*T. japonicum*	1,00,000	30 DAT accordingly	Adult release or staple parasitized egg card bits (Tricho cards)
Nilaparvata lugens (Rice brown plant hopper)	*Cyrtorhinus lividipennis*	100 adults Or 50–70 nymphs/sq.m	On noticing incidence	Released on hopper infested spots
Bollworm complex (cotton borers)	*T. chilonis*	1,50,000	@ weekly interval, 6 times initiating from 40 DAT	Staple parasitized egg card bits on lower leaf (Tricho cards)
Opisina arenosella (coconut black headed caterpillar)	*Goniozus nephantidis*	5000 adults	Need based or for each generation	Adult release
Eriosoma lanigerum (apple wooly aphid)	*Aphelinus mali*	1000 adults/ infested tree	Once, Infestation noticed	Adult release
Cydia pomonella (apple codling moth)	*T. embryophagum*	2000 adults/ tree	@weekly interval	Adult release
Icerya purchase (citrus scale)	*Rodolia cardinalis*	2000 adults/ tree	@weekly interval	Adult release
Maconellicoccus hirsutus (grape mealy bug)	*Cryptolaemus montrouzieri*	2500–3000 beetles/ha or beetles/ infested vine	Once, when noticed	Adult release
Eichornia crassipes (water hyacinth)	*Neochetina eichorniae* or *N. bruchii*	1000–5000 adults/ water body	Once, onset monsoon in waterbody	Adult release
Heteropsylla cubana (Subabulpsyllid)	*Curinus coeruleus*	20–25 adults/ tree	Twice during July and October	Adult release

Conservation of Released Natural Enemies after Augmentation

Pesticide use should be reduced because most natural adversaries are highly vulnerable to them. The idea behind Integrated Pest Management (IPM) was to reduce the need for both biological and chemical methods of pest control using the least amount of pesticides necessary for commercial harvesting and proper pesticide application that causes the least amount of disruption to the organisms that naturally control pest populations. Reducing reliance on pesticides is possible through the deployment of resistant cultivars, cultural practices that limit pest abundance or harm, behavioral manipulation of pest mating and host-finding, and even direct physical intervention (Ballal, 2022). However, many IPM initiatives have not been able to advance past the preliminary step of creating survey procedures and economic thresholds for pesticide application. Some of the important conservation methods include:

(i) **Reduction in pesticide usage:** The vast majority of natural enemies are extremely vulnerable to pesticides, which is why the use of pesticides is one of the primary factors limiting their effectiveness in the field. The original concept that gave rise to IPM was to combine biological and chemical control of pests. This was accomplished by lowering the amount of pesticides used to the bare minimum necessary for economically viable production and applying the necessary pesticides in a way that would cause the least amount of disruption to the biological control agents. The use of resistant cultivars, cultural approaches that minimize pest abundance or damage, methods of altering insect mating or host-finding behavior, and, in some situations, physical methods of control can help lessen the need for pesticides (Collier and Steenwyk, 2004). However, many IPM initiatives have not been able to advance past the first stage, which involves the development of sampling methods and economic criteria for the application of pesticides.

(ii) **Selection of choice of insecticides to protect natural enemies:** The effect that a pesticide has on populations of natural enemies depends not only on the physiological effect that the chemical has but also on how the pesticide is utilized, such as when and how it is administered. Herbicides and fungicides, in addition to insecticides and acaricides, can sometimes be toxic to the natural enemies of insects and mites. The likelihood of toxicity is highest with insecticides and acaricides. A database containing information on the impacts of pesticides on spiders, mites, and beneficial insects has been compiled and summarized

(Croft, 1990; Benbrook, 1996). This database provides a comparison of the toxicity of various pesticides as well as the "selectivity ratio," which is the dose required to kill 50% of the target pest divided by the dose that kills 50% of the impacted natural enemy species. Synthetic pyrethroids are among the insecticides that are among the most toxic to beneficial insects. On the other hand, *Bacillus thuringiensis* and insect growth regulators are among the insecticides that are among the least hazardous. In general, systemic insecticides, which can only be exposed to insects by digesting plant material, and insecticides that must be swallowed to exert their harmful effects have a far smaller impact on natural enemies than they do on pests. Correct timing and application methods can lessen the impact of pesticides on natural enemies by reducing the amount of pesticide that comes into contact with the beneficial organism. The use of less tenacious pesticides reduces exposure, especially when combined with an understanding of the natural enemy's biology to prevent exposure during vulnerable life phases. It is possible that natural enemies in neighboring regions will not be affected by treatments of alternating strips within a field or spot applications in regions of high pest density. As such, the natural enemy's and the pest's movement may determine how successful area reduction strategies are.

(iii) **Habitat provision and resources for released natural enemies:** The released natural enemies require acclimatization with the new surroundings for their establishment and subsequent perpetuation. In the summer, pollen, nectar, and honeydew (made by aphids) may be essential or helpful for the adult stages of numerous predators and parasitoids. Since many agricultural plants only bloom evenly for a brief period, supplementary sources of pollen and nectar may need to be found among flowering plants growing at the field's boundaries or even within the field itself. The proliferation of several species of plants in the same area, however, can impede host-finding efforts, especially for parasitoids that have specialized in one particular plant. While pollen and other food sources can help keep generalist predator populations stable, the success of these predators ultimately rests on their capacity to promptly respond to outbreaks of the target pest by either aggregating or multiplying. Therefore, understanding the biology and behavior of the natural enemy and pest is essential before making any decisions on plant diversification or other techniques of boosting the nutrition of natural enemies (Collier and Steenwyk, 2004).

Success achieved from Augmentation of bio-control agents (BCA)

- After the introduction of a Metaseiulus occidentalis (a mite) is cultured in a lab and used to control almond mites. orchards, California growers saw yearly pesticide cost savings of $24 to $44 per acre (Headley and Hoy, 1987).
- In 1999, the lepidopteran bio-agent, *Epiricania melanoleuca* was used to control a *Pyrilla perpusilla* outbreak across 6 lakh acres in Northern India by releasing between 8000 and 10,000 cocoons or 8000 and 100,000 eggs per hectare at regular intervals. Within 20 days, the insect infestation was eradicated, and the cost of spraying insecticides was reduced by almost ₹480 million (Satyagopal et al., 2014).
- In India, tiller damage caused by rice yellow stem borer (*Scirpophaga incertulas*) was reduced by 50.1–61.3% when T. japonicum and T. chilonis were released at a rate of 50,000 / hectare/week for 11 weeks (Mohammad, 2018), and leaf damage caused by leaf folder (*Cnaphalocrocis medinalis*) was reduced by 63.8–75.5%. These results were obtained using the rice variety Swarna (Kumar and Khan, 2005).

Augmentation Biological Control in Greenhouses

For the commercialization of a native natural enemy, it must first be proved that it can be used in greenhouses; must be successful in commercial pest control of target pests in greenhouses, and must be amenable for mass-production. Pre-evaluation is also necessary to demonstrate its capability as a biological control agent. In terms of ecological factors like as reproduction, predation, and dispersion, the occurrence of diapause, and host range should be considered and evaluated (Hussey and Bravenboer. 1964). Interactions with native natural enemies that were released at the same time need to be researched as required (Hunter, 1997). It is also necessary to demonstrate its capability as a biological control agent. Since the purpose of augmentative bio-control is typically not long-term, it is often costly to use natural enemies as a temporary measure, particularly when compared with the costs involved in traditional pesticide application (Huffaker, 1971a; Collier and Steenwyk, 2004). The costs are incurred in the rearing, storage, and deployment of the enemies of nature. Therefore, the use of natural enemies for increase has been generally limited to circumstances where released natural enemies (NE) could be restricted. These situations also come up against crops of high value grown in greenhouses. For instance, whiteflies (*Trialeurodes variorum*) the most efficient supplementary biological control targets in greenhouses are vegetable crops. The use of *Encarsia formosa*@ 3000 adults/acre at weekly intervals against this insect pest has been very well established. Host feeding is also a major contributor to host mortality in whiteflies by *Encarsia formosa*, besides parasitism. An essential pest is the two-spotted spider mite (Tetranychusurticae) in the greenhouse. *Phytoseiulus persimilis* releases for this mite control are effective @ 6–8 adult mites/plant. And the use of Braconid wasp, *Aphidius colemani@* 1–2 adults per sq. m is found to be effective against the cotton aphid, *Aphis gossypii* (Bezark, 1997).

In Europe, greenhouses have found great success with the inoculation of insects with parasitoids and predators at specific times of the year as a method of biological control. Growers shifted to this approach because of the high cost of chemical pest control and the widespread insecticide resistance among greenhouse pests. Parasitic mite, *Encarsia formosa* was originally incorporated into the program as a means of combating mite, *Phytoseiulus persimilis*, was employed to regulate populations of two-spotted spider mites and greenhouse whitefly (Hussey and Scopes, 1985). Pests such as whiteflies, thrips, leaf miners, aphids, and caterpillars have had supplementary natural enemies introduced over time. Biological pest treatment in Europe has far lower costs than chemical pest control. Farmers can benefit from this network of experts by subscribing to specialized publications and participating in grower-focused research groups, farmers are kept up-to-date on the program's implementation details, new developments, and new natural enemies.

The application of augmentative biological control has proven to be more effective in the development of greenhouse-grown vegetables when compared to field crops. This is due to the distinctive environmental conditions operating in greenhouses, which make it conducive for the natural enemies to flourish successfully, as well as due to the years of experience that gardeners have gained while tending to these plants. Conditions such as heat, humidity, air circulation, light intensity, daylight, and carbon dioxide concentration in certain contemporary greenhouses are all precisely regulated and monitored by computers, which allows for an optimal growing environment for the plants housed within. The crops are shielded from the wind and rain and are typically grown as a monoculture, which reduces the amount of pests that can be associated with them. The levels of competition between different species are often quite low at each trophic level (Glenister, 1991).

Innovative Technological Advancements in Augmentation for Improving the Effectiveness of Biological Control Agents

When there are fewer pests, it is the right time to release the biological control agents. Generally, these agents frequently fail to establish themselves in greenhouses since there are not many prey or hosts available. However, there are few different approaches that can be used to build natural enemies that have been released. The usage of artificial rearing units within greenhouses is one such approach. In Europe, the rearing sachet method was designed with the intention of releasing predatory mites such as *Amblyseius swirskii* and *Neoseiulus californicus* (Messelink et al., 2014). The use of sachet systems for rearing natural enemies was introduced in Japan through importation. However, after being utilized in greenhouses, few disadvantages were discovered in using them, i.e., they were susceptible to the effects of water or pesticides, and the emergence rate of predatory mites from the rearing system was frequently insufficient to adequately manage the infestation (Shimoda et al., 2017). Food additives, both artificial and natural, designed for spray application have been created to encourage the proliferation of generalist predators in greenhouses across Europe (Messelink et al., 2014). For instance, the pollen of maize was sprayed as a means of improving *A. swirskii* populations, and for the purpose of providing mired bugs and predatory mites with food, sterile eggs of *Ephestia kuehniella* species were sprinkled (Shimoda et al., 2017).Instead of releasing only one species as a biological control agent, it is more effective and advantageous to release different kinds of natural enemies, where each of which has a unique host range and can therefore manage a wider variety of pest species (Heimpel and Mills, 2017).

In a manner analogous to crop breeding, three potential genetic manipulation strategies are being employed to boost the efficacy of entomophages in field conditions: Artificial selection, hybridization or, usage of heterosis, and biotechnology (recombinant DNA or rDNA) techniques. It has been suggested that natural enemy arthropods could be made more effective in IPM systems by the application of artificial selection for resistance to pesticides (Roush and Hoy, 1981). Some techniques used for improved entomophages (parasitoids and predators) are as follows:

(i) ***Trichogramma* strain tolerant to endosulfan insecticide**: A strain of *Trichogramma* sp. was developed at the National Bureau of Agricultural Insect Resources (NBAIR), Bangalore: *T. chilonis* strains with increased resistance to endosulfan have been commercialized under the brand name "Endogram" by a private company. Several states have adopted and implemented the "Endogram" system. At NBAIR, Bangalore, this strain was also improved to have several tolerances to other insecticides such as monocrotophos and fenvalerate (Jalali et al., 2006).

(ii) ***Trichogramma* strain tolerant to multiple pesticides**: A strain of *T. chilonis* that is resistant to numerous insecticides, including endosulfan, monocrotophos, fenvalerate, spinosad, and lambda cyhalo-thrin, was developed and then tested in the field against *Helicoverpa armigera*, and the results of this test were found to be positive (NAIP, 2012).

(iii) ***Trichogramma* strain tolerant to high temperatures**: It has been found that temperatures above 32°C reduce the survival and parasitizing capabilities of *T. chilonis* (Jalali and Venkatesan, 2005). Two different strains of *T. chilonis* have been shown to be resistant to heat (32–40°C) and to have 65% females and 116 egg-laying capacity indicating a strong biotic potential. Under high temperatures, these strains could be employed to combat sugarcane borer infestations (NAIP, 2012).

(iv) **Green lace wing, *Chrysoperla carnea* resistant to Monocrotophos**: Monocrotophos-resistant *C. carnea* strain was first reported by Patel and Yadav (2000). Moreover, cross-resistance to additional insecticides, including dimethoate, acephate, phosphamidon, and methyl-odometer, has been demonstrated by the same *Chrysoperla* strain.

(v) **Carbaryl and Permethrin resistant predatory mite, *Metaseialus occidentalis***: Breeding resistant to carbaryl and permethrin is an example of artificial selection. has been implemented successfully in the predatory mite, *M. occidentalis* in California. There have also been developments in *M. occidentalis* that produce strains with multiple resistances. *via* hybridization and then further selection under laboratory conditions (Hoy, 2009).

(vi) **Parasitoid, *Aphytismelinus* resistant to Carbaryl insecticide**: *Aphytis melinus* is a major commonly used parasitoid used against numerous species of armored scales, notably the California red scale and the San Jose scale (Spollen and Hoy, 1992a). The parasitoid is often utilized in citrus groves as a method of preventing the spread of California red scale. For this reason, it is crucial to do pesticide residue testing before releasing the parasitoid into the wild. Spollen and Hoy (1992b) reported carbaryl-resistant *A. melinus* from California orange groves.

(vii) **Hybrid Trichogammatidwasps development through hybridization**: A hybrid strain of *T. chilis* was produced to demonstrate tolerance to high temperatures as well as repeated applications of pesticides. Temperature tolerance (32°C–38°C) was induced in a *T. chilonis* strain resistant to various pesticides for the purpose of increasing the strain's capacity to parasitize its main rearing host, *C. cephalonica* eggs (Jalali and Venkatcsan 2005).

(viii) **Improved natural enemies through rDNA technology**: In this method, the desirable gene (or genes) is cloned, and introduced into a vector. This gene must also be heritable, meaning that it can be reliably inserted into their genome, expressed in the right tissues at the right time, and passed on to future generations (Routray et al., 2016). After that, the modified strain of the natural enemies must be healthy and able to function well in agricultural field conditions. Rapid progress in improving natural enemies may be possible through the application of genetic engineering techniques replacing the need for conventional rearing for a large number of generations. Furthermore, genes from any species could be utilized, as opposed to selecting them solely from the pool of genes contained in arthropods' natural predators. For example, development of a mutant strain of *Bacillus thuringiensis*, which can be cultured continuously; the toxin is encapsulated within a cell that increases its persistence under field conditions.

Advantages of Augmentative Biological Control

The utilization of native natural resources has two distinct advantages, i.e., natural enemies have adapted to the environment of the home country situations, and they do not have any supplementary detrimental impacts on hybridization with native species. The use of biological management as a strategy for the eradication of pest arthropods in forestry, rangelands, and agricultural conditions is not being maximized to its full potential (Huffaker, 1971b). Further and ongoing work in the classical, augmentation and conservation methods of biological control is warranted in light of the well-documented benefits of these approaches. This is because it is possible to further enhance the benefits already linked with these methods. Furthermore, efforts should be made to combine and create novel methods of biological control (Tauber et al., 1985). Methods such as genetically engineering pests, manipulating their behavior with semiochemicals and improving artificial diets and rearing techniques have all been developed to make augmentative releases more economically feasible on a large scale.

Conclusion

Even though there is no negative environmental impact from using predator and parasitoid organisms, their effectiveness is constrained by the conditions under which they are most useful. As their efficacy increases, they may eventually replace the more hazardous chemical approach to pest management. Classical biological control has a very favorable cost-benefit ratio (1:250), while augmentative control is competitive with pesticides (1:2–1:5) (Bale et al., 2008), but requires much less investment in research and development. Since biological control is the most long-lasting, cost-effective, and ecologically safe method of pest management, it will play a pivotal role in the field in the future. It is possible to increase the efficiency of natural enemies by manipulating their biology in various ways, such as by decreasing the developmental period, increasing the number of offspring produced, changing the sex ratio, increasing their tolerance to heat and humidity, or shifting their preference for a particular host or habitat. An improvement in genetics may be useful when the natural enemy is known to be a promising biocontrol agent; the limiting trait is influenced primarily by a single major gene; the gene can be obtained through selection, mutagenesis, or cloning; the manipulated strain is fit and effective; and the released strain can be maintained on some form of reproductive isolation.

References

Bale, J.S., vanLenteren, J.C and Bigler, F. Biological control and sustainable food production. *Biological Science*. 363, (2008): 761-76.

Ballal, C.R. Success Stories in Biological Control: Lessons Learnt. Vantage: *Journal of Thematic Analysis*. 3, no. 1, (2002): 7-30.

Benbrook, C. 1996. Pest Management at the Crossroads. Consumers Union, Yonkers NY.

Bezark, L.G. 1997. Suppliers of Beneficial Organisms in North America. State of California, Deptt. of Food and Agriculture, Biological Control Services Program, Div. of Pest Management, Environmental Protection and Worker Safety.

Collier, T. and Steenwyk, R.V. A critical evaluation of augmentative biological control. *Biological Control*. 31, no. 2, (2004): 245-56. DOI: https://doi.org/10.1016/j.biocontrol.2004.05.001

Croft, B. 1990. Arthropod Biological Control Agents and Pesticides. New York; John Wiley and Sons.

Day, R., Haggblade, S., Moephuli, S., Mwang'ombe, A. and Nouala, S. Institutional and policy bottlenecks to IPM. *Current Opinion in Insect Science*. 52, (2022): 100946. DOI: https://doi.org/10.1016/j.cois.2022.100946.

DeBach P., Rosen, D., Kennett, C.E. 1971. Biological Control of Coccids by Introduced Natural Enemies. In: *Huffaker C.B. (eds.) Biological Control. Springer, Boston, MA*

Dhaliwal, G.S. and Arora, R. 2004. Integrated Pest Management: Concepts and Approaches. Chennai: Kalyani Publishers.

Glenister, C.S. 1991. Biological control: making it work. Part 5: Supporting adoption of biological control, pp. 243-250. In: *Macdonald J. F. (ed.). Agricultural Biotechnology at the Crossroads. Biological, Social and Institutional Concerns*. National Agricultural Biotechnology Council Report 3, Ithaca, New York, USA.

Hale, M.A. and Elliott, D. 2002. Successes and challenges in augmentative biological control in outdoor agricultural applications: A Producer's Perspective. In: *1st International Symposium on Biological Control of Arthropods, Augmentative biological control in outdoor agricultural applications*. Honolulu, Hawaii, USA, January 14-18. Pp. 185-88.

Headley, J.C. and Hoy, M.A. Benefit/cost analysis of an integrated mite management program for almonds. *Journal of Economic Entomology*. 80, (1987): 555-59.

Heimpel, G.E. and Mills, N.J. 2017. Biological control: Ecology and Applications. Cambridge, UK: Cambridge University Press.

Hoffmann, M. P., R. L. Ridgeway, E. D. Show, and J. Matteoni. 1998. Practical application of mass reared natural enemies: Selected case histories, pp. 268-293. In: *Ridgeway, R. L., M. P. Hoffmann, M. N. Inscoe, and C. S. Glenister (eds.). Mass-Reared Natural Enemies: Application, Regulation, and Needs*. Thomas Say Publications in Entomology, Entomological Society of America, Lanham, Maryland, USA.

Hoy, M.A. Biological control of arthropod pests: Traditional and emerging technologies. *American Journal of Alternative Agriculture*. 3, no. 2–3, (1988): 63-68.

Hoodle, M.S., van Driesche, R.G. and Sanderson, J.P. Biology and use of the whitefly parasitoid, Encarsia Formosa. *Annual Review of Entomology*. 43, (1998): 645-69. https://doi.org/10.1146/annurev.ento.43.1.645

Huffaker, C.B., P. S. Messenger, Paul DeBach. 1971a. The Natural Enemy Component in Natural Control and the Theory of Biological Control of Pests in Glasshouse.

Huffaker, C.B. 1971b. Biological Control. p 511. New York, New York: Plenum Press,

Hunter, C.D. 1997. Suppliers of beneficial organisms in North America. California Environmental Protection Agency Department of Pesticide Regulation, Environmental Monitoring and Pest Management Branch. Sacramento, California. https://www.agresearch.montana.edu/wtarc/producerinfo/entomology-insect-ecology/General/SuppliersBeneficialOrganismsNA.pdf

Hussey, N.W. and L. Bravenboer. 1964. Culture by the Introduction of Natural Enemies. New York: Cornell University Press.

Hussey, N.W. and N. Scopes. 1985. Biological Pest Control, The Glasshouse Experience. p. 240. New York: Cornell University Press.

Jalali, S.K. and Venkatesan, T. 2005. Annual report of the ICAR Cess Fund Project: Development of a starain of *Trichogrammachilonis* Ishii tolerant to newer insecticides and high temperature. *Project Directoriate of Biological Control, Bangalore*, 45pp.

Jalali, S.K., Singh, S.P., Venkatesan, T., Murthy, K.S and Lalitha, Y. Development of endosulfan tolerant strain of egg parasitoid *Trichogrammachilonis* Ishii (Hymenoptera: Trichogrammatidae). *Indian Journal of Experimental Biology*. 44, (2006): 584-90.

Kumar, S. and Khan, M.A. Bio-efficacy of *Trichogramma* spp. against yellow stem borer and leaf folder in rice ecosystem. *AnnalsofPlant ProtectionSciences*. 13, no. 1, (2005): 97-9.

Lacey, L.A. and Georgis, R. Entomopathogenic Nematodes for Control of Insect Pests Above and Below Ground with Comments on Commercial Production. *The Journal of Nematology*. Jun; 44, no. 2, (2012): 218–25.

Messelink, G.L., Bennison, J., Alomar, O., Ingegno, B.L., Tavella, L. and Shipp, L. Approaches to conserving natural enemy populations in greenhouse crops: Current Methods and Future Prospects. *BioControl*. 59, (2014): 377–393.

Mohammad, Y. Effectiveness of Trichogramma*japonicum* Utilization for Biological Control Agents on *Scirpophagaincertulas* in Indonesia. *Asian Journal of Crop Science*.10, (2018): 31-39. DOI: 10.3923/ajcs.2018.31.39

Myrick, S., Norton, G.W., Selvaraj, K.N., Natarajan, K. and Muniappan, B. Economic impact of classical biological control of papaya mealybug in India. *Crop Protection*. 56, (2014): 82-6.

NAIP, 2012. Annual report 2011-12, National Agricultural innovation Project (NAIP)-ICAR, New Delhi. Effect of abiotic stresses on the natural enemies of crop pests: *Trichogramma, Chrysoperla, Trichoderma* and *Pseudomonas* and mechanism of tolerance to these stresses. www.nbair.res.in/pdbc-naip/home.html, 1-6p

Patel, I.S and Yadav, D.N. Standardization of bioassay technique for developing genetically improved monocrotophos resistant strain of green lacewing, *Chrysoperla scelestes* Banks. *Pest Management and Economic Zoology*. 8, (2000): 47-51.

Rabb, R.L. and R. E. Stinner, and R. van den Bosch. 1976. Conservation and Augmentation of Natural Enemies. P.233-254. In C.B. Huffaker and P.S Messenger (Eds.). Theory and Practice of Biological Control. Academic, New York.

Roush, R.T and Hoy, M.A. Genetic improvement of *Metaseiulus occidentalis*: Selection with methomyl, dimethoate and carbaryl and genetic analysis of carbaryl resistance. *Journal of Economic Entomology*. 74, (1981): 138-41.

Routray, S., Dey, D., Baral, S., Das, A.P. and Mahantheshwara, B. 2016. Genetic improvement of natural enemies: A review. *Agricultural Reviews*. 37, no. 4, (2016): 325-32. DOI: 10.18805/ag.v37i4.6463

Satyagopal, K., S.N. Sushil, P. Jeyakumar, G. Shankar, O.P. Sharma, D.R. Boina, S.K. Sain, et al. 2014. AESA based IPM package for sugarcane. Pp 56.

Shimoda, T., Kagawa, Y., Mori, K., Hinomoto. N., Hiraoka, T. and Nakajima, T. A novel method for protecting slow-release sachets of predatory mites against environmental stresses and increasing predator release to crops. *BioControl*. 62, (2017): 495-503.

Spollen, K.M and Hoy, M.A. 1992a. Genetic improvement of an arthropod natural enemy: Relative fitness of a carbaryl resistant strain of the California red scale parasite, *Aphytismelinus DeBach. Biological Control*. 2, (1992): 87-94.

Spollen, K.M and Hoy, M.A. 1992b. Resistance to carbaryl in populations of the parasite *Aphytismelinus deBach* (Hymenoptera: Aphelinidae): Implications for California citrus IPM. *Biological Contro*l. 2, no. 3, (1992): 203-10. DOI: https://doi.org/10.1016/1049-9644(92)90060-Q

Srivastava, K.P. A Textbook of Applied Entomology: By Augmentation of Natural Enemies – Book Chapter.

Tanwar, R.K., Jeyakumar, P. and Vennila, S. 2010. Papaya mealybug (*Paracoccusmarginatus*) and its management strategies. Technical Bulletin 22. National Centre for Integrated Pest Management (NCIPM), New Delhi. DOI: https://icar.org.in/files/Papaya%20 MealybyugNCIPM.pdf

Weicker, F. and Company Management Consultants. 2000. Pre-feasibility assessment of a commercial research and production facility. Final Report, *unpublished.*

10

Milli-Composting as an Alternative to Modern Composting

Devina Seram[1] and Haobijam James Watt[2]

[1]Department of Entomology, School of Agriculture, Lovely Professional University, Phagwara, Punjab, India

[2]Department of Agricultural Economics and Extension, School of Agriculture, Lovely Professional University, Phagwara, Punjab, India

Introduction

One of the most important concerns confronting both developed and developing nations in this age of rapid industrialization is how to deal with the ever-increasing volume of solid waste. One aspect of the environmental crisis that occurs alongside the development of the global economy is the rapid increase in waste volume (Zibres et al., 2011). The threat that organic waste poses to both the natural world and people's health in today's modern society makes it a primary topic of investigation in the modern community. The annual production of agricultural waste across the globe is greater than hundreds of megatons, and a significant portion of this waste is either improperly disposed of or burned directly, which contributes further to the acceleration of both global warming and air pollution (Zhang et al., 2016).

The compost that is produced through the biological process of composting can be used as an organic fertilizer in agriculture, vegetable gardens, and other types of gardens. Composting is a biological process that acts on the transformation of the organic matter that is present in the residues into material that is humidified (Antunes et al., 2021). In addition to having low production costs, this method is also environmentally friendly, making it a viable option for the long-term processing of this waste (Lopez-Gonzalez et al., 2015). Composting processes that are mediated by invertebrates from the soil fauna, such as vermicomposting and milli-composting, are proposed along with a variety of mixtures, practices, and management to improve the efficiency of composting as well as the quality of the compost produced. Traditional composting relies heavily on the activity of microorganisms. Vermicomposting and milli-composting are two examples of composting processes that use invertebrates from the soil fauna. Milli-composting is a relatively unknown biotechnology that uses diplopods, more commonly known as millipedes, to increase the biotransformation of vegetable wastes into stable organic matter. It not only aids in climate change mitigation but also aids in carbon sequestration, so milli-composting can be used as an alternative to modern composting as it is crucial for preventing pollution and promoting lush plant life.

Milli-composting is a relatively new biotechnology that is also kind to the environment. It facilitates the transformation of plant residues into stable organic matter through the activity of diplopods, which are more commonly known as millipedes. The species of diplopods known as *Trigoniulis corallinus* is suitable for milli-composting because it is found throughout the tropics, can be found in a variety of agricultural settings and has a coloration that makes it easily distinguishable from other species. Diplopods are found all over the world, including in tropical, subtropical, and temperate regions, and they are crucial to the process of increasing the fertility of the soil. In microcosm experiments, these organisms can mobilize nutrients that are bound up in litter and enrich the soil with nitrogen (N), carbon (C), calcium (Ca), magnesium (Mg), phosphorus (P), and potassium (K). This is because diplopod feces contain a high level of microbial activity, which contributes to their ability to consume large amounts of litter. This material is broken up while traveling through the digestive tract, which increases the specific surface area of the material. Additionally, the material becomes wet and is subsequently enriched with microorganisms (Aquino and Correia, 2005). Diplopods can metabolize between 0.3 and 7% of the material that they ingest, and when they excrete microbial activity, it continues in their fecal pellets. This results in an increase in the bioconversion of plant residues (Ambarish and Sridhar, 2013).

Millipedes are not earthworms, even though many people think of them in that way. Millipedes are classified as members of the class Diplopoda, while earthworms are members of the Oligochaeta class. Millipedes are important members of the soil and litter fauna in temperate and tropical parts of the world, where they and other invertebrates aid in the breakdown of plant organic matter. Millipedes, as primary destructors of plant debris, play highly important roles in soil formation as well as enhance the physiochemical properties of the compost. In the present scenario, the adverse effects of inorganic fertilizers on soil health have resulted in the practice of organic farming for sustainable agriculture, where organic waste recycling will be of immense importance. Attempts are made to reduce the use of chemical fertilizer through vermicompost and other biostimulants. Composting is a biological process in which organic biodegradable wastes are converted into hygienic, humus-rich products (compost) for use as a soil conditioner and an organic fertilizer (Popkin, 1995). Nowadays, solid waste compost when applied to agricultural soils has shown beneficial effects on crop development and yields as well as improved soil ecosystem.

Millipedes have an advantage over earthworms in this respect because earthworms cannot crush fibrous and recalcitrant residues found in nature. As a result, millipedes have a greater competitive advantage over earthworms. It is also important to note that there are regions that do not have access to manure for the production of vermicompost, which makes the production of milli-compost a more viable option. This is something that should be taken into consideration. Milli-composting is a viable activity in rural and peri-urban areas because lignocellulosic residues and millipedes are more easily found in these areas. It is known that millipedes are macro detrivores, which are terrestrial arthropods that feed on decomposing vegetable matter and mineral soil.

Milli-Composting Production Process and Obtaining the Milli-Compost

A concrete ring with a height of 0.5 m, a width of 1 m, and the capacity to receive 400 L of waste can be used in the milli-composting process so that it can be processed (Figure 10.1). The first step of the process consists of counting the residues and depositing them inside the ring. The process of milli-composting is developed from the combination of waste products derived from Bauhinia species (leaves of cow paw), Paspalum species (grass clippings), Musa species (leaves of banana), and chopped cardboard. The volume of each component, in liters, is as follows: 160 L, 120 L, 80 L, and 40 L, respectively. After that, they are mixed to create a very uniform mixture, after which they are added to the ring and wetted (Figure 10.1). The percentage of waste is determined based on a preliminary study that involves observing the consumption of various types of waste by millipedes. The wastes consist of both urban and agricultural debris. During the second stage, the ring is provided with an approximate volume of 2.2 L of millipedes (Trigoniuluscorallinus), which is equivalent to a population of approximately 3960 adult individuals. These millipedes are manually collected from earthworm beds, compost pans, and grass that contains fresh grass cuttings. The ring is kept covered with a protective screen that has two purposes: The first is to stop the millipedes from climbing the wall of the ring and escaping, and the second is to stop anything unwanted from entering the ring while the milli-composting process is taking place.

Throughout the milli-composting process, it is necessary to monitor the amount of moisture that is present in the materials contained in the rings. In order to keep the moisture level between 50% and 60%, regular additions of water from a house can (about 4L worth) are made once a week or as needed. This is done to maintain the moisture level and the objective of which is to create an environment that is conducive for the millipedes' continued existence and the milli-composting process as a whole. After the process described above, the milli-compost is finally obtained 180 days later. After collection, the leftovers are sorted through a sieve with a mesh size of 2 mm (Figure 10.1), placed in plastic bags, and later used in the propagation of vegetable seedlings.

Figure 10.1: Schematic steps involved in milli-composting process using millipedes of the species *Trigoniuluscorallinus* to create a substrate for growing vegetable seedlings. (Antunes et al., 2022)

Properties of Milli-Compost

The macroporosity, microporosity, total porosity, water retention capacity, and volumetric density are the physical characteristics that can be evaluated in the substrates (Brasil, 2008; Teixeira et al., 2017). These characteristics are obtained by the table method of tension, using metal rings of 100 mL and tension of 60 cm of water column (Brasil, 2008; Teixeira et al., 2017). To evaluate the physical-chemical characteristics of the substrates, pH measurements are taken in a solution of distilled water with a 5:1 volume-to-volume ratio, and electrical conductivity is measured in the same aqueous extract that is obtained for pH measurement. This is done in accordance with the method described by Brasil (2006). As for the chemical properties, samples of each substrate can be checked in order to determine the total levels of P, K, C, and Mg levels. These levels are determined in an aqueous acid extract following the substrates' own nitric-perchloric digestion (Teixeira et al., 2017). The elementary analyzer (CHN) is used to conduct the analysis which determines N and C concentrations (Nelson and Sommers, 2015). For each of the aforementioned analyses, three sets of data can be collected and analyzed.

Conclusion

The addition of milli-compost results in an increase in the nutritional content of the soil that is approximately 201%, 138%, 85%, 293%, and 29% higher, respectively, for nitrogen, phosphorus, potassium, calcium, and magnesium. It demonstrates a higher ability for water retention while exhibiting a lower volumetric density. The composting method is an alternative and a potentially fruitful way to get rid of environmental threats that affect people's health as well as the environment. When compared to composting with vermicompost and traditional compost, milli-compost contains a greater quantity of plant nutrients that are exchangeable with one another. It is necessary to spread knowledge among farmers about the significance of milli-composting and to encourage additional study of the process of milli-composting as a different way of recycling waste materials. The milli-composting method makes it easier to recycle garbage from nearby farms and cities by transforming it into an organic compost known as milli-compost. This milli-compost combines the physical, physicochemical, and chemical qualities that are desirable in a good substrate. When organic matter is added to the soil, millipedes are recognized to play a significant part in the breakdown and decomposition of the substance. In a similar vein, millipede compost is known to bring the pH closer to neutral, raise the levels of macro and micronutrients, and reduce the ratio of carbon to nitrogen. This suggests that millipede compost could play an important part in enriching the soil in terms of both the nutrition it provides and the microflora that it fosters. However, more research needs to be done on the generation of compost by millipedes feeding on agricultural residues, the enrichment of soil, and their impact on the growth and yield of crops. In a similar vein, the application of millipede compost in proper proportions with farmyard manure or any other organic manure could play a crucial part in organic farming as well as an environmentally responsible approach toward sustainable agriculture. It is vital to do research in order to have an understanding of the rate at which millipede compost can be added to different plant species in order to boost growth, dry matter, and yields.

References

Antunes, L.F.S., Fernandes, L.S., Vaz, A.F.S., Silva, M.S.R.A., Ferreira, D.M.T.S., Correia, M.E.F. Millicomposting: Sustainable technique for obtaining organic compost for the cultivation of broccoli seedlings. *Cleaner Engineering and Technology.* 7, (2022): 1-7. DOI: https://doi.org/10.1016/j.clet.2022.100442

Antunes, L.F.S., Souza, R.G., Vaz, A.F.S., Ferreira, T.S., Correia, M.E.F., 2021. Evaluation of millicomposts from different vegetable residues and production systems in the lettuce seedling development. *Org. Agr.* 11, (2021): 367-78. https://doi.org/10.1007/s13165-020-00342-y.

Brasil. 2008. Ministry of Agriculture, Livestock and Supply (MAPA). Technical Regulation of the National Program for combating and eradicating brucellosis and tuberculosis.

Lopez-Gonzalez, J.A., Suarez-Estrella, F., Vargas-García, M.C., Lopez, M.J., Jurado, M. M., Moreno, J. Dynamics of bacterial microbiota during lignocellulosic waste composting: Studies upon its structure, functionality and biodiversity. *Bioresour. Technol.* 175, (2015): 406-16. DOI: https://doi.org/10.1016/j.biortech.2014.10.123.

Popkin, R. Good news for waste watchers: Recycling, composting show results for the future. *Environ. Prot. Agncy J.* 21, no. 1, (1995): 188-90.

Senthilkumar, N., Lakhmidevi, R., Sumanthi, R., Sathiskumar, R., Divya, G. R., Lenora, L. M. D. Millicompost: An Alternative Biocompost for Forest Nurseries. *International Journal of Current Research.* 10, no. 6, (2018):70971-74.

Silva Júnior, J.V. da, Beckmann-Cavalcante, M.Z., Brito, L.P. da S., Avelino, R.C., Cavalcante, ´I.H.L. Aproveitamento de materiais alternativos naproduçao ˜ de mudas de tomateiro sob adubaçao ˜ foliar. *Rev. Cienc. Agron.* 45, (2014): 528-36. DOI: https://doi.org/10.1590/S1806-66902014000300013.

Teixeira, P. C., Donagemma, G. K., Fontana, A., Teixeira, W. G. 2017. Manual of soil analysis methods. (3rd ed., p. 573). Embrapa Solos-Livro técnico (INFOTECA-E). DOI: https://www.infoteca.cnptia.embrapa.br/handle/doc/1085209

Zhang, L., Zhang, H., Wang, Z., Chen, G., Wang, L. Dynamic changes of the dominant functioning microbial community in the compost of a 90-m^3 aerobic solid state fermentor revealed by integrated meta-omics. *Bioresour. Technol.* 203, (2016): 1-10. DOI: https://doi.org/10.1016/j.biortech.2015.12.040.

Zibres, L., Renard, Q., Dufey, J., Duyet, Tu. K.P., Lebaily, H.N., Francis, P., Hanbruge, E. Valorization of water hyacinth in vermicomposting using epigeic earthworm Perionyx excavates in central Vietnam. *Biotechnology, Agronomy, Society and Environment,* 15, (2011): 85-93.

11

Detection and Management of Plant Diseases for Regenerative Agriculture

***Kartik Sawant*[1], *Meenakshi Rana*[2], *Seweta Srivastava*[3] and *Devendra Kumar*[4]**

[1–4]*Department of Plant Pathology, School of Agriculture, Lovely Professional University, Phagwara, Punjab, India*

Introduction

In the last two decades, sustainable agriculture has grown to become one of the most pressing issues in agriculture. Although there are numerous definitions of sustainable agriculture, most agree on three fundamental overlapping components: Ecological, economic, and social sustainability (Kaur 2013; Pilgeram 2013). Meanwhile, the prevalence of plant diseases is a major impediment to sustainable agricultural production systems worldwide, particularly in the tropics and subtropics. Farmers frequently apply agrochemicals (i.e., pesticides, insecticides, fungicides, herbicides, etc.) over their recommended dose to manage plant diseases, raising serious concerns about food safety, soil, environmental quality, and pesticide resistance (Dordas 2008; Kaur 2013).

Disease detection on plants is critical for sustainable agriculture, and manually monitoring plant diseases is difficult. It usually requires a significant amount of work, knowledge of plant diseases, and extended processing time. Early detection is the base for efficient plant disease prevention and control, and it is critical in agricultural production management and decision-making. Plant disease identification has become increasingly important in recent years (Li et al., 2021). Diseased plants usually have visible marks or lesions on their leaves, stems, flowers, or fruits. In general, each disease or pest condition has a distinct visible pattern that can be used to identify abnormalities. Typically, the leaves of plants are the primary source for identifying plant diseases, and most disease symptoms may first appear on the leaves (Ebrahimi *et* al., 2017). Rapid advances in biotechnology have resulted in the development of various molecular diagnostic tools over the last decade. These tools, which are based on the properties of nucleic acids (DNA or RNA) or proteins of the target agents, have improved the efficacy, accuracy, and speed of detecting and identifying disease-causing agents, as well as the characterization of a wide range of pathogens and pests. Various techniques are used in biotechnology for disease diagnosis and management, but molecular markers are used in particular for the detection, identification, quantification, and characterization of plant pathogens that cause diseases in plants (Kumar *et al.,* 2015).

Despite the availability of many technologies in previous years, many challenges remained to identify uncultivable fungi, detect cryptic species, and differentiate fungal communities' diversity involved in causing plant disease. Nobody knew how many fungal species existed at the time. However, more and more diagnostic laboratories are now using molecular techniques to detect and identify plant-pathogenic fungi-caused diseases (Ahmed et al., 2014; Kim et al., 2010). Disease management in agriculture is the technique of reducing disease in crops to maximize the quantity or quality of harvest production. Fungi, oomycetes, bacteria, viruses, viroids, virus-like organisms, phytoplasmas, protozoa, nematodes, and parasitic plants are among the species that cause infectious diseases in crops. Resistance genes, fungicides, nematicides, quarantine, and other methods can be used to control infections (Lvet et al., 2020).

Plant

A living thing that can grow on land, in water, or even on other plants; it typically consists of a stem, leaves, roots, and flowers; and it bears fruit in the form of seeds.

Disease

A specific abnormal condition that adversely affects the structure or function of all or part of an organism and that is not immediately due to any external injury is referred to as a disease. Diseases can be caused by both internal and external factors. Diseases are typically understood to be distinct medical conditions that can each be identified by a unique set of signs and symptoms.

Sustainable Agriculture

The goal of sustainable agricultural practices is to preserve and improve the fertility of the soil while also protecting the natural environment and expanding the Earth's base of natural resources. Also, increases the amount of money that can be made from farming and fosters environmental stewardship and responsibility, which are also the objectives.

Biotechnology

Biotechnology can be defined as the application of organisms, cells, parts thereof, and molecular analogs for the production of goods and services and it integrates natural sciences and engineering sciences to achieve this goal.

Classification of Pathogens and Plant Diseases in Agriculture

Plant disease is an impairment in a plant's vital activities that causes disruption or alteration and affects all plant species, whether wild or farmed. Plant disease is defined as anything that inhibits a plant from reaching its full potential or output. Plant disease will reduce agricultural output. Pathogens (Figure 11.1) are the primary cause of plant illnesses (Vidhula and Sruthy, 2021) and can be classified based on the causal agent (Figure 11.2) and are as follows.

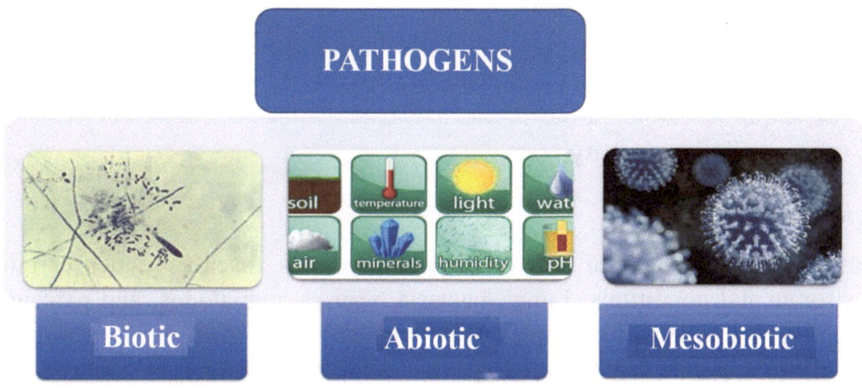

Figure 11.1: Classification of pathogens.
Source: Centers for disease control and prevention (CDC)

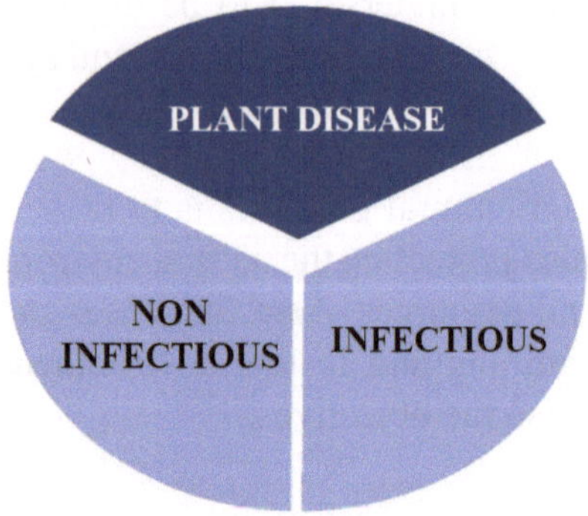

Figure 11.2: Classification based on causal agent.
Source: Centers for disease control and prevention (CDC)

Non-Infectious

Non-infectious pathogens are those that are inanimate or abiotic. As a result, they are not contagious. These are generated by perturbations in the plant body induced by a lack of certain natural traits caused by poor soil, air, and mechanical conditions. Examples are

- Low/high temperature
- Unfavorable oxygen levels
- unfavorable water levels
- Hail
- Wind
- Air pollution toxicity

Infectious

Pathogenic organisms or viruses cause infectious diseases under certain environmental conditions. Fungi, bacteria, viruses, nematodes, and parasitic blooming plants are all examples of pathogens. They obtain nutrition, water, and everything else they require to proliferate from their host. The majority of plant illnesses are caused by fungal and viral pathogens, with a few caused by bacterial and nematode pathogens. Some viruses can infect a variety of plants, but others require a specific host. The ability of pathogens such as fungi and bacteria to survive, disseminate, and reproduce varies.

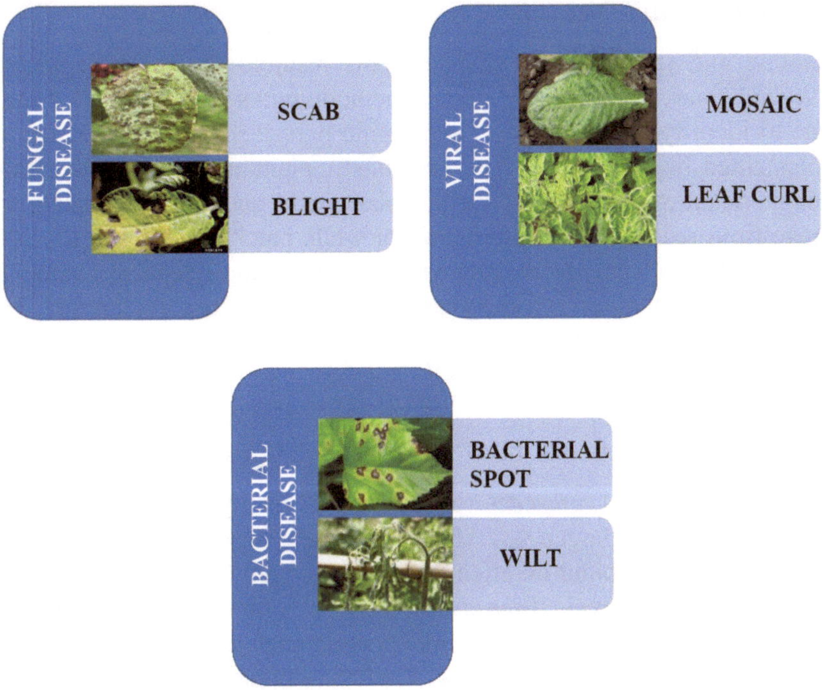

Figure 11.3: Infectious plant diseases.

Biotic

These infectious agents are neither living nor nonliving, for example, viruses and viroids.

Factors Affecting Plant Diseases

Biotic Factors

Pathogens of living nature are divided into the following categories. Fungi, Bacteria, Phytoplasma, Nematode, Algae, Phanerogams, and Organisms similar to Rickettsia and Protozoa.

Abiotic Factors

Instead of creating illness, these factors destabilize the plants. The primary causes are as follows: Nutrient deficiency or surplus, light, moisture, oxygen deficiency, improper cultural practices, temperature, pesticide toxicity, air pollutants, a deviation in soil conditions (acidity, alkalinity), etc.

Detection of Plant Diseases

Scientific and technological advancements resulted in the implementation of several innovative plant-treatment methodologies (Figure 11.4). When it came to identifying distinct cryptogrammic parasites, new approaches were implemented beginning in 1765. Previously, plant diseases were identified through visual examination of various plant parts, such as leaf curling, color change from red to yellow, and brown or white patches (Lopez et al., 2003). With scientific advancement, microscopic examination became available to see and research diverse pathogens involved in disease production (Belet Chane and Boyraz 2019).

Visual Observations

Effective initial detection is mostly dependent on signs and symptoms.

The visible physical presence of either the pathogen or the structures generated by the pathogen is a sign. The fungal mycelia and spore masses of downy mildews found on affected leaves, as well as the bacterial ooze of *Xanthomonas* leaf streak disease on rice, are common examples of easily identifiable symptoms. Symptoms are apparent changes that occur in the host plant as a result of pathogen infection. There is a characteristic manifestation of

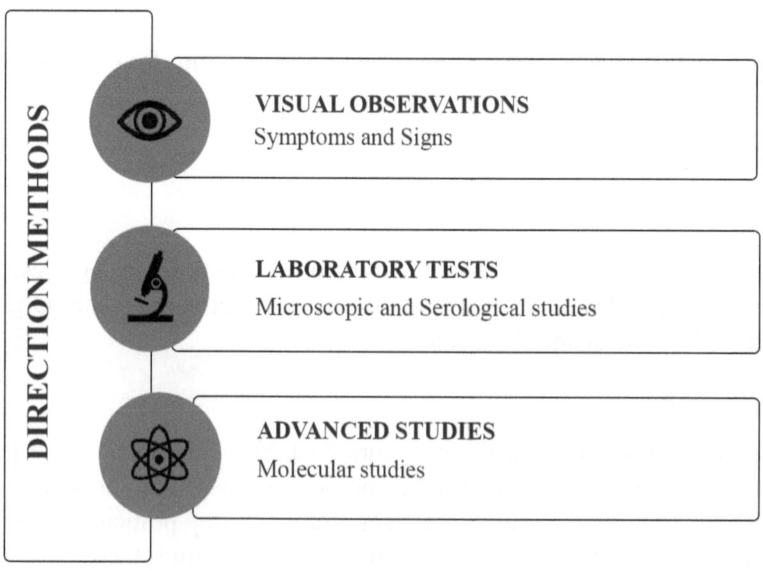

Figure 11.4: Methods of detection for plant disease.
Source: Kartik

symptoms for any illness in a given plant, which usually occurs in a sequential series during the duration of the disease. The disease syndrome is a collection of symptoms that depicts the disease image. The entire plant or any organ of the plant may exhibit morphological signs. These have been divided into many groups for ease of study. Plant disease morphological symptoms are primarily classified into six groups.

- Necroses
- Growth abnormalities
- Metaplastic symptoms
- Proleptic symptoms
- Color changes
- Wilts

Laboratory Tests

Conventional fungal pathogen detection methods depend on isolation of the fungal pathogen or an appropriate nutritive medium, microscopic study of spores and spore-bearing structures, and taxonomic identification (Table 11.1). These processes are time-consuming and labor-intensive, produce uneven results, and necessitate extensive knowledge of fungal taxonomy (Narayanasamy 2002).

Table 11.1: Typical symptoms of plant diseases and how to visualize them

Cause of diseases	Description	Instrument	Magnification	Examples
Fungi	Small filaments that resemble threads; from these filaments, huge fruiting structures may develop.	Microscope	20–250	Rusts, smuts, leaf spots, wilts, powdery mildews, cankers.
Bacteria	Tiny single-celled organism	Microscope	400–1000	Bacterial blight, fire blight, cankers.
Viruses	Very small RNA/DNA particles with a protein covering that are either spherical or rod-shaped.	Electron Microscope	20,000–100,000	Potato virus, TMV, tomato leaf curl, Cucumber mosaic virus.
Phytoplasmas	Very small organisms with no cell walls and an unclear form.	Electron Microscope	20,000–50,000	Little leaf, aster yellows

Source: Chandrashekara et al., 2012

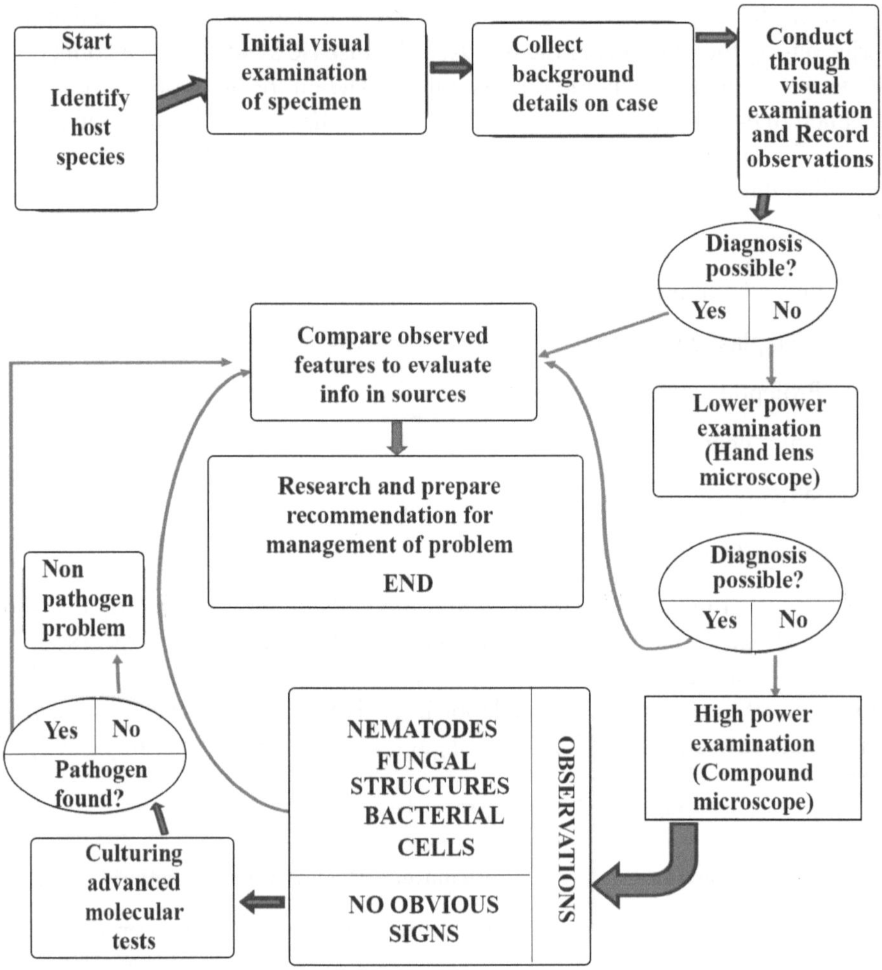

Figure 11.5: Detection and diagnosis of plant diseases.
Source: V. J. Galea

Advanced studies

Due to the limits of traditional approaches, molecular techniques were developed for the research of identification and classification difficulties. A wide range of molecular approaches are becoming more important tools in many aspects of fungal diagnoses. Immunological approaches, nucleic acid-based probe technology, and polymerase chain reaction (PCR) technology are among these techniques. The former methods depend on phenotypic characteristics, but the latter, which is based on genotypic characteristics, provides rapid, highly specific, effective, and perhaps more accurate findings. In contrast to the fundamental procedures, organism isolation does not need culturing (Badali and Nabili 2012; Spring and Thines, 2010). When morphological characteristics are not evident, an immunological approach called enzyme-linked immunosorbent assay (ELISA) was used to quantify and detect fungus. These immunological approaches, which rely on an antigen-antibody response in conjunction with a fluorescent dye or enzyme, are now accessible for easy and quick results (Mostafa et al., 2012; Peruski and Peruski, 2003).

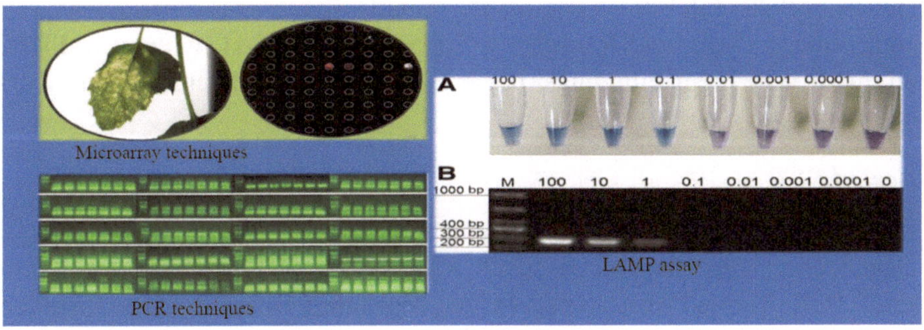

Figure 11.6: Various molecular detection techniques (images).

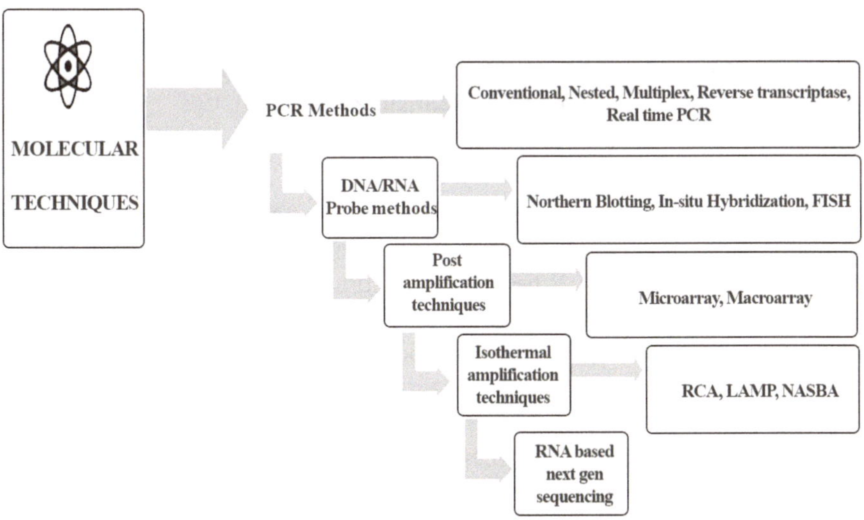

Figure 11.7: Different types of molecular techniques.
Source: Kartik

PCR (Polymerase Chain Reaction)-Based Methods

Conventional PCR

The development of PCR resulted in a massive change in the field of plant pathology. Using precise primers, this approach allows for the replication of a specific segment of DNA in millions of copies via alternating cycles of denaturation, annealing, and elongation. PCR was initially quite specific for detecting illnesses caused by bacteria and viruses. It is now commonly used for detecting plant pathogens as well. The efficiency of DNA extraction and the concentration of deoxynucleoside triphosphate (DNTs) are important factors in PCR. Inhibitors contained in sample assays can sometimes impair performance (Fang and Ramasamy, 2015). This method, without a doubt, produces very precise findings, but it is prohibitively expensive and labor-intensive.

Nested PCR

To obtain high specificity and sensitivity, nested PCR is performed. For fungus identification, for example, nested PCR sensitivity is 1000 times better than single PCR (Yeo and Wong, 2002). In this procedure, two rounds of amplification are performed in which a single pair of primers is used to amplify a broad tract of DNA, and this amplified sequence of DNA serves as a target for the second round, which is performed using two internal primers. Because two cycles of amplification need to be done in separate tubes, there is a significant danger of contamination with this method of PCR. As a result, the possibility of false-positive findings owing to contamination, as well as the rigorous effort required, is a key downside of this approach (Rahman et al., 2013).

Multiplex PCR

Multiplex PCR can reduce time and money by employing several pairs of primers in the same reaction, allowing for the simultaneous detection of multiple targeted DNA sequences. When plants become infected with more than one pathogen, this approach becomes extremely important in plant pathology. On the agarose gels, different fragments unique to target pathogenic fungus were amplified and recognized based on their molecular sizes. Amplicon size has a significant impact on DNA-synthesizing accuracy and to eliminate this issue, primers must be properly constructed, as well as their relative concentration and annealing temperature (Dasmahapatra and Mallet, 2006).

RT PCR

A vast number of alterations have been made to the Standard PCR technique since its inception and some of these modifications have increased the usability and diagnostic capability of PCR in a variety of biological and medical disciplines (Tang et al., 1997). RNA is reversely transcribed into cDNA using random primers and RT enzyme and then amplified using any PCR-based technique. As a result, RT-PCR is primarily employed to identify and diagnose infections caused by RNA-containing viruses (such as retroviruses). The detection of RNA-containing viruses can aid in the development or testing of antimicrobial vaccines or treatment. *Fusarium graminearum* that cause Fusarium ear blight (FEB) disease in cereals such as wheat, rye, barley, oat, and maize have been quantified using RT-PCR (Brown et al., 2011).

Real-Time PCR

Conventional PCR technologies only provide amplification of target DNA, which is insufficient for quick illness diagnosis. A novel technology, RT-PCR, has been introduced in recent years. This technique is an upgraded version of traditional PCR in which the DNA may be measured while it is being amplified (Mackay, 2004). Real-time PCR has been used to identify and quantify pathogenic fungi such as *Aspergillus versicolor, Cladosporium cladosporoiodes, Stachybotrys chartarum,* and *Alternaria alternate* (Black 2009).

DNA/RNA Probe-Based Methods

With exceptional sensitivity and speed, DNA/RNA probe techniques employ the probe to identify plant diseases caused by microorganisms such as fungi. For the identification of fungi, DNA probes are typically utilized as an alternative to PCR. However, in current approaches, they are typically employed in combination with PCR (McCartney et al., 2003).

Northen Blotting

The Northern blot, also known as the RNA blot, is used to identify pathogenic fungi by transferring RNA onto a carrier and is also useful for studying gene expression. First, the RNA size is determined by the location of the bands on the blot. If the size of the RNA is known, it may be used to approximate the coding capacity of transcripts and the size of the protein that it encodes. To begin, one major downside of the blotting approach is the danger of mRNA destruction during electrophoresis. Second, detection with several probes is challenging. When compared to RT-PCR, the sensitivity of northern blotting is rather poor. *Magnaporthe grisea* was found in rice plants using real-time PCR and Northern Blotting (Qi and Yang, 2002).

In-situ hybridization

In situ hybridization histochemistry is another name for RNA blotting. It contains useful information for recognizing and counting fungus. This approach employs single-stranded RNA probes, commonly known as riboprobes. These probes are marked with the letter 35S. Northern blots are quite similar to in situ hybridization. Both of these rely on the hybridization of tagged DNA/RNA probes to mRNA homologous sequences. The initial material used in these two procedures differs. In the case of a northern blot, tissue digest is utilized as the starting material, whereas a histological slice is used in *in situ* hybridization. Regardless of whether direct hybridization is used, signal hybridization identifications are most efficiently obtained following fungal growth or biological amplification (Jensen, 2014).

The fundamental advantage of this method is that it makes the best usc of tissue that is in short supply, such as clinical biopsies and cultured cells. On the same tissue, hundreds of different hybridizations can be performed. Tissue libraries can be created and kept in the freezer for future use. In situ hybridization can be performed in a variety of methods, including probing with synthetic oligonucleotides, cDNAs, and cRNAs. However, riboprobe probing produces the most sensitive and efficient findings. Radioactive or non-radioactive nucleotides can be used to label probes. 35S riboprobes are the most sensitive approach for identifying mRNA among these radioactive probes (Hayden *et al.*, 2002).

Post Amplification Technique

Microarray

The two technological advancements resulted in the concept of microarray technology. First, there are attempts to sequence DNA, and then there is a focus on expressed genome components. Microarray DNA chip technology allows for the simultaneous study of thousands of mRNAs and is used to monitor changes in gene expression. This approach differs from the others in that it offers measurements of gene expression on specific groups of genes (Eshaque and Dixon, 2006). This technology allows for a thorough comprehension of the fungal cell in a single array. Because it does not need large-scale DNA sequencing, the microarray approach is relatively simple to utilize. Although this approach can detect worldwide changes in gene expression, it has severe limitations. To begin, this approach needs a substantial amount of mRNA. Second, the expense and accessibility of this study are constraints (Singh and Kumar, 2013).

The Macroarray plot is also known as the DNA array hybridization plot or the Reverse dot plot. This approach relies on the sensitivity of DNA amplification rather than radioisotopes (Singh and Kumar, 2013). In an increasing number of laboratories throughout the world, Macroarray is quickly becoming a standard molecular tool for diagnostic and epidemiological investigations. It is based on the simultaneous amplification of related species by PCR. In addition, it analyzes many amplified sequences at the same time in a single hybridization procedure. It's a more sensitive technology than just PCR. This technique combines PCR amplification with hybridization, which boosts sensitivity up to 1000-fold or more than PCR alone (Taoufik*et* al., 2004). *Alternaria alternata, Aspergillus fumigatus, Fusarium solani, Candida albicans,* and *Cladosporium herbarum* have all been identified using this assay (Sato et al., 2010).

Isothermal Amplification-Based methods

Rolling Circle Amplification (RCA)

RCA technology is a reliable and straightforward approach that provides a universal platform for the localization of a wide range of compounds, including nucleic acid sequences (Gusev et al., 2001). Rolling replication of small single-stranded DNA circular molecules underpins RCA. The RCA method, also known as rolling circle replication, is isothermal (RCR). This procedure necessitates the use of DNA polymerase, a primer to trigger replication, DNTPs, and DNA binding and unwinding proteins. The RCA reaction is used by the majority of the organizations for signal amplification, with tiny circular probes serving as templates. Primers in RCA reactions serve two functions: They operate as RCA signal amplifiers and discriminators by being complementary to the DNA-targeted sequence (Kuhn et al., 2002).

Loop-Mediated Isothermal Amplification (LAMP)

LAMP is a robust and new nucleic acid amplification technology that may be used instead of PCR. It amplifies the targeted nucleic acid with great specificity under isothermal conditions. LAMP does not need a thermal cycler to generate temperature variations; instead, it just needs a single temperature for DNA amplification (Tsui et al., 2011). The LAMP technique does not require expensive equipment to achieve high levels of accuracy, and it requires fewer preparation steps than traditional and qPCR methods. Another benefit of LAMP over PCR in many circumstances is that inhibition reactions occur less frequently in LAMP than in PCR. The thermos-dependent dimorphic fungus, *Paracoccidioides brasiliensis*, has been effectively detected using LAMP (Endo et al., 2004).

RNA-Sequence-Based Next-Generation Sequencing

RNA-Seq is a relatively recent deep-sequencing technique. In most cases, a huge population of RNA is converted to a cDNA library with adaptors attached

Table 11.2: Techniques for identifying phytopathogenic fungus using molecular diagnostics

Methods	**Advantages**	**Disadvantages**
Conventional PCR Methods for detecting phytopathogenic fungi through molecular diagnostic techniques.	PCR produces quick and precise results. when using a particular species primer	Species Required much labor and cost
Nested PCR	The use of two sets of primers in PCR increases the productivity and specificity of target DNA amplification.	Contamination risk due to two amplification cycles
Multiplex PCR	Using many pairs of primers in the same reaction saves time and money.	Primer and probe interference reduces sensitivity
Reverse transcriptase (RT) PCR	PCR provides quantitative data on pathogens and is more sensitive than traditional PCR.	Formation of each assay is time-consuming and requires expensive equipment and the reagents
Real-time PCR (q PCR)	Automated and no need of post-amplification analysis	Cost and complexity as a result of concurrent heat cycling and fluorescence detection
In situ hybridization	Maximum use of the short supply of tissue	Difficulty in identifying targets that have low DNA and RNA copies
Microarray	It is simple to utilize since it does not need large-scale DNA sequencing.	Large amount of mRNA is required
LAMP	LAMP Rapid, sensitive and highly specific	The primer design is sophisticated; it detects just one pathogen; there is a possibility of sample contamination.
Northern blotting	Detection of RNA size	Applied only on a small sample of the genes.

Source: Aslam *et al.*, 2017

to one or both ends. Following that, each fragment with or without amplification, is high-throughput sequenced to produce short sequences from one end, as in single-end sequencing, or both ends, as in pair-end sequencing. Depending on the DNA sequencing technology employed, reads are typically between 30 and 400 bp in length. The library is constructed to test how closely the RNA sequencing findings show the original RNA transcripts. To construct an RNA-seq library, either the RNA or the cDNA must be fragmented to enable processing by next-generation sequencing. mRNA should be primed for RT reactions using either random primers or oligo primers. The advantage of utilizing oligo (dT) is that the bulk of the cDNA generated should be poly-adenylated mRNA, resulting in more informative sequences (Mortazavi et al., 2008).

Management of Plant Diseases

The procedures intended to exclude, decrease, or destroy inoculums are part of pathogen management. The host must be managed through strategies that increase plant vigor, encourage resistance through nutrition, introduce genetic resistance through breeding, and provide need-based protection through chemical measures. Environment management is the technique of altering the environment such that disease or pathogen growth is inhibited and the host is not more vulnerable to damage. Each disease prevention method falls short of being effective on its own. But when many of these tactics are applied together, a tolerable level of control may be attained. The management of the entire farm must be combined with effective disease management (Rasool et al., 2018).

Components of Integrated Disease Management

The IDM can be defined as "a sustainable strategy to controlling disease by integrating biological, cultural, physical, and chemical instruments in a way that reduces economic, health, and environmental hazards" (Rasool et al., 2018). As a consequence, the three main components of disease management are listed below: Cultural control, biological control, and chemical control. Even though these components will be addressed individually, it should be noted that the various components are frequently complementary to one another, with strong interaction among and between them and the environment, and that it is critical to move away from relying on a single technology and toward a more ecological approach based on a fundamental understanding of population biology at the local farm level, as well as reliance on the integration of control components (Thomas, 1999).

Cultural control: This refers to the modification of the agricultural environment to make it less favorable for pathogens, such as modifying planting time, plant spacing, season, irrigation, green manuring, crop rotation, and crop mix, either alone or in conjunction with other strategies. Cultural management measures not only promote crop health, but they are also beneficial in lowering inoculum potential and increasing the biological activities of antagonists in the soil (Rasool et al., 2018). Crop rotation, intercropping, and multiple cropping have long been viewed as effective methods of lowering disease and pathogen populations (Brust and King, 1994). In capital-scarce, labor-intensive, pest- and disease-prone agriculture systems, intercropping yields greater and more consistent returns than monocropping (Perrin, 1977). To improve output, agricultural methods must be used effectively, assuring higher productivity and the sustainability of the production system. One approach to ensure this occurs is the adoption of a sustainable practice, one of which is cultural control (Costa et al., 2018).

To promote growth and development of the crop from the pathogen, agricultural pest, and weed, effective agricultural practices are used to manipulate the preplanting and plant development circumstances (Carvalho *et al.,* 2015). Crop rotation, fertilization, scavenging of infested plant parts, harrowing, bait plants, weeding, destruction of crop residues, well-managed irrigation, pruning, spacing between rows, planting density, appropriate planting time, use of varieties adapted to the regions, use of mulch, among other good agricultural practices for pest, disease, and weed control stand out (Carvalho *et al.,* 2015).

Biological control: Biological control may be used as an effective way to manage plant diseases, boost output, protect the environment and biological resources, and move toward a sustainable agricultural system. The phrase "biological control" and its shortened synonym "bio-control" have been used in a variety of biological domains, most notably entomology and plant pathology. In plant pathology, the word refers to the employment of microbial antagonists as well as host-specific pathogens to suppress diseases. Biological control agents (BCAs) are those organisms that are used for controlling pests or diseases. Biological management of plant diseases by microorganisms is thought to be a more natural and ecologically friendly alternative to present chemical treatment approaches (Prajapati et al., 2020).

Advantages of using Bio-control agents (BCAs) in pest and disease management

(i) Saves the environment from pollution

(ii) Don't have a negative impact on beneficial organisms

(iii) Cheaper than pesticides and does not run into resistance issues

(iv) BCAs require just a single treatment to sustain themselves, whereas fungicide requires many applications

(v) BCAs are strong, resilient, and long-lasting

(vi) BCAs contributed to the development of induced systemic resistance in the crop species

(vii) Not a problem with water contamination

(viii) Biofertilizers and biocontrol agents can be used together

(ix) Biocontrol agents are relatively easy to handle and apply to the target

(x) By promoting good soil microflora, biocontrol agents not only control the disease but also promote healthy root and plant growth

Chemical control: The R and D-based chemical firms have had great success innovating new products, which has helped them bring a wide range of solutions to market for farmers all over the world to employ in their initiatives to boost agricultural production and safeguard their harvests and livelihoods. This chemical diversity has been proven to be crucial for several reasons, and without the ongoing research efforts of the top companies, farmers would be dependent on aging, less effective products supplied by generic manufacturers, with their problems of unsustainable unavailability due to outdated product safety as well as losses of efficacy due to resistance. Thankfully, in today's environment, this range of options has so far been preserved (Leadbeater and Gisi, 2010).

Chemicals are a crucial component of successful integrated pest management (IPM) systems for controlling plant diseases. In the middle of the 1800s, lime sulfur and Bordeaux combinations were introduced, and in the first half of the 1900s, fungicides with various sites of action and protective and contact qualities against several target sites in fungal metabolism played a significant role. In the 1960s, fungicides that block a particular target site were developed. Numerous particular fungicides were widely used up until recently because of their wide application windows, systemic activity, and protective and curative characteristics (Knight et al., 1997). To preserve green leaf areas and boost grain output, chemical management is frequently utilized. When the infection level is visible to the naked eye to be greater than 5% of the leaf area, fungicide treatment has been proven to be effective (Cook et al., 1999). One part of crop management is seed treatment. It is a cutting-edge and affordable delivery technology that partially replaces the traditional foliar spray and safeguards the genetic potential of the seed against diseases from the moment of sowing (Smiley et al., 2002). Fungicide seed treatments can control root rots and other plant diseases because they have antagonistic action against pathogenic fungi on seed-borne and root rot diseases (Bailey and Lazarovits, 2003).

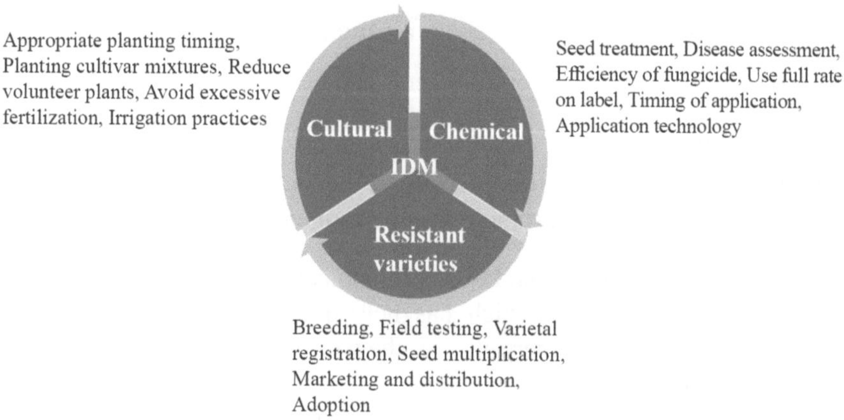

Figure 11.8: Integrated disease management practices.
Source: Kartik

Control of Diseases Through Sustainable Nutrient Management

Numerous researchers have reported on the suppression or control of disease using nutritional modification, which was accomplished by either altering the nutrient availability or the nutrient absorption (Huber and Graham, 1999). The nutritional state of the plant and the factors that might affect the development of the illness, such as thick stands, changes in light interception, and humidity inside the crop stand, are directly impacted by fertilizer application on the development of plant diseases under field settings (Dordas, 2008). The consensus is that healthy plants have lower disease susceptibility and infection rates because they have balanced nutrients. Therefore, it's critical to offer a balanced diet during the period when the nutrient may be utilized to prevent disease the most.

In addition to fertilization, any management technique that modifies the soil's environment such as tillage, seedbed firmness, nursery location, moisture control via irrigation, and manures can influence the development of diseases. For instance, the club root of crucifers is not directly affected by liming, but club root spore germination is reduced at pH > 7.0. Alternately, liming soils to near-neutral pH raises the prevalence of diseases including wheat take-all infection and potato scab (Havlin et al., 2009). The prevalence of potato scab, which is brought on by *Streptomyces scabies*, might thus be reduced either by reducing soil pH or by adding Manganese (Mn). The enhanced resilience of the tuber tissue to the pathogen and the reduction of *S. scabies'* vegetative development prior to infection may be the causes of the suppressive action on Mn (Gupta et al., 2017).

Nutrient management is a key factor in controlling plant diseases since disease resistance in any plant is primarily genetically determined but is closely related to the nutritional state of the plants or pathogens. Because there is a dynamic relationship between the nutritional condition of pathogenic plants and their abiotic environment, good nutrient management in cultivated crops can effectively reduce the severity of the majority of diseases. Furthermore, when the crops are undernourished, nutrient management results in a more noticeable reduction in the severity of diseases. The nutritional state of the host plants also affects their morphological or histological traits, which in turn controls pathogen entrance, penetration, and infection dissemination to unaffected plant sections. When plants are healthy and well-nourished, diseases can be suppressed to an acceptable level or to a level that can be further managed by pesticides or other traditional methods that are more effective, efficient, and environmentally friendly. Although not the sole strategy, plant nutrients should be a part of disease control plans since they are effective in reducing the incidence of disease (Gupta et al., 2017).

This increased tolerance or resistance mechanism of the host plant may be caused by modifications in the physiological or biochemical processes as influenced by the plant nutrients. The usage of plant nutrients to treat plant diseases in comparison to improving soil fertility that can sustain agricultural yield for our future generations needs to be the subject of further in-depth research. Understanding the physiological or biochemical processes involved in the use of nutrition to prevent or treat disease is crucial. Genetically modified crops require less nutritional demand when compared to the conventional crops grown.

Conclusion

Despite the availability of numerous technologies in past years, several obstacles remained in identifying unculturable fungi, detecting cryptic species, and discriminating fungal communities' diversity involved in causing plant disease. Nobody knew how many fungal species existed at the time. However, more and more diagnostic laboratories are now adopting molecular approaches to detect and identify plant pathogenic fungi-caused diseases. IDM entails scouting and the timely use of a variety of techniques and tactics.

These may include site selection and preparation, using resistant cultivars, changing planting procedures, adjusting the environment through drainage, irrigation, pruning, thinning, shading, and so on, and, if necessary, spraying pesticides. Monitoring environmental parameters (temperature, moisture, soil pH, nutrients, and so forth), disease forecasting, and determining economic thresholds are all significant components of the management plan.

References

Ahmed, S. A., Van Den Ende., B. H. G., Fahal., A. H., Van De Sande., W. W., De Hoog, G. S. (2014). Rapid identification of black grain eumycetoma causative agents using rolling circle amplification. *PLoS neglected tropical diseases*, 8, no. 1, (2014): e3368.

Aslam, S., Tahir, A., Aslam, M. F., Alam, M. W., Shedayi, A. A., Sadia, S. Recent advances in molecular techniques for the identification of phytopathogenic fungi–a mini review. *Journal of Plant Interactions*, 12, no. 1, (2017): 493-504.

Badali, H., Nabili, M. Molecular tools in medical mycology; Where We Are! *Jundishapur Journal of Microbiology*, 6, no. 1, (2012): 1-3.

Bailey, K. L., Lazarovits, G. Suppressing soil-borne diseases with residue management and organic amendments. *Soil and tillage research*, 72, no. 2, (2003): 169-80.

Belet Chane, T., Boyraz, N. Biotechnological tools for detection, identification and management of plant diseases. *Afr J Biotechnol,* 18, (2019): 798-805.

Black, J. 2009. Quantitative real-time polymerase chain reaction (qPCR) of filamentous fungi in carpet. Research Triangle Park, NC: RTI Press.

Brown, N. A., Bass, C., Baldwin, T. K., et al. Characterisation of the Fusarium graminearum-wheat floral interaction. *Journal of pathogens*, (2011): pp. 626345.

Brust, G. E., King, L. R. Effects of crop rotation and reduced chemical inputs on pests and predators in maize agroecosystems. *Agriculture, ecosystems & environment*, 48, no. 1, (1994): 77-89.

Carvalho, A. M. D., Coser, T. R., Rein, T. A., Dantas, R. D. A., Silva, R. R., Souza, K. W. Management of cover crops at flowering and physiological maturity and its effect on maize yield. *Pesquisa Agropecuária Brasileira*, 50, (2015): 551-61.

Chandrashekara, K. N., Raju, S. R., Chandrashekara, C. 2012. Molecular and Biotechnological Approaches in Plant Disease Management. *Ecofriendly Innovative Approaches in Plant Disease Management.* International Book Distributors.

Cook, R. J., Hims, M. J., Vaughan, T. B. Effects of fungicide spray timing on winter wheat disease control. *Plant Pathology*, 48, no. 1, (1999): 33-50.

Costa, F., De, KD., Menezes, JFS., et al. Agronomic performance of conventional soybean cultivated with organomineral and mineral fertilizers. *Nucleus Magazine*,15, no. 2, (2028): 301-09

Dasmahapatra, K. K., Mallet, J. DNA barcodes: Recent successes and future prospects. *Heredity*, 97, no. 4, (2006): 254-55.

Dordas, C. Role of nutrients in controlling plant diseases in sustainable agriculture. A review. *Agronomy for sustainable development*, *28*, no. 1, (2018), 33-46.

Ebrahimi, M. A., Khoshtaghaza, M. H., Minaei, S., Jamshidi, B. (2017). Vision-based pest detection based on SVM classification method. *Computers and Electronics in Agriculture*, 137, (2017): 52-8.

Endo, S., Komori, T., Ricci, G., et al. Detection of gp43 of Paracoccidioides brasiliensis by the loop-mediated isothermal amplification (LAMP) method. *FEMS microbiology letters*, 234, no. 1, (2004): 93-97.

Eshaque, B., Dixon, B. Technology platforms for molecular diagnosis of cystic fibrosis. *Biotechnology advances*, 24, no. 1, (2006): 86-93.

Fang, Y., Ramasamy, R. P. Current and prospective methods for plant disease detection. *Biosensors*, 5, no. 3, (2015): 537-61.

Gupta, N., Debnath, S., Sharma, S., Sharma, P., Purohit, J. 2017. Role of nutrients in controlling the plant diseases in sustainable agriculture. In *Agriculturally important microbes for sustainable agriculture*. pp. 217-62. Singapore: Springer.

Gusev, Y., Sparkowski, J., Raghunathan, A., et al. Rolling circle amplification: A new approach to increase sensitivity for immunohistochemistry and flow cytometry. *The American journal of pathology*, 159, no. 1, (2001): 63-69.

Jensen, E. Technical review: In situ hybridization. *The Anatomical Record*, 297, no. 8, (2014): 1349-53.

Hayden, R. T., Qian, X., Procop, G. W., Roberts, G. D., & Lloyd, R. V. (2002). In situ hybridization for the identification of filamentous fungi in tissue section. *Diagnostic Molecular Pathology*, 1, no. 2, (2002): 119-26.

Huber, D. M., Graham, R. D. (1999). The role of nutrition in crop resistance. *Mineral nutrition of crops: Fundamental mechanisms and implications*, 18, no. 12, (1999): 169.

Kaur, G. Sustainable development in agriculture and green farming in India. *OIDA International Journal of Sustainable Development*, 6, no. 12, (2013): 61-4.

Kim, S. W., Li, Z., Moore, P. S., Monaghan, A. P., Chang, Y., Nichols, M., John, B. A sensitive non-radioactive northern blot method to detect small RNAs. *Nucleic acids research*, 38, no. 7, (2010): e98-e98.

Kumar, A., Singh, R., Singh, R., Sengar, R. S. Molecular Approach for Detection of Plant Pathogen. *Biotech Today*, 5, no. 2, (2015): 14-9.

Knight, S. C., Anthony, V. M., Brady, A. M., et al. Rationale and perspectives on the development of fungicides. *Annual review of phytopathology*, 35, no. 1, (1997): 349-72.

Kuhn, H., Demidov, V. V., Frank-Kamenetskii, M. D. Rolling-circle amplification under topological constraints. *Nucleic acids research*, 30, no. 2, (2002): 574-80.

Leadbeater, A., Gisi, U. 2010. The challenges of chemical control of plant diseases. In *Recent developments in management of plant diseases*. pp. 3-17. Dordrecht: Springer

Li, L., Zhang, S., Wang, B. (2021). Plant disease detection and classification by deep learning—a review. *IEEE Access*, 9, (2021): 56683-98.

López, M. M., Bertolini, E., Olmos, et al. (2003). Innovative tools for detection of plant pathogenic viruses and bacteria. *International Microbiology*, 6, no. 4, (2003): 233-43.

Lv, H., Fang, Z., Yang, L., Zhang, Y., Wang, Y. An update on the arsenal: Mining resistance genes for disease management of Brassica crops in the genomic era. *Horticulture research*, 34, no. 7, (2020).

McCartney, H. A., Foster, S. J., Fraaije, B. A., Ward, E. Molecular diagnostics for fungal plant pathogens. *Pest Management Science: Formerly Pesticide Science*, 59, no. 2, (2003): 129-42.

Mortazavi, A., Williams, B. A., McCue, K., Schaeffer, L., Wold, B. Mapping and quantifying mammalian transcriptomes by RNA-Seq. *Nature methods*, 5, no. 7, (2008): 621-28.

Mostafa, A. T., Armin, A., Hamid, P., Reza, A. M. Rapid detection methods for analysis of fungi and mycotoxins in agriculture products. *Res J Recent Sci*, 1, no. 7, (2012): 90-8.

Narayanasamy, P. 2002. *Microbial plant pathogens and crop disease management*. CRC Press. *Science Publishers.*

Perrin, R. M. Pest management in multiple cropping systems. *Agro-Ecosystems*, 3, (1977): 93-118.

Peruski, A. H., Peruski Jr, L. F. Immunological methods for detection and identification of infectious disease and biological warfare agents. *Clinical and vaccine immunology*, 10, no. 4 (2003): 506-13.

Pilgeram, R. The political and economic consequences of defining sustainable agriculture in the US. *Sociology Compass*, 7, no. 2, (2013): 123-34.

Prajapati, S., Kumar, N., Kumar, S., Maurya, S. Biological control a sustainable approach for plant diseases management: A review. *Journal of Pharmacognosy and Phytochemistry*, 9, no. 2, (2020): 1514-23.

Qi, M., Yang, Y. Quantification of Magnaporthe grisea during infection of rice plants using real-time polymerase chain reaction and northern blot/ phosphoimaginganalyses. *Phytopathology*, 92, no. 8, (2002): 870-76.

Rahman, M. T., Uddin, M. S., Sultana, R., Moue, A., Setu, M. Polymerase chain reaction (PCR): A short review. *Anwer Khan Modern Medical College Journal*, 4, no. 3, (2013): 30-6.

Rasool, K., Rasool, F., Bhat, N.A., et al. IDM: A new approach to disease management. *International Journal of Chemical Science,* 2, no. 2, (2018): 7-11.

Sato, T., Takayanagi, A., Nagao, K., et al. Simple PCR-based DNA microarray system to identify human pathogenic fungi in skin. *Journal of clinical microbiology*, 48, no. 7, (2010): 2357-64.

Singh, A., Kumar, N. (2013). A review on DNA microarray technology. *International Journal of Current Research and Review*, 5, no. 22, (2023): 1.

Smiley, R. W., Cook, R. J., Paulitz, T. 2002. Seed treatments for small grain cereals. Oregon State University, USA, 8.

Spring, O., Thines, M. 2010. Molecular techniques for classification and diagnosis of plant pathogenic Oomycota. In Molecular identification of fungi . pp. 35-50. Berlin, Heidelberg: Springer.

Tang, Y. W., Procop, G. W., &Persing, D. H. Molecular diagnostics of infectious diseases. *Clinical chemistry*, 43, no. 11, (1997): 2021-38.

Taoufik, A., Nijhof, A., Hamidjaja, R., Jongejan, F., Pillay, V., Sonnevelt, M., De Boer, M. 2004. Reverse line blot hybridisation in the detection of tick-borne diseases. *Biotech Magazine Online as published in BTi September*.

Thomas, M. B.. Ecological approaches and the development of "truly integrated" pest management. *Proceedings of the National Academy of Sciences*, 96, no. 11, (1999): 5944-51.

Tsui, C. K., Woodhall, J., Chen, W., et al. Molecular techniques for pathogen identification and fungus detection in the environment. *IMA fungus*, 2, no. 2, (2011): 177-89

Vidhula T., &Sruthy, K B. 2021. Plant Disease Identification System-A Review. *International Research Journal of Engineering and Technology* (IRJET).

Yeo, S. F., & Wong, B. Current status of nonculture methods for diagnosis of invasive fungal infections. *Clinical microbiology reviews*, 15, no. 3, (2002): 465-84.

12

Economics of Regenerative Agriculture Benefits and Drawbacks

Harshal Avinashe[1], Sharad Sachan[2], Sneha Choudhary[3], and Nidhi Dubey[4]

[1,4]*Department of Genetics and Plant Breeding, Lovely Professional University, Phagwara, Punjab, India*

[2]*Department of Agricultural Economics and Extension, Lovely Professional University, Phagwara, Punjab, India*

[3]*Department of Plant Pathology, Lovely Professional University, Phagwara, Punjab, India*

Introduction

Regenerative agriculture is an approach to farming and land management that focuses on restoring and improving the health of ecosystems and the soil (Latacz and Nuppenau. 2019). It goes beyond sustainable agriculture by actively seeking to regenerate and rejuvenate the natural resources utilized in agricultural practices (Lal, 2018). The goal of regenerative agriculture is to create a resilient and sustainable farming system that enhances biodiversity, improves soil health, sequesters carbon, and promotes overall ecosystem health (Paustian, et.al., 2018). Regenerative agriculture encompasses a range of practices and principles that vary depending on the specific context and agricultural systems (Smith *et. al.*, 2015).

Regenerative Agriculture often include:

- Conservation tillage: Minimizing soil disturbance to maintain soil structure, prevent erosion, and preserve beneficial microorganisms.

- Cover cropping: Planting cover crops in between cash crops to protect the soil from erosion, improve nutrient cycling, and enhance soil fertility.
- Crop rotation and diversification: Alternating crops and incorporating diverse plant species to improve soil health, break pest cycles, and optimize nutrient use.
- Composting and organic matter management: Adding organic materials to the soil to improve its fertility, water-holding capacity, and nutrient availability.
- Managed grazing: Employing rotational grazing strategies to mimic natural grazing patterns, improve soil health, and promote plant growth.
- Agroforestry: Integrating trees, shrubs, or other perennial plants into agricultural systems to provide additional ecosystem services, such as shade, windbreaks, and habitat for beneficial organisms.
- Water management: Implementing practices that optimize water use efficiency, such as rainwater harvesting, irrigation management, and soil moisture monitoring.

Regenerative agriculture plays a crucial role in sustainable food production by addressing the environmental, social, and economic challenges faced by modern agricultural systems (Ponisio et al., 2015). The following are important reasons why regenerative agriculture is important:

1. **Soil Health and Fertility**: Regenerative agriculture focuses on building and preserving healthy soils. By using practices such as cover cropping, crop rotation, and organic matter management, regenerative agriculture improves soil structure, increases nutrient availability, and enhances microbial activity. Healthy soils support robust plant growth, reduce the need for synthetic fertilizers, and contribute to long-term agricultural productivity (White and Hall, 2017).
2. **Climate Change Mitigation**: Regenerative agriculture has the potential to sequester significant amounts of carbon dioxide from the atmosphere and mitigate climate change. Practices like cover cropping, agroforestry, and holistic grazing enhance carbon sequestration in soils, plants, and trees, which helps to offset greenhouse gas emissions and reduce the carbon footprint of food production (West *et. al.*, 2014).
3. **Biodiversity Conservation**: Regenerative agriculture promotes biodiversity by creating habitats for beneficial insects, birds, and other wildlife. Regenerative practices support a wide range of organisms and

help protect endangered species by incorporating diverse plant species, preserving natural ecosystems, and minimizing chemical inputs, Biodiversity conservation is essential for maintaining resilient and sustainable food systems (Chappell *et. al.,* 2013).

4. **Water Management**: Regenerative agriculture employs strategies to optimize water use and enhance water quality. Practices such as conservation tillage and cover cropping reduce soil erosion and improve water infiltration, helping to conserve water resources. By promoting healthy soils with good water-holding capacity, regenerative agriculture contributes to better drought resistance and water efficiency in food production (Albrecht *et. al.,* 2018).
5. **Resilience and Adaptation**: Regenerative agriculture practices enhance the resilience of farming systems to climate variability and extreme weather events. By diversifying crops, improving soil health, and implementing agro-ecological principles, regenerative farming systems are better equipped to withstand pests, diseases, and other environmental challenges. This resilience reduces the risk of crop failure and helps farmers adapt to changing conditions. (Albrecht *et. al.*, 2018).
6. **Economic Sustainability**: Regenerative agriculture provides economic benefits to farmers. By reducing reliance on expensive inputs like synthetic fertilizers and pesticides, regenerative practices can lower production costs. Additionally, regenerative products often command premium prices in markets where consumers prioritize environmentally friendly and sustainable food choices, which can improve farmers' profitability and long-term economic viability (Chappell *et. al.*, 2013).

Overall, regenerative agriculture offers a holistic and sustainable approach to food production that addresses the environmental challenges of modern agriculture while ensuring the long-term productivity and resilience of farming systems. It promotes a harmonious relationship between agriculture and the natural environment, supporting the well-being of farmers, communities, and future generations.

Benefits of Regenerative Agriculture

Regenerative agriculture offers a wide range of benefits, including improved soil health, increased carbon sequestration, enhanced biodiversity, water management and conservation, climate resilience, reduced erosion, decreased reliance on synthetic inputs, improved nutrient cycling and water quality, drought resistance, increased farm profitability, market demand for sustainable products, job creation, preservation of traditional farming knowledge, and community resilience and engagement (Altieri and Nicholls, 2013). These benefits contribute to more sustainable, resilient, and environmentally friendly food production systems.

1. **Improved Soil Health**: Regenerative practices prioritize soil health by focusing on organic matter management, minimizing soil disturbance, and promoting diverse cover crops. These practices enhance soil structure, increase water-holding capacity, improve nutrient cycling, and promote beneficial soil microorganisms. Healthy soils support robust plant growth, increase crop resilience, and reduce the need for synthetic fertilizers and pesticides.
2. **Increased Carbon Sequestration**: Regenerative agriculture plays a critical role in mitigating climate change by sequestering carbon dioxide from the atmosphere. Practices such as cover cropping, agroforestry, and rotational grazing enhance carbon capture and storage in soils, plants, and trees. Healthy soils with high organic matter content act as a carbon sink, reducing greenhouse gas emissions and helping to offset the carbon footprint of food production.
3. **Enhanced Biodiversity**: Regenerative practices create habitats for beneficial insects, birds, and other wildlife, promoting biodiversity conservation. By incorporating diverse crops, preserving natural ecosystems, and minimizing chemical inputs, regenerative agriculture supports a wide range of plant and animal species. This biodiversity contributes to ecosystem resilience, enhances pollination services, and promotes natural pest control, reducing the reliance on synthetic pesticides.
4. **Water Management and Conservation**: Regenerative agriculture employs strategies that optimize water use and improve water quality. Practices such as conservation tillage, cover cropping, and agroforestry reduce soil erosion, increase water infiltration, and enhance water-holding capacity in soils. These practices help conserve water resources, improve water efficiency in agricultural systems, and reduce the impacts of droughts and floods.
5. **Climate Resilience**: Regenerative practices make farming systems more resilient to climate variability and extreme weather events. By improving soil health, increasing water-holding capacity, and diversifying crops, regenerative agriculture enhances the ability of plants to withstand droughts, floods, and temperature fluctuations. These resilient systems

are better able to adapt to changing environmental conditions, reducing the risk of crop failures and production losses.

6. **Reduced Erosion and Soil Loss**: Conventional agriculture practices often contribute to soil erosion, which leads to the loss of valuable topsoil and nutrient depletion. Regenerative practices, such as conservation tillage and cover cropping, minimize soil disturbance and protect the soil surface, reducing erosion and soil loss. This helps to maintain soil fertility, preserve valuable agricultural land, and prevent sediment runoff into water bodies.

7. **Decreased Dependency on Synthetic Inputs**: Regenerative agriculture minimizes the use of synthetic fertilizers and pesticides, reducing the environmental impacts associated with their production and application. By prioritizing soil health, nutrient cycling, and pest management through ecological means, regenerative practices decrease the reliance on chemical inputs. This reduces input costs for farmers and minimizes potential negative impacts on human health and the environment.

8. **Improved Nutrient Cycling**: Regenerative practices optimize nutrient cycling in agricultural systems. By incorporating cover crops, rotating crops, and utilizing livestock integration, regenerative agriculture improves nutrient availability and retention in the soil. This reduces nutrient runoff into water bodies, minimizes nutrient losses, and improves nutrient use efficiency. As a result, farmers can achieve higher yields and reduce the need for synthetic fertilizers.

9. **Enhanced Water Quality**: Regenerative practices contribute to improved water quality in rivers, lakes, and groundwater. By minimizing the use of chemical inputs, such as pesticides and synthetic fertilizers, regenerative agriculture reduces the risk of water contamination. Additionally, practices like riparian buffer zones and conservation practices along waterways help filter and reduce the impacts of agricultural runoff, enhancing overall water quality.

10. **Drought Resistance**: Healthy soils in regenerative systems have improved water-holding capacity, enabling crops to better withstand drought conditions. The increased organic matter content and soil structure in regenerative systems enhance water infiltration and retention, providing a buffer against drought stress. This resilience to droughts reduces the risk of yield losses and improves the overall productivity and sustainability of agricultural systems.

11. **Increased Farm Profitability**: Regenerative practices can lead to increased farm profitability through various mechanisms. By reducing input costs associated with synthetic fertilizers and pesticides, regenerative agriculture offers potential cost savings for farmers. Moreover, regenerative practices often provide opportunities for diversification, such as integrating livestock or cultivating multiple crops. These diversification strategies can generate additional revenue streams, improve market opportunities, and reduce the financial risk associated with a single crop.
12. **Market Demand for Sustainable Products**: Consumers are increasingly seeking sustainably produced food, driving market demand for regenerative products. Regenerative agriculture aligns with consumer preferences for environmentally friendly and socially responsible food choices. Farmers who adopt regenerative practices can tap into this growing market demand, differentiate their products, and potentially command premium prices. Certifications and labeling schemes for regenerative products further enhance consumer trust and marketability.
13. **Job Creation**: Regenerative agriculture often requires more labor-intensive practices compared to conventional agriculture. As a result, the adoption of regenerative practices can lead to job creation in rural communities. From implementing cover cropping to managing rotational grazing systems, regenerative practices offer opportunities for skilled labor in areas such as farming, agroecology, land management, and ecological restoration.
14. **Preservation of Traditional Farming Knowledge**: Regenerative agriculture supports the preservation of traditional farming practices and cultural heritage. By emphasizing agroecological principles, regenerative practices draw on traditional knowledge and wisdom passed down through generations. This helps maintain and revitalize traditional farming practices, fostering cultural diversity and promoting sustainable agricultural systems.
15. **Community Resilience and Engagement**: Regenerative agriculture fosters community engagement and collaboration. Farmers, stakeholders, and communities can work together to implement regenerative practices, share knowledge, and collectively address sustainability challenges. This community-centered approach strengthens social connections, empowers local farmers, and builds resilience at the community level.

Economic Benefits of Regenerative Agriculture

Regenerative agriculture offers a wide range of economic benefits, including cost savings on inputs, increased farm profitability, diversification of revenue streams, premium pricing for sustainable products, market access to niche markets, risk reduction and resilience, improved soil health and fertility, carbon sequestration and climate change mitigation, water management and conservation, enhanced biodiversity and< ecosystem services, improved water quality, reduction in soil erosion and land degradation, preservation of genetic diversity, job creation and rural development, localized economic benefits, improved food security, long-term sustainability and succession planning, reduced health risks and healthcare costs, positive externalities and public goods, and improved resilience and stability in rural communities (Foster and Clark, 2018). These economic advantages contribute to a more sustainable, resilient, and prosperous future for agriculture and society as a whole.

1. **Cost Savings on Inputs**: Regenerative practices minimize the need for expensive synthetic fertilizers, pesticides, and other chemical inputs, reducing input costs for farmers. By focusing on soil health, nutrient cycling, and ecological pest management, regenerative agriculture offers cost-effective alternatives.

2. **Increased Farm Profitability**: Regenerative practices can lead to improved farm profitability. By reducing input costs and optimizing resource use, farmers can achieve comparable or even higher yields while minimizing expenses. Additionally, regenerative agriculture provides opportunities for premium pricing and market differentiation, allowing farmers to capture additional value from sustainably produced products.

3. **Diversification of Revenue Streams**: Regenerative agriculture often encourages diversification through the integration of livestock, agroforestry, or multiple crops. This diversification can provide additional revenue streams, improve market opportunities, and reduce the financial risk associated with relying on a single crop or commodity.

4. **Premium Pricing for Sustainable Products**: Consumers increasingly value sustainably produced food, and regenerative agriculture aligns with their preferences. Farmers who adopt regenerative practices can tap into the growing market demand for sustainable products and potentially command premium prices. Certifications and labeling schemes for regenerative products further enhance consumer trust and marketability.

5. **Market Access and Niche Markets**: Regenerative agriculture offers opportunities for farmers to access niche markets and supply chains that prioritize sustainably produced food. These markets often provide higher margins and greater stability, as they are less susceptible to price fluctuations compared to commodity markets.
6. **Risk Reduction and Resilience**: Regenerative agriculture enhances farm resilience by diversifying production systems and reducing reliance on external inputs. This reduces the vulnerability of farmers to price volatility, input supply disruptions, and extreme weather events. Regenerative practices also improve the ability of agricultural systems to adapt to climate change, enhancing long-term resilience.
7. **Improved Soil Health and Fertility**: Regenerative practices focus on building healthy soils through practices such as cover cropping, crop rotation, and organic matter management. Healthy soils improve nutrient availability, water-holding capacity, and microbial activity, resulting in improved crop yields and reduced fertilizer requirements.
8. **Carbon Sequestration and Climate Change Mitigation**: Regenerative agriculture plays a vital role in mitigating climate change by sequestering carbon dioxide from the atmosphere. Practices such as cover cropping, agroforestry, and rotational grazing enhance carbon capture and storage in soils, plants, and trees. Carbon sequestration can also provide additional revenue opportunities through carbon market programs and incentives.
9. **Water Management and Conservation**: Regenerative practices improve water infiltration and reduce runoff, resulting in more efficient water use and reduced irrigation needs. This can lead to cost savings on water expenses and contribute to water conservation efforts, particularly in regions prone to water scarcity.
10. **Enhanced Biodiversity and Ecosystem Services**: Regenerative agriculture promotes biodiversity conservation through practices that create habitats for beneficial organisms. These practices support pollination, natural pest control, and other ecosystem services, reducing the reliance on synthetic inputs and enhancing the overall ecological balance of agricultural systems.
11. **Improved Water Quality**: Regenerative practices, such as reduced chemical inputs and minimized soil erosion, contribute to improved water quality in rivers, lakes, and groundwater. This has economic benefits by

reducing costs associated with water treatment and purification, as well as supporting the health of aquatic ecosystems.

12. **Reduction in Soil Erosion and Land Degradation**: Regenerative agriculture practices, such as conservation tillage and cover cropping, minimize soil erosion and land degradation. This helps protect valuable topsoil, maintain land productivity, and reduce the need for costly measures to combat erosion.

13. **Preservation of Genetic Diversity**: Regenerative agriculture promotes crop and livestock diversity, which helps preserve genetic resources. Preserving genetic diversity is important for adapting to changing environmental conditions, developing resilient crop varieties, reducing the risk of crop failures, and enhancing long-term agricultural productivity.

14. **Job Creation and Rural Development**: Regenerative agriculture often requires more labor-intensive practices compared to conventional agriculture. The adoption of regenerative practices can create employment opportunities in rural communities, stimulating local economies and supporting rural livelihoods.

15. **Localized Economic Benefits**: Regenerative agriculture encourages local sourcing and supports regional food systems. This helps to stimulate local economies by keeping food production, processing, and distribution within the community, creating jobs and supporting local businesses.

16. **Improved Food Security**: Regenerative agriculture enhances food security by improving crop yields, reducing dependence on external inputs, and building resilience to climate variability. By diversifying production systems and optimizing resource use, regenerative practices help ensure a stable and sustainable food supply.

17. **Long-Term Sustainability and Succession Planning**: Regenerative agriculture emphasizes long-term sustainability by prioritizing soil health, biodiversity conservation, and ecological balance. This approach supports the viability and profitability of farming operations over the long term, enabling successful inter-generational transfer and succession planning.

18. **Reduced Health Risks and Healthcare Costs**: By minimizing the use of synthetic pesticides and fertilizers, regenerative agriculture reduces exposure to harmful chemicals for farmers, farmworkers, and consumers.

This can result in improved health outcomes, reduced healthcare costs, and a safer working environment for agricultural workers.

19. **Positive Externalities and Public Goods**: Regenerative agriculture generates positive externalities and public goods, such as improved air and water quality, carbon sequestration, and wildlife habitat conservation. These benefits contribute to overall societal well-being and have economic value, even if they are not directly captured in market transactions.

20. **Improved Resilience and Stability in Rural Communities**: Regenerative agriculture fosters community resilience by strengthening social connections, promoting knowledge sharing, and engaging local stakeholders in sustainable farming practices. This resilience helps rural communities withstand economic shocks, maintain cultural heritage, and contribute to the overall stability and well-being of rural areas.

Drawbacks and Challenges of Regenerative Agriculture

Regenerative agriculture offers significant economic and environmental benefits, but some challenges and drawbacks need to be addressed for its widespread adoption. These challenges include transition costs, scalability concerns, market barriers, measurement complexities, initial yield reductions, knowledge gaps, limited access to capital and financing, policy support, input availability, and mindset shifts (Landers *et. al.*, 2018). Overcoming these challenges requires collaborative efforts from farmers, policymakers, researchers, and other stakeholders to create a supportive environment and provide the necessary resources and knowledge for the successful implementation of regenerative agriculture (Tittonell and Giller, 2013)

1. **Transition Costs and Learning Curve**: Adopting regenerative agriculture often involves a transition period and initial investment. Farmers need to learn new practices, acquire knowledge about soil health, biodiversity, and ecosystem dynamics, and implement changes on their farms. This learning curve can be time-consuming and may require financial resources for training, experimentation, and infrastructure modifications.

2. **Limited Scalability and Productivity Concerns**: Regenerative agriculture, with its emphasis on ecological processes and holistic systems, may face challenges in terms of scalability and productivity. Some regenerative practices, such as conservation tillage or cover cropping, require more labor and careful management compared to conventional practices. Scaling up regenerative systems without

compromising their ecological benefits can be demanding. Moreover, concerns may arise regarding productivity trade-offs, as regenerative practices prioritize soil health and ecosystem services over maximizing yields. While regenerative systems can achieve comparable yields in the long run, short-term productivity fluctuations can be a deterrent for some farmers.

3. **Market Barriers and Consumer Awareness**: Despite the growing demand for regenerative products, market barriers and limited consumer awareness can hinder the economic viability of regenerative agriculture. Consumers may not always be familiar with the benefits of regenerative practices or may have limited access to regenerative products due to supply chain constraints. Education and marketing efforts are needed to raise consumer awareness and foster strong market demand for regenerative products.

4. **Measurement and Certification Challenges**: Assessing and quantifying the impact of regenerative practices can be complex. Metrics for measuring soil health, carbon sequestration, or biodiversity conservation may require specialized knowledge and monitoring techniques. Developing reliable and cost-effective measurement tools is essential to demonstrate the economic benefits and environmental outcomes of regenerative agriculture. Similarly, establishing certification standards that accurately reflect regenerative practices can be challenging. Ensuring transparency and credibility in labeling and certification schemes is crucial to maintaining consumer trust and enabling fair market competition.

5. **Initial Yield Reductions and Transition Period**: Transitioning to regenerative agriculture may initially result in yield reductions as the soil undergoes the process of regeneration. Farmers may experience lower yields during the transition period until the soil health improves and the ecological processes are restored. This can be financially challenging for farmers who depend on consistent yields for their livelihoods.

6. **Knowledge and Training Gaps**: The adoption of regenerative practices requires a deep understanding of ecological processes, soil health, and farm management principles. However, there may be knowledge and training gaps among farmers, agronomists, and extension services regarding regenerative agriculture. Providing accessible and comprehensive education and training programs is crucial to support the adoption and successful implementation of regenerative practices.

7. **Access to Capital and Financing**: The initial investment required for transitioning to regenerative agriculture can be a barrier for many farmers, especially those with limited access to capital and financing options. Financial support mechanisms such as loans, grants, and subsidies can facilitate the adoption of regenerative practices and help overcome the financial challenges associated with the transition.
8. **Policy and Institutional Support**: The lack of supportive policies and institutional frameworks can hinder the widespread adoption of regenerative agriculture. Policy incentives, research funding, technical assistance, and extension services tailored to regenerative practices are needed to facilitate knowledge dissemination, create a favorable economic environment, and support the long-term viability of regenerative agriculture.
9. **Limited Availability of Inputs and Services**: In some regions, the availability of regenerative inputs, such as organic seeds, compost, or cover crop seeds, may be limited. Additionally, the availability of specialized services, such as agronomic advice and technical support for regenerative practices, may be scarce. Addressing these supply chain and service gaps is essential to facilitate the adoption and implementation of regenerative agriculture.
10. **Adoption Barriers and Mindset Shift**: Shifting from conventional agricultural practices to regenerative agriculture requires a fundamental mindset shift. The cultural and social norms ingrained in the agricultural sector, as well as resistance to change, can act as barriers to adoption. Overcoming these barriers requires education, awareness-building, and showcasing successful case studies to demonstrate the economic and environmental benefits of regenerative agriculture.

Case Studies and Success Stories

The studies and success stories from India showcase the diverse applications of regenerative agriculture across different regions and farming systems. They highlight the economic benefits of improved soil health, increased productivity, market demand for organic and sustainably produced food, and the empowerment of local communities. Furthermore, these examples demonstrate the positive environmental outcomes of regenerative agriculture, such as biodiversity conservation, water management, and ecological restoration. By adopting regenerative practices, these initiatives have not only improved the livelihoods of farmers but also contributed to the resilience and sustainability of India's agricultural systems.

1. **Araku Tribal Coffee, Andhra Pradesh**: Araku Tribal Coffee is a cooperative located in the Eastern Ghats region of Andhra Pradesh, India. The cooperative has adopted regenerative agricultural practices, including shade-grown coffee and agroforestry systems. By integrating coffee cultivation with diverse tree species, farmers have improved soil health, increased biodiversity, and enhanced water conservation. The cooperative has seen economic benefits through increased coffee yields, premium pricing for sustainably produced coffee, and improved livelihoods for tribal farmers. The success of Araku Tribal Coffee has demonstrated the potential of regenerative agriculture to promote sustainable livelihoods and environmental conservation in tribal communities (Balaji, 2020).
2. **Sahaja Samrudha, Karnataka**: Sahaja Samrudha is an organic farmers' collective based in Karnataka, India. The collective promotes regenerative agriculture practices among smallholder farmers, focusing on seed conservation, agroecology, and traditional farming techniques. Through farmer training programs and seed banks, Sahaja Samrudha has helped farmers transition to regenerative practices and maintain local seed varieties. This has resulted in improved soil health, increased crop diversity, and reduced input costs. Farmers involved in Sahaja Samrudha have experienced improved incomes, reduced dependence on external inputs, and strengthened community networks. (Pal, 2017)
3. **Prakruti Organic Farm, Maharashtra**: Prakruti Organic Farm, located in Maharashtra, India, is an example of regenerative agriculture in a peri-urban setting. The farm follows organic and regenerative practices, including composting, vermiculture, and agroforestry. These practices have improved soil fertility, increased water-holding capacity, and reduced the need for external inputs. Prakruti Organic Farm supplies organic produce to local markets, catering to the growing demand for healthy and sustainably produced food. The success of the farm has inspired neighboring farmers to adopt regenerative practices and create a local network of sustainable agriculture practitioners. (Anonymous, 2022)
4. **Navdanya, Uttarakhand**: Navdanya is a non-governmental organization (NGO) and seed bank based in Uttarakhand, India. The organization promotes regenerative agriculture and seed sovereignty through seed conservation, organic farming, and agroecology. Navdanya has established community seed banks, trained farmers in sustainable farming practices, and advocated for policies that support regenerative agriculture. Their efforts have revitalized traditional farming knowledge,

conserved indigenous seed varieties, and empowered farmers to adopt regenerative practices. Navdanya's work has had a significant impact on the livelihoods of farmers, biodiversity conservation, and the promotion of sustainable food systems. (Wikipedia, n.a.)

5. **Sadhana Forest, Tamil Nadu**: Sadhana Forest is an ecological restoration project located in Tamil Nadu, India. The project aims to restore degraded land by implementing regenerative practices such as reforestation, water conservation, and sustainable agriculture. Sadhana Forest follows agroforestry techniques, including the planting of indigenous tree species alongside food crops. This approach has led to increased soil fertility, improved water management, and the regeneration of native flora and fauna. The project has not only restored degraded land but has also created a sustainable living model that demonstrates the economic viability of regenerative agriculture in arid regions (Anonymous n.a).

Conclusion

In conclusion, regenerative agriculture offers a transformative and sustainable approach to farming and land management. Its emphasis on restoring and improving ecosystem health, soil fertility, and biodiversity makes it a crucial tool in addressing the environmental challenges faced by modern agricultural systems. By implementing practices such as conservation tillage, cover cropping, crop rotation, and organic matter management, regenerative agriculture can enhance soil health, increase carbon sequestration, promote biodiversity conservation, improve water management, and build resilience to climate variability. The economic benefits of regenerative agriculture are also significant. Farmers can reduce input costs, increase farm profitability, access niche markets, and diversify revenue streams by adopting regenerative practices. Additionally, regenerative agriculture can create employment opportunities, stimulate local economies, and contribute to rural development. By improving soil health, reducing erosion, and conserving water resources, regenerative agriculture helps ensure long-term sustainability, food security, and economic stability. While regenerative agriculture offers numerous benefits, there are challenges to its widespread adoption. Transition costs, scalability concerns, market barriers, measurement complexities, initial yield reductions, knowledge gaps, limited access to capital, policy support, input availability, and mindset shifts all need to be addressed. Collaborative efforts from farmers, policymakers, researchers, and stakeholders are necessary to overcome these challenges and create a supportive environment for regenerative agriculture. Overall, regenerative agriculture provides a holistic and resilient approach to food production that integrates ecological principles, economic viability, and societal well-being. By prioritizing the restoration and improvement of natural resources, regenerative agriculture offers a path toward a more sustainable and prosperous future for agriculture and society as a whole.

References

Albrecht, H., Timler, C. J., Klawitter, R. Agricultural landscapes: The major driver of spatial patterns in soil microbial communities of grassland ecosystems. *Land Degradation and Development,* 29, no. 10, (2018): 3473-83.

Altieri, M. A., Nicholls, C. I. The adaptation and mitigation potential of traditional agriculture in a changing climate. *Climatic Change*, 140, no. 1, (2013): 33-45.

Anonymous. 2022. Prakrutik Farm Organic Natural Health. Retrieved April 11, 2023 from https://prakrutikfarm.com/about-us/

Anonymous (n.a). Sadhana Forest started its ecological revival and sustainable living work on December 19th, 2003. Retrieved April 10, 2023 from https://sadhanaforest.org/about-us/

Balaji, R. 2020. From seed to Brew: How a coffee project in Andra Pradesh's Araku changed the lives of Adivasi farmers. Retrieved April 10, 2023 from https://yourstory.com/socialstory/2020/08/coffee-project-araku-valley-tribals-naandi-foundation

Chappell, M. J., Wittman, H., Bacon, C. M., Ferguson, B. G., Barrios, R. G. Food sovereignty: An alternative paradigm for poverty reduction and biodiversity conservation in Latin America. *F1000Research*, 2, (2013): 235.

Foster, J. B., Clark, B. The planetary emergency. *Monthly Review,* 70, no. 3, (2018): 1-19.

Lal, R. (2018). Digging deeper: A holistic perspective of factors affecting soil organic carbon sequestration in agroecosystems. *Global Change Biology*, 24, no. 8, (2018): 32853301.

Landers, J. N., Anderson, C. J., Wallander, S., Marquart-Pyatt, S. The price of progress: Health and ecosystem costs of the Green Revolution. *Social Science Research*, 73, (2018): 94-108.

Latacz-Lohmann, U., Nuppenau, E. A. Farmer heterogeneity and the adoption of sustainable soil management practices: A microeconomic analysis. Journal of Environmental Management, 235, (2019): 100-11.

Pal, S. 2017. Over 5000 organic farmers are reviving traditional crop varieties. Thanks to One Organization. Retrieved April 10, 2023 from https://www.thebetterindia.com/81830/traditional-farming-organic-food-sahaja-samruddha-karnataka/

Paustian, K., Lehmann, J., Ogle, S., Reay, D., Robertson, G. P., Smith, P. (2016). Climate-smart soils. *Nature*, 532, no. 7597, (20160): 49-57.

Ponisio, L. C., M'Gonigle, L. K., Mace, K. C., Palomino, J., de Valpine, P., and Kremen, C. (2015). Diversification practices reduce organic to conventional yield gap. Proceedings of the Royal Society B: Biological Sciences, 282, no. 1799, (2015): 57-65

Smith, P., Cotrufo, M. F., Rumpel, C., et al. Biogeochemical cycles and biodiversity as key drivers of ecosystem services provided by soils. *Soil*, 1, no. 2, (2015): 665-85.

Tittonell, P., Giller, K. E. When yield gaps are poverty traps: The paradigm of ecological intensification in African smallholder agriculture. *Field Crops Research,* 143, (2013): 76-90.

West, P. C., Gerber, J. S., Engstrom, P. M., Mueller, N. D., Brauman, K. A., Carlson, K. M., Siebert, S. (2014). Leverage points for improving global food security and the environment. Science, 345, 6194, (2014): 325-28.

White, R., and Hall, A. The economics of Agroecology: From ideology to practice. *Journal of Rural Studies*, 57, (2017): 1-10.

Wikipedia (n.a.). Navdanya. Retrieved April 10, 2023 from https://en.wikipedia.org/wiki/Navdanya

13

Integrated Management of Plant Diseases Toward Regenerative Agriculture

***Dechan Choskit*[1] *and Meenakshi Anand*[2]**

[1]*CSIR-Indian Institute of Integrative Medicine, Jammu, Jammu & Kashmir, India*

[2]*Department of Home Science (Human Development), Lakshmibai College, University of Delhi, Delhi, India*

Introduction

The growing interest in agricultural production over the past few decades has resulted in more farming systems that rely on chemical pesticides or chemical fertilizers, which has seriously harmed the soil, water, air, and climate while paying little attention to quality or concern for the agricultural system's negative effects on ecosystems. But as time went on, scientific research began to focus on sustainable agriculture, organic farming, clean agriculture, and other ideas that eventually evolved into agricultural systems that maintain the balance in all areas of the agricultural and production process, as well as consumer and environmental concerns. These agricultural approaches have attracted the attention of numerous farmers, and many large farms are now interested in organic farming in the context of sustainable agriculture. The idea of sustainable agriculture is simply the production of food, fiber, other plant products, or even animal products using agricultural techniques that protect the environment, public health, and human societies from the detrimental effects arising from carrying out that process using the standard format as well as technology that also takes into account animal welfare, such that this type of agriculture will enable us to produce healthy foods without compromise (Bahlai et al., 2010; Muller et al., 2017, and Palmgren et al., 2015).

The Future of Sustainable Agriculture

The adoption of sustainable agriculture was made to raise awareness of their significance and to support the governments of plantations and businesses that operate within the parameters of sustainability, in particular organic agriculture, which is clean and recently devoted a lot of studies and research as well as major scientific conferences on sustainable agriculture, as were a lot of major exhibitions for organic goods and food made from organic ingredients around the world. Specifically, the potential role that organic agriculture could play in reducing the adverse effects of climate change, reducing deforestation rates, maintaining biological risk, and also arranging the steady increase in the world's population, attracted attention to the level of international policy.

Sustainable agriculture is being incorporated into national plans and development by agricultural and environmental policies all over the world. For a shift in agricultural production practices and an uptick in both quantity and quality of output. This significantly boosts investments in sustainable agriculture, which is reflected in higher budgets for scientific research, reclaimed land use, and financial incentives, as well as better infrastructure (Rundlöf et al., 2016 and Seufert, 2019).

Integrated Disease Management

In the 1960s, when chemicals, particularly fungicides and insecticides, came under fire from environmentalists for being overused and causing environmental pollution, chemical residues in food, land, water, and the air, and the resulting health risks, integrated disease management (IDM) came into focus.

- It focused on the other methods of disease control.
- It involved cultural, biological, epidemiological, and alternative means to achieve disease control.
- Nowadays, there is an emphasis on disease "management" rather than on "Control."

Definition of IDM

"Disease management system that in the context of the associated environment and population dynamics of microorganisms, utilizes all suitable techniques and methods in a manner as compatible as possible and maintains the disease below economic level."

- In general, it is the incorporation of all appropriate management strategies for the prevention and treatment of diseases.
- In IDM, it's important to avoid using fungicides carelessly, cultivating in monocultures, and using susceptible cultivars.
- IDM makes ensuring that soil health is properly managed, that healthy seeds and planting material are used, that fields are kept clean, that cultural practices are used to suppress the disease, that bio-control agents are used, and that resistant plant genotypes are grown.

Different Approaches of IDM System

1. The combined control approach: This method combines several different control techniques, such as adjusting the sowing time, treating the seeds, using resistant varieties, scheduling chemical sprays, etc. This sort of IDM is frequently used as a set of procedures where the likelihood of a disease occurring is high.
2. The surveillance-based approach: This sophisticated IDM strategy considers the economic threshold levels or economic damage levels and is based on crop health monitoring and surveillance.
3. Advanced IDM system: This system uses high-input technology such as forecasting supported by computers, remote sensing, scouting, multiple pathogen thresholds, information on the life cycles of pathogens, epidemiology of diseases, environmental factors, and knowledge-based decision-making.

Main Components of IDM

1. Host resistance
2. Induced systemic resistance
3. Genetically improved plants
4. Cultural practices
5. Physical methods
6. Plant nutrition
7. Biological control
8. Use of pesticides of plant origin
9. Judicious use of chemicals

Host Resistance

A tactic that makes use of cultivars created through conventional breeding or genetic engineering that are both pest-resistant and pest-tolerant (Nelson et al., 2018). These cultivars have physical, morphological, or biochemical traits that make them less appealing to pests or less conducive to their growth, development, or effective reproduction. These cultivars are resistant to or tolerant of insect damage, which lowers production losses. The IPM's first line of defense is this choice.

Resistant cultivars can ensure protection against diseases, save time, money, and energy spent on alternative control methods, and prevent environmental pollution with chemicals. They can also be a simple, practical, effective, and affordable form of plant disease control. In low-value crops, where other methods are frequently too expensive, the development of varieties resistant to common and important diseases can be an acceptable recommendation for the farmers. They are the only practical method of controlling such diseases as wilts, rusts, and others caused by viruses, where chemical control is very expensive and impractical. Plants' ability to withstand disease is also influenced by their genetic makeup, which might be monogenic, oligogenic, or polygenic.

Advantages of host plant resistance

Host plant involves no additional expense to the farmers and does not require inputs or application expertise; instead, the resistant cultivars have a steady and cumulative effect on the disease.

Disadvantages of host plant resistance

It takes 5 to 10 years to generate a pathogen-resistant type. Host plant resistance can put a pathogen under such strong selection pressure that new pathogen biotypes may evolve as a result. Because the main pathogen is no longer competing for resources, the introduction of varieties that are resistant to one pathogen causes the emergence of a new pathogen problem.

Induction of Host Resistance

Plants actively react to a range of environmental cues, including gravity, light, temperature, physical stress, and the availability of water and nutrients. Additionally, plants respond to a range of chemical cues made by microbes linked with the soil and plants. Such stimuli can either condition or trigger a plant's host defense through biochemical changes that increase resistance to several diseases later infecting the plant. Depending on the kind, source, and quantity of the stimuli, the induction of host defense may be local or systemic in character. Salicylic acid (SA), a substance that is regularly created during pathogen infection and normally induces the expression of pathogenesis-related (PR) proteins, mediates systemic acquired resistance (SAR). While the induced systemic resistance (ISR) is mediated by jasmonic acid (JA) and/or ethylene, which is produced after applications of some nonpathogenic rhizobacteria, some of these PR proteins include a variety of enzymes that may act directly to lyse the invading cells, reinforce cell wall boundaries to resist infections or induce localized cell death. It's interesting to note that the SA- and JA-dependent defense mechanisms can compete with one another, and some bacterial pathogens use this to their advantage to bypass the SAR. Because different microbes and insect feeding can activate different host-resistance pathways to varying degrees, it is plausible that multiple stimuli are constantly being received and processed by the plant. As a result, the magnitude and duration of host defense induction will likely change over time.

Genetically Improved Plants

Transgenic creatures are those that contain genes from different species inserted into their genomes. These organisms can be created by combining and introducing genes from plants, bacteria, and animals into the live cells of other organisms. By inserting specific genes that confer resistance to diseases like viruses, fungi, and insects into plant genomes, it is possible to produce disease-resistant transgenic plants. These transgenic plants lower the need for pesticides, benefiting the environment while saving farmers money. In order to combat viral illnesses, genetically modified plants are typically used. For example, in the United States, the transgenic papaya cultivar "Rainbow" has been produced to be resistant to papaya ring spot virus.

Integration of Different Cultural Practices

While individual applications of various cultural practices, such as crop rotation, mulching, tillage, various soil amendments, soil solarization, soil sterilization, changing the date of sowing, plant spacing, etc., can control diseases to a certain extent, but when these practices are combined, they not only control the diseases but also increase crop yield. The pepper veinal mottle virus is propagated less by intercropping maize and sorghum with peppers, which act as barriers against the virus's aphid vectors. Compared to the unsolarized control, soil solarization (for 40 days) alone reduced the gladiolus wilt by 67.3%. However, soil solarization (for 40 days) combined with the addition of cabbage, cauliflower, broccoli, and sarson leaf residues reduced the gladiolus wilt by 74.6%.

Physical Methods of Disease Control

In May-June, water-soaked wheat seeds are treated with solar heat for 5–6 hours to effectively control loose wheat smut. By using controlled atmosphere storage, refrigeration, and irradiation, the majority of post-harvest illnesses can be prevented. Soil solarization has been used to control fungi that are otherwise difficult to control, such as *Rhizoctonia solani*, *Fusarium spp.*, *Sclerotium*, etc. that cause soil-borne diseases. During April, May, and June, this method involves first irrigating the soil beds before covering them with a thin (20 mm) layer of translucent mulch. It increased soil temperatures in some cases by up to 50°C, which is harmful to many soil-borne plant diseases. It has been utilized in tropical and subtropical climates to raise disease-free nurseries. It also does a great job of controlling weeds. Black rot disease can be controlled by soaking cabbage seeds in hot water for 15-20 minutes at 52°C. (caused by *Xanthomonas campestris pv. campestris*).

Plant Nutrition

An essential part of integrated disease control is crop plant nutrition, which directly influences disease (IDM). Plants that are undernourished or overfed are more likely to contract illnesses and generate lower-quality, lower-yielding food and goods. Plant metabolism and, consequently, the frequency and severity of illness are impacted by fertilizer application factors such as quantity, proportion, timing, and method. Wheat that has received fertilization with both Phosphorus (P) and Potassium (K) has much less leaf rust damage and powdery mildew infection. The incidence of various diseases may also be impacted by macronutrient deficiencies. Potassium has a direct impact on the number, development, multiplication, survival, vigor, and length of the life cycles of pathogens, which helps it play a critical role in the survival of crop plants under environmental stress circumstances.

Biological Control

The essential element of an IDM system is the employment of biocontrol agents. One of the key elements of disease management strategies that are benign to the environment is the science and art of employing live creatures as biocontrol agents. These biocontrol agents frequently take the place of fungicides in IDM for sustainable agriculture, where they are of great utility. The biocontrol agents either produce antibiotics, cause hyperparasitism, or compete with the pathogen to slow its growth. Different biocontrol agents like *Bacillus subtilis*, *Pseudomonas fluorescens*, *Gliocladium spp.*, *Trichoderma* spp., *Chaetomium globosum*, *Pseudomonas cepacia*, *Bacillus cereus*, *Agrobacterium radiobacter*, etc. are employed to control different diseases. *Trichoderma viride* is the most significant and adaptable biocontrol agent used to manage a variety of plant diseases, including *Rhizoctonia solani* and *Sclerotium rolfsii*, which are otherwise challenging to manage by other means. In a similar vein, *Fusarium lateriticum* has been applied to apricot tree primary wounds to prevent the canker disease brought on by *Eutypa armeniacae* and *Heterobasidion annosus*, a root rot disease, which was effectively controlled by applying *Peniophora gigantea oidia* paste to pine stumps, which is a disease that spreads through uncovered stumps left behind after falling. *Ampelomyces quisqualis* and *Darluca* spp. are used to manage rust fungus and powdery mildew, respectively, by hyperparasitizing them. The K-84 strain of *Agrobacterium radiobacter* has been utilized to treat crown gall disease all over the world.

Use of pesticides of plant origin

Plant-based pesticides are made from components of plants, and their genes are also utilized to genetically modify crops to express resistance to viral, fungal, and insect attacks. The control of plant diseases depends heavily on plant components and their antifungal extracts. In India's Rig Veda, plants having pest-killing abilities are mentioned. Garlic (*Allium sativum*) has a long history of being valued for its medicinal, antibacterial, and pesticidal characteristics as well as being used for these purposes. The ethanolic leaf extracts of Eucalyptus sp., *Chenopodium ambrosioides*, *Lippia alba*, *Aegle marmelos*, and *Cestrum diurnum* can inhibit the growth of *Rhizoctonia solani*. The seed extract of Piper nigrum was discovered to be efficient against *R. bataticola*.

Judicious Use of Fungicides

For many years, chemicals have been employed successfully to fend off the effects of various illnesses. To manage a wide range of plant diseases destroying different crops, fungicides with varied modes of action, such as protective (broad spectrum fungicides), post-infection activity (EBI), pre-symptom and post-symptom (benzimidazoles and triazoles), may be utilized. As a result of the excessive use of these chemicals, which led to water pollution, residues on food and fruit crops, effects on non-target organisms, and the development of chemical resistance in pathogens, attention has been focused on the rational use of fungicides, which includes carefully observed control methods and cultural practices.

Types of Integrated Disease Management

Integration Of Cultural And Chemical Control

A constant supply of fresh watermelons and a decline in diseases brought on by *Colletotrichum lagenarium*, *Pseudomonas syringae pv. lachrymans*, and *Pseudoperonospora cubensis* have been made possible by the integration of chemicals and cultural practices (including improved cultivars). The Tomato Leaf Curl Virus was delayed from spreading for 3–5 weeks and tomato yields were increased by covering tomato nursery seedlings with nylon net for 25–30 days and applying four sprays of monocrotophos at intervals of 10 days after transplanting.

Integration Of Chemical And Biological Control

For the successful control of some illnesses, bio-control agents such as *Pseudomonas fluorescens*, *Trichoderma viride*, *T. harzianum*, *Bacillus subtilis*, *Pseudomonas putida*, *P. cepacia*, *Talaromyces flavus*, and *Agrobacterium radiobacter* strain K 84 can be used in conjunction with pesticides.

Integration of Resistance, Cultural, Biological, and Chemical Control

Club root (*Plasmodiophora brassicae*) of vegetables is controlled using a combination of cultural practices (crop rotation, good farm hygiene procedures, quarantine), fertilizers, soil fumigation and solarization, pesticides (fungicide transplant dips, soil drench, soil incorporations, seed treatments, trace elements, and surfactants), resistant cultivars, and biocontrol agents.

IPM and Plant Diseases

The integration of various control techniques has historically been stressed in plant pathology research and extension activity, and this approach meets both traditional and contemporary definitions of IPM. Applicators can pick up the signs of common plant diseases and pests. To ensure that a pest is appropriately diagnosed, it is important to have a sample as well as pathogenic organisms. It might not always be able to do this. The harm or symptoms that some plant diseases produce are frequently used to identify them. Additionally, to identify the majority of plant disease causative agents, laboratory testing and inspection are required.

IPM and Sustainable Development in Terms of Plant Diseases

IPM is an approach that all pest control programs have mostly followed. The term IPM has only sometimes been used in the realm of phytopathology, as was already indicated. IPM is widely utilized in the disciplines of crop protection, and it is generally regarded as one of the core principles of biointensive pest control (Jacobsen, 1997). IPM programs, particularly those that focus on plant disease control, outline the tools and decision-making needed for specific crop pests. To support the needs of growers in agriculture for sustainability as well as to achieve the knowledge and implementation structure in potential of IPM, IPM that is related to all types of pathogens with the crops and plant species in all agroecological systems requires a lot of achievements in modern technology applications, such as crop growth loss models (Bale et al, 2008).

Principles of Plant Disease Management

Protection: Use of biocontrol and fungicides, practical disease avoidance, adequate plant spacing, grouping plants according to their nutrient and water requirements, avoidance of planting times that favor causal agents, care taken when harvesting, and handling of plants and plant parts carefully during transplanting and harvesting are all important practices.

Exclusion: Government controls, avoid introductions, and prevent the transit and introduction of plant infections. Quarantine facilities, purchasing disease-free stock and clean seeds, treating seeds (fungicides on seed coatings), eradicating infestations, and altering practices are all ways to stop the spread of illness into uncontaminated areas.

Resistance: The most dependable, efficient, and cost-effective method of eradicating plant diseases; includes resistance genes inside the plant, physical barriers and horizontal resistance, as well as the production of toxins and plant systemic immune responses such salicyclic acid. "Vertical resistance: One gene produces one protein, preventing infection; resistant crop can withstand pathogen attack" (Penetration, no colonization).

Therapy: This is accomplished by introducing a chemical control agent into the physiological functions of the plant in order to stop the progression of disease development after infection has already taken place.

Avoidance: Like the cultural techniques that reduce the risk of infection, including the time of planting, how the seedbed is prepared, how water is managed, how much space each plant gets, and fertilizer, avoiding damaging plants, etc.

Eradication and Reduction of Inoculums: Reduces inoculum density, improves sanitation, removes infected plant debris, prunes diseased plants and infected wood, rotates crops, gets rid of weeds and other potential hosts, employs methods that hinder the movement of vectors, sterilizes the soil, and uses biocontrol to stop the spread of introduced diseases.

Impact of IPM on Plant Health, Pesticide Use and Yield

IPM's plant health management program is linked to achieving sustainability. The quantity of illnesses that are treated using environmentally friendly methods often requires the most management efforts for plant health. For instance, the majority of earlier studies on the control of root infections focused more on the use of crop rotation, soil fumigants, and other similar techniques. The attempts to control plant infections have led to the development of improved solutions for the management of plant diseases. The most significant recent contribution of plant pathology to wifehood is genetic research, including the gene-for-gene model and other sophisticated applications based on molecular levels. IPM's major goal is to protect crops from all pests while maintaining the natural equilibrium. The 20th century saw the development of several important methods for controlling plant diseases, and current biotechnology has helped this field evolve to become the most important recently. With biocontrol and IPM, plant health management is a shifting target (Cook, 2000). There is a connection between IPM programs and pesticide usage whereby it may be decreased and pest control techniques updated without producing any negative effects (Kenmore et al., 1984; Gallagher et al., 2005 and Hassanali et al., 2008).

Conclusion

Protecting ecosystems and natural resources that depend on or are influenced by agriculture is part of sustainable agriculture. Using all of the knowledge, resources, and techniques at our disposal, IPM comprises an integrated strategy for the prevention and/or control of organisms detrimental to plants. In order to eliminate or minimize hazards to human health and the environment, IPM tries to limit the use of pesticides and other types of intervention to levels that are both economically and environmentally viable. IPM's plant health management program is linked to achieving sustainability. The quantity of illnesses that are treated using environmentally friendly methods often requires the most management efforts for plant health. If IPM is used, there are four possible outcomes: pesticide usage falls but yields rise; both pesticide use and yields rise; pesticide use rises but yields decrease; and pesticide use and yields rise.

References

Bahlai, C.A, Xue, Y., McCreary, C.M., Schaafsma, A.W., Hallett, R.H. Choosing organic pesticides over synthetic pesticides may not effectively mitigate environmental risk in soybeans. *PloS one.* 5, (2010): 11250.

Bale JS, van Lenteren JC, Bigler F. 2008. Biological control and sustainable food production. *Phil Trans R Soc B.* 363, (2008): 761- 76.

Cook RJ. Advances in Plant Health Management in the Twentieth Century. *Annual Review of Phytopathology.* 38, (2000): 95-116.

Gallagher K, Ooi P, Mew T, et al. 2005. Ecological basis for low-toxicity integrated pest management (IPM) in rice and vegetables. In: Pretty J, editor. The pesticide detox. London, UK: Earthscan

Hassanali A, Herren H, Khan ZR, Pickett JA, Woodcock CM. Integrated pest management: The push-pull approach for controlling insect pests and weeds of cereals, and its potential for other agricultural systems including animal husbandry. Phil Trans R Soc B. 363, (2008): 611-21.

Jacobsen BJ. Role of plant pathology in integrated pest management. *Annual Review of Phytopathology.* 35, (1997): 373-91.

Kenmore PE, Carino FO, Perez CA, et al. Population regulation of the brown planthopper within rice fields in The Philippines. *J Plant Prot Tropics.* 1, (1984): 19-37.

Lechenet M, Bretagnolle V, Bockstaller C, Boissinot F, Petit MS, Petit S, Jolain NMM. Reconciling Pesticide Reduction with Economic and Environmental Sustainability in Arable Farming (RNC Guedes, Ed.). *PLoS ONE.* 9, (2014): 97922.

Muller A, Schader C, El-Hage Scialabba N, et al. Strategies for feeding the world more sustainably with organic agriculture. *Nature Communications.* 8, (2017): 1290.

Nelson R, Wiesner-Hanks T, Wisser R., Kurti PB. Navigating complexity to breed disease-resistant crops. *Nat Rev Genet.* 19, (2018): 21-33.

Pacini, C., Wossink A.,abGiesen G., aVazzana C., Huirne R. Evaluation of Sustainability of Organic, Integrated and Conventional Farming Systems: A Farm and Field-scale Analysis, Agriculture, Ecosystems & Environment. 95, (2003): 273-88.

Palmgren MG, Edenbrandt AK, Vedel SE, et al. Are we ready for back-to-nature crop breeding? *Trends in Plant Science.* 20, (2015): 155-64.

Pretty J.. Agricultural sustainability: Concepts, principles and evidence. *Philos Trans R Soc Lond B Biol Sc*i. 363, (2008): 447-65.

Rundlöf M, Smith HG, Birkhofer K. 2016. Effects of Organic Farming on Biodiversity. eLS. Pp 1-7. Chichester, UK: John Wiley & Sons, Ltd.

Sánchez-Bayo F, Wyckhuys KAG. Worldwide decline of the entomofauna: A review of its drivers. *Biological Conservation.* 232 (2019): 8-27.

Seufert V. Comparing Yields: Organic Versus Conventional Agriculture. Encyclopedia of Food Security and Sustainability. *Elsevier.* (2019), Pp 196-208.

14

Modern Agricultural Practices for Soil Conservation

Rodrick Lepcha[1], Sona Jaison[2], and Princy Thakur[3]

[1,3]*Soil Science and Agricultural Chemistry, School of Agriculture Lovely Professional University, Phagwara, Punjab, India*

[2]*School of Agriculture, Lovely Professional University, Phagwara Punjab, India*

Introduction

The soil is an invaluable resource that sustains life in innumerable ways. It provides a dynamic foundation for agriculture, supporting growing crops, and serving as a habitat for organisms. It plays a crucial role in maintaining biodiversity. Additionally, soil acts as a filter, purifying and storing water, and it reserves essential plant nutrients. however, soil resources face various threats that endanger their health and productivity. Soil erosion, the removal of the top layer of soil through natural or human-induced processes, poses a significant challenge. It not only leads to a loss of fertile soil but also causes widespread environmental degradation, and reduced agricultural productivity in response to this difficulty, soil conservation emerges as a principal priority. Soil conservation incorporates a range of practices and strategies to prevent soil erosion, and promote sustainable land use. By adopting and implementing effective soil conservation measures, we can safeguard this irreplaceable resource for future peers and ensure food security and environmental sustainability.

The main highlight of the study is to shed light on the causes and consequences of soil erosion, exploring different conservation practices and techniques;

also to understand the vital processes and factors contributing to soil erosion, including wind and water erosion. Additionally, to study its impact on soil and crop production and strategies for different conservation practices to mitigate erosion, preserve soil fertility, and ensure the long-term sustainability of our agricultural systems and ecosystems.

The Agent that Degrades the Soil

Soils are typically degraded by water and wind erosion. Approximately 56% of all degraded land is due to water erosion, whereas 28% is due to wind erosion (Munyanziza & Oldeman, 1994). This amounts to nearly 1100 million hectares (Mha) worldwide. Soil particles on sloping and naked areas are washed away by runoff, whereas soil particles on flat and unprotected lands are blown away by the wind (Balasubramanian, 2017). Often neglected is the fact that plowing causes tillage erosion, which progressively pushes soil downslope in plowed areas and has negative consequences on crop yield right there. Depending on the soil's parent material, climate, topography, and water management practices, it may become compacted, drained poorly, acidified, alkalinized, or salinized, all leading to soil degradation (Blanco & Lal, 2008).

Soil Erosion

Soil erosion is the process through which the soil's upper layer is worn away. Erosion can be split into two categories. Natural and Enhanced Erosion. When rocks and sediments are eroded and carried away by factors such as water, wind, ice, and gravity, this phenomenon is called geologic erosion. It is an essential part of the rock cycle and has a major impact on the long-term development of the Earth's surface. Human activities have increased the erosion rate, a phenomenon known as accelerated erosion or land degradation. Erosion processes can be accelerated by human-induced factors beyond their natural rates, with serious environmental and economic repercussions. Deforestation, agriculture, construction and urbanization, mining, poor land management, etc. are just a few examples of the human activities that accelerate erosion. Natural soil formation processes, including geologic erosion, follow a consistent pattern of weathering that typically occurs at modest rates in all soils. It takes place gradually and is unaffected by human interference (Bashir et al., 2017). Erosion from natural causes can shape the landscape in unexpected ways over long periods. Water erosion, wind erosion, glacial erosion, and gravity erosion are the main causes of geological erosion.

The loss of topsoil, which is the only part of the soil that contains the most vital plant nutrient, makes erosion control a critical issue. The negative impact on productivity can be reduced significantly if erosion is reduced to a tolerable level. Soil profile, soil management, climatic circumstances, etc. all play a role in determining the severity and consequences of soil erosion (Lal & Moldenhauer, 1987). In India, a default T value of 11.2 Mg ha1 yr1 (5.0 t ac1 yr1) is being followed (Mandal et al., 2006), which refers to soil loss tolerance. The impact of the T value shifts depending on soil type. While modest soil erosion would not have a major impact on yield in deeply rooted crops, it might have far-reaching consequences in shallow, sloped soils. Therefore, it's important to identify erosion's critical limitations for various soils, ecozones, land uses, and agricultural systems.

Water Erosion

Water erosion is when precipitation occurs and soil infiltration capacity gets saturated the excess water starts to flow which leads to the washing off of detached soil and transport to another place. It is dominated in humid and sub-humid regions due to frequent rainstorms. There is also a problem with water erosion in arid and semi-arid regions where limited precipitation occurs mostly in storms when the soil is bare and devoid of vegetative cover. A particularly spectacular example is concentrated gully erosion, which can have devastating consequences leading to landscape aesthetics being drastically altered, and the removal of sediments in large amounts.

Types of Water Erosion

Splash erosion: When raindrops impact the soil directly, they can cause splash erosion by dislodging and splashing soil particles. Erosion from wind and water can cause soil particles to become detached and carried away by surface runoff, and it is most common on bare or thinly vegetated land.

Sheet erosion: A homogeneous thin layer of dirt or sediment is removed from the surface in this way. Rainwater or overland flow are the most common culprits in the occurrence of sheet erosion, which is most prevalent on sloping terrain with sparse vegetation or disturbed soil.

Rill erosion: When rills, or little channels, form on the surface of the soil, erosion occurs. These waterways often have a depth of less than 30 cm (12 in) and are somewhat small. Rill erosion, induced by concentrated runoff, is a familiar sight in agricultural fields and construction sites.

Gully erosion: Gully erosion is more destructive than rill erosion because it results in the development of deep channels. Gullies are more destructive than rills because of their greater width and depth. They thrive in regions that receive heavy or consistent rainfall.

Streambank erosion: Erosion of riverbanks and other shorelines is known as streambank erosion. It occurs when water erosion removes soil and sediment from riverbanks, causing them to collapse and recede. The land is lost as waterways expand due to streambank erosion.

Tunnel erosion: When water permeates the earth and erodes soil layers, it results in tunnel erosion, also called piping erosion. When water flows through more permeable soil layers, it can erode and transport away finer particles, a major cause of erosion.

Wind Erosion

Soil, silt, and other materials can be moved and eroded by the wind, which is why it is considered a natural process. The process of surface wear caused by wind erosion takes place when the speed of the wind is great enough to dislodge and transport away loose soil particles. In arid and semiarid environments, when plant cover is sparse and soil moisture is low, wind erosion is a typical problem. It poses a serious threat to the economy and the environment.

Processes of Wind Erosion

Detachment: Particles are dislodged and carried upward by the wind when it exerts a force on the surface. Fine-grained and weakly bonded particles, like silt and sand, are the most likely to detach.

Transportation: The wind now picks up the loosened soil and transports it, the distance and altitude to which it is carried depends on the particle size and the strength of the wind.

Surface creep: Rolling across the surface, larger soil particles (0.5 to 2 mm in diameter) collide with other soil particles, knocking them loose. This process is known as surface creep. Large particles in this process travel barely a few meters.

Saltation: Silt and clay, which are between 0.05 and 0.5 mm in diameter, salt together to form larger particles. These dirt particles are just light enough to be swept off the ground without becoming airborne. These particles scratch the soil's surface as they move through a sequence of low bounces and short hops, leading to abrasion and the subsequent breakdown of particles.

Suspension: Suspension takes place between sand and clay particles of soil that are less than 0.1 mm in size. For long periods, soil particles can float in the air, although bigger dust particles (0.05 to 0.1 mm) may fall to the ground within a couple of miles, while particles of (0.01–0.001mm) may travel hundreds to thousands of kilometers before settling down.

***Deposition*:** Deposition commonly occurs when the wind speed decreases.

Types of Wind Erosion

***Sweeping drift*:** It is characteristically used to describe very mild wind erosion where some fine soil from loose soil becomes suspended.

***Active drift*:** It is a more severe form of wind erosion that includes all three types of particle movement along with the formation of drift banks

***Deflation wind erosion*:** Deflation erosion is the removal of loose material by the wind from flat expanses of dry, uncemented sediments found in deserts, dry lake beds, floodplains, and glacial outwash plains.

***Abrasion/Corrasion wind erosion*:** Abrasion/ Corrasion wind erosion is the process of erosion produced by the suspended particles that impact solid objects.

Soil Conservation Practices

Protecting the soil's top layer from erosion requires identifying the most effective means of doing so. There are two approaches exist, one of which is dependent on the percentage of slope Mechanical or engineering techniques, and agronomic conservation practices.

Agronomical Method

The agronomic method is superior as it saves money in the long run while still producing desirable results. Rainfall can be absorbed by using agronomic conservation strategies, which are more cost-effective, long-lasting, and efficient than traditional methods. It can be used where the proportion of slope is less than 2%.

Various Agronomical Measures

Conservation Tillage

With the help of plants and fewer tillings, conservation tillage can protect soil from wind and water erosion. Planting and tillage are used in conservation production techniques to lessen the amount of water and soil lost during cultivation. When compared to plowing, improved cropping systems that employ fewer intensive forms of cultivation can cut airborne pollution by as much as 50–95%. The ability of farm managers to integrate safe agricultural practices with reliable pest control is crucial for the success of conservation tillage.

No-Till Farming

Because it leaves most of the residue behind and keeps the soil surface undisturbed, no-till is an efficient conservation approach for lowering wind erosion. Soil moisture is preserved, evaporation is cut down, and drying is slowed as a result. Erosion is reduced on moist soils. With no-till, you don't need to till the soil before planting your seeds in tiny furrows. Because the cover crop residue is left on the soil surface, no-till prevents the soil from drying out as quickly in hot weather. About 80% of wind erosion can be prevented when plow till is replaced by no-till. The residue layer not only aids in water infiltration but also boosts the soil's organic matter and microbial richness.

Stubble-Mulch Tillage

By reducing the amount of soil inversion and leaving crop residue on the surface, stubble-mulch tillage is a conservation tillage practice. Subsurface tillage is a method that does not include burying crop leftovers. Water losses from no-till fallow can be larger than those from stubble-mulch tillage (Schillinger and Bolton, 1993), but heavy residue cover in the no-till systems reduces evaporation and enhances water storage in the spring. In places with low residue production, stubble mulch tillage may be preferable to no-till systems with complicated crop rotations for enhancing water retention and soil aggregation.

Conservation Agriculture

Minimal soil disturbance, perennial ground cover, and crop rotation or association are the cornerstones of the conservation agriculture practice. Minimal soil disturbance, permanent soil cover, and crop rotation or association are the cornerstones of conservation agriculture, which is neither a prescription nor a conventional technology (FAO, 2001). The goal of conservation agriculture is to increase crop yields while simultaneously preserving and increasing the soil resources upon which crop growth depends. Modern technologies are encouraged in conservation agriculture to boost soil quality and biodiversity. The foundations of conservation agriculture (CA) include no-till farming and other soil conservation methods (Dumanski et al., 2006). Cover crops, crop wastes, and mulch are used to keep the soil always covered. Crop sequences, intercropping, and relay cropping/mixed cropping are all examples of crop associations that can be used as part of a larger crop rotation. Widespread grain and oilseed crops, pastures and rangelands, and high-value intensive crops like horticulture and agroforestry can all benefit from applying conservation agriculture concepts with the right adjustments. The purpose of conservation agriculture is to enhance the soil's physical, hydrological, and biological qualities to boost soil productivity and yield stability while decreasing production costs and manpower needs for soil preparation and management. The goal of conservation farming is to strike a good economic, ecological, and agricultural balance by maximizing yields and profits. Following is a description of the guiding concepts of Conservation agriculture and the actions to be encouraged: (Dumanski, et al., 2006)

- Using a no-till method, ensure that the soil is always covered and that as little mechanical disturbance as possible is made to the soil.
- Improve soil quality through rotating crops, using cover crops, and employing IPM techniques.
- Solicit fertilizers, insecticides, herbicides, and fungicides in proportion to crop-related nutrient losses. The idea behind this is to improve soil quality rather than plant nutrition.
- Encourage correct application of crop protection products to cut down on waste, boost productivity, and shield the environment.
- Encourage legume fallows, composting, and the use of organic fertilizers and nutrients for the soil.
- Encourage the manufacture of food, fiber, fruit, energy, and medications with added value. Manufacturing goods with added value is manufacturing for sale.

Sustainable Land Management

The concept of sustainable land management expands the scope of conservation agriculture to include economics, markets, profitability, and long-term viability. Improved crop and livestock output in response to market opportunities contributes to value-added production, food sufficiency, and poverty alleviation within a land management system that prioritizes sustainability. Enhanced crop management, agroforestry, specialized crops, and permanent crops are all viable options for getting the job done. Sustainable land management is defined by the Food and Agriculture Organisation of the United Nations (FAO, 2005) as "the adoption of land use systems that, through appropriate management practices, enable land users to maximize economic and social benefits from the land while maintaining or enhancing ecological support functions of the land resources." Smyth and Dumanski (1993) were the first to talk about and explain the idea of sustainable land management. They said that sustainable land management is a set of technologies, policies, and activities that try to balance social and economic issues with environmental ones.

1. Keep up with demand and increase output (productivity)
2. lessen production risks and strengthen soil's ability to absorb shocks from degradation (stability/resilience);
3. Keep soil and water from being contaminated (preservation) and preserve natural resources (preservation).
4. Ensure social acceptability and equitable access to the advantages of better land management (acceptability) while remaining economically viable (viability).

Contour Farming

Soil erosion is reduced with contour farming because the soil is worked into gentle slopes that follow the natural shape of the ground. Furrows that don't match the natural curves of the land provide water a simple path downward, leading to quick runoff when it rains. By plowing the land in a direction perpendicular to the slope, farmers can increase water infiltration into the soil and reduce the loss of valuable nutrients, organisms, and plants that would otherwise wash down the slope. This method of plowing helps prevent soil erosion on sloped and curved terrain by diverting water flow. Water can more easily seep into the soil when contour lines act as a break to prevent the formation of gullies and rivulets during periods of high rainfall. Contour plowing is an effective method for preserving soil quality. Reducing soil fertility due to misuse, salinization, acidification, or other chemical soil pollutants is a major goal of soil conservation efforts.

Crop Rotation

It is best practice in crop rotation to rotate between different plant species rather than growing the same crop year after year. There are many advantages for farmers that implement this strategy of soil protection. Legumes, often known as nitrogen-fixing plants, are rotated in to enhance soil structure through their varied root systems, reduce insect populations, and increase soil nitrogen levels. By alternating deep-rooted and shallow-rooted crops, crop rotation can enhance soil structure and fertility. As a result, erosion can be slowed and infiltration improved, lowering the potential for flooding farther down the road. There are several positive effects on the soil. Crop rotation is preferable to monoculture because it helps maintain favorable growing conditions by retaining soil fertility.

Wind breaks

Reducing the destructive impact of wind is an important part of this soil conservation practice. Planting hedges in rows protects from snow and wind for your crops.

Cover Crops

Green manure, water retention, a more even distribution of nitrogen, and a more natural habitat for microorganisms are all benefits of using cover crops. By strengthening aggregates and encouraging biological activity, they prevent soil from being eroded by the wind and boost its overall quality. To prevent soil erosion during the time between harvesting and planting, farmers frequently employ cover crops. Cover crops planted between rows of primary crops shield delicate plants from damaging winds. Cover crops, however, are best suited to humid and temperate climates with high rainfall, and in arid and semi-arid regions, they can compete with main crops for scarce water. Cover crops are especially useful in areas where agricultural residual yields are low.

Strip Cropping

In strip cropping, crops are planted in alternating rows of high and low density. In order to prevent erosion, erodible landscapes can be protected by alternating strips of row crops with annual or perennial grass. The strips should be as wide as necessary to protect the crops from sand drifts. Sensitive crops are better protected by narrower strips. The strips should be wide enough to accommodate the width of the machinery used in fertilizing and turning the soil.

Engineering Methods

This conservation technique will be implemented only in areas with a slope percentage of more than 2%; in this method, bunds, terracing, and grass waterways are used to slow down the speed of surface runoff water and so control erosion.

Various engineering methods

1. **Contour Bunding**: Water can be retained in the catchment for an extended period through the use of bunding or the construction of a small embankment. It is mostly done to shorten the length of the slope

and reduce the speed of runoff water. Therefore, there is less runoff and soil erosion, and more water infiltrates into the ground. As the name implies, contour bunding is building a succession of smaller bunds across the slope of the land along a contour to redirect the flow of water and force it to walk rather than run down the slope. In addition, water can be pushed against it to increase soil moisture. It cuts the slope in half, which lessens the amount of water that can run off the slope and does damage to the soil. It is built on slopes between 2% and 6% in locations with an annual rainfall of fewer than 600 millimeters.

2. **Side bunds** are built parallel to the contour of the bund on either side.
3. **Lateral bunds** are built parallel to the slope between two side bunds to shorten the contour bund's total length. The amount of water that runs off to one side is diminished as a result. **Supplemental bunds** are built to restrict the horizontal distance between two contour bunds.
4. **Marginal bund:** To clearly mark the edge of a field, road, river, etc., a bund is built. Shoulder bunds are constructed at the outside edge of the terrace to collect runoff water and reinforce the bench's foundation.
5. **Graded Bunding**: By bundling or building little embankments, we may shorten the slope and slow the flow of runoff water. Therefore, there is less runoff and soil erosion, and more water infiltrates into the ground. Bunds are used in watersheds for a variety of purposes, including preventing erosion and retaining water. It is placed across the slope rather than following the natural lines of the soil. The area between two parallel bunds functions as a channel, with a gentle longitudinal slope along the bund to direct runoff away from the gap. Recommended for a slope of 2–8%, it is typically built in regions with annual precipitation of more than 700 mm.
6. **Grassed waterways**: Drainage and diversion of catchment runoff is accomplished by the use of grassed streams and diversions. Surface runoff water from the bunded/terraced region can be disposed of in grassed waterways, which can be either natural or man-made watercourses lined with erosion-resistant grasses. Grass plants planted along a waterway's edge operate as a protective barrier against soil erosion brought on by the channel's turbulence. The grassy canals are built within the terrain's natural gradient. Sites for grassed rivers are chosen based on crucial criteria including a slope of less than 20% and a water flow velocity of less than 1.8 m/s.

7. **Bench terraces:** To follow the natural slope of a piece of land, bench terraces are built in the shape of platforms. They are built on sloping terrain with an incline of 10–33%. By using a cut-and-fill technique, we can transform hilly land into terraces by creating a series of steps that catch water as it moves across the soil's surface. Slope determines the many bench terrace varieties.

   (a) **Level bench terraces**: Irrigated bench terraces, tabletop terraces, and paddy terraces are all names for the same type of terrace. Optimal for regions with a medium annual rainfall and very permeable soils.

   (b) **Bench terrace sloping outward**: The durability of the terrace's outer edge is dependent on the construction of a shoulder bund, which is ideal in low-rainfall areas with permeable soils. Orchard-style bench terraces are another name for these tiers.

   (c) **Bench terraces sloping inward**: These terraces, also known as hill-type bench terraces, are favored in high-rainfall areas with impermeable soils because of the drainage channel provided on the inner side of the terrace. Favored for cultivating plants that are particularly vulnerable to flooding.

Soil Erosion Models: A Modern Approach to Soil Conservation

Soil erosion is a major environmental problem because it threatens agricultural output and biodiversity around the world. By simulating runoff and erosion scenarios with and without protective measures, models enable optimal land use planning and identification of vulnerable locations. It is possible to reliably anticipate runoff, soil erosion, and sediment output across a variety of agroclimatic zones and management strategies. Soil loss and runoff rates from agricultural regions, land use plans, relative soil loss indices, and more can all be estimated with the use of soil erosion models, which the government can use to inform soil and water conservation policy and practice. Using refined and tested models, the likelihood of soil erosion can be calculated. To model soil erosion, researchers from several fields, including soil science, plant science, hydrology, sedimentology, meteorology, and others, must collaborate. It is essential for any model to be able to encompass processes, factors, and causes across multiple spatial and temporal scales. Models provide a quantitative and consistent means of calculating soil erosion and sediment output across a variety of scenarios, making them essential for achieving full control of soil erosion. Many different models of soil erosion have been developed over the years. Effective use of these models depends on regional soil and climate variables due to their varying degrees of complexity and data requirements.

Inclusion of Geographical Information System (GIS) and Remote Sensing

Soil erosion models have been substantially aided by the development of cutting-edge technologies like remote sensing and GIS. The forecasting power of soil erosion models has been boosted by combining GIS and remote sensing with empirical and process-based approaches. In order to simulate erosion and create maps of eroded areas, crucial data is stored in a GIS. Land cover estimation over extensive areas is an important input for erosion modeling, and remote sensing can help with that. Modeled data can be expanded from small plots using remote sensing and GIS techniques.

Water Erosion Models

Empirical Models

Because of their ease of application in scenarios with sparse data and input parameters, empirical models are a good starting point for tracing the origins of sediment and nutrient deposition. The Universal Soil Loss Equation (USLE), the Revised Universal Soil Loss Equation (RUSLE), and the Modified Universal Soil Loss Equation (MUSLE) are all examples of empirical models.

1. **Universal Soil Loss Equation (USLE):**

   The United States Loss Equation (USLE) was created in the United States and is currently the most popular empirical model for assessing soil losses worldwide (Wischmeier and Smith, 1965).

   USLE is employed in the conservation planning and design process. Under certain situations, the USLE may accurately anticipate soil loss from agriculturally used soils. In contrast to complex models, the USLE is advantageous due to its low barrier to entry, ease of usage, and lack of dependence on either a large training set or a large number of input parameters. Because of the equation's simplicity, not all aspects of soil erosion are taken into account when it is used in practice.

   The average annual soil loss is estimated as

   $$\mathbf{A = R \times K \times LS \times C \times P}$$

   where **A** is the average annual soil loss (Mg ha^{-1}),

   R is the rainfall and runoff erosivity index for the location of interest,

K is erodibility factor,

LS is topographic factor,

C is cover and management factor,

P is supporting practice factor,

2. **Modified USLE (MUSLE):**

   The MUSLE is essentially a USLE with some tweaks. In contrast to USLE, which bases its sediment output forecasts on precipitation, MUSLE uses a runoff factor that takes into consideration the moisture content of the soil before it rains. With this adjustment, USLE can be utilized to foretell sediment yield in response to storms.

3. **Revised USLE (RUSLE):**

   This model is founded on empirical and process-based methodologies, making it more extensive and thorough than USLE (Renard et al., 1997). Soil erosion from farmland and pasture may now be more precisely estimated thanks to updates made to the RUSLE. The RUSLE is an updated computerized version of the USLE that keeps its core components while also taking into account new findings and insights.

   The equation is: **A = R.K.L.S.C.P**

   where **A** is the computed soil loss,

   R is the rainfall-run-off erosivity factor,

   K is the soil erodibility factor,

   L is the slope length factor,

   S is the slope steepness factor,

   C is the cover management factor,

   P is the supporting practices factor

Process-Based or Physically-Based models

EUROSEM (European Soil Erosion Model)

Scientists from seven different European nations work together on EUROSEM, and runoff plots specifically constructed to verify the models are located in England and Spain (Morgan et al., 1998). Erosion on both the field and watershed scales, as well as sediment and dissolved material pollution of water bodies, are evaluated by EUROSEM. It can be used as a planning tool in the selection of soil conservation measures and functions on an event-by-event basis.

CREAMS (Chemicals, Runoff, and Erosion from Agricultural Management Systems)

It has been used extensively to estimate field-scale pollution emissions from agricultural land and is the most extensively validated process-based model available. This framework is equally applicable to agricultural land. It was designed to function in areas up to 40 ha, although this can be increased to 400 ha if necessary. Daily estimates of runoff volume and peak rate, infiltration, evapotranspiration, and percolation are generated by the hydrologic module. The erosion factor is partially grounded in physics and partially in experience.

ANSWERS (Areal Non-point Source Watershed Environment Response Simulation)

This model was developed to analyze the financial viability of potential land-use decisions within the watershed by evaluating the sediment yield of the entire watershed as opposed to that of field-sized regions. It is executable on a mainframe because it uses a large database consisting of data from sources such as topographical maps, soil surveys, crop and management surveys, and other similar sources.

WEPP (Water Erosion Prediction Project)

The WEPP is an example of the next generation of prediction technology; it is a process- and computer-based model (Flanagan et al., 1995). Using the fundamental principles of overland flow dynamics, infiltration, evaporation, evapotranspiration, erosion mechanics, percolation, drainage, surface ponding, interception of rainfall and runoff by plants, decomposition of residues, soil consolidation, tillage, and land management, WEPP is a process- and computer-based model for hillslopes and watersheds. When used with different tillage and cropping systems, WEPP can reliably anticipate soil erosion both during storm occurrences and on a continuous basis. As a function of surface cover, soil characteristics, surface roughness, and tillage, it models erosion across and between hills, as well as sediment movement and deposition in channels and reservoirs.

EGEM (Ephemeral Gully Erosion Model)

Based on fundamental concepts of gully bed and side-wall dynamics, the EGEM was designed to provide predictions about gully development and erosion (Foster and Lane, 1983). When it comes to channel incision and head cut development, the EGEM takes into account the dynamic processes of focused flow. When it comes to predicting channel erosion, the EGEM is one of the few process-based models available.

Wind Erosion Models

Empirical Models

Wind erosion equation: The Wind Erosion Prediction Equation (WEQ) is the classical equation of wind erosion prediction (Woodruff and Siddoway, 1965).

$$\mathbf{E = f\,(I, K, C, L, V)}$$

where **E** is the average annual soil loss (Mg ha−1 yr−1),

I is soil erodibility index (Mg ha−1 yr−1),

K is soil ridge roughness factor,

C is climate factor,

L is width of the unsheltered field (m),

V is equivalent vegetative cover factor.

The WEQ is the most popular model and an empirical model that simplifies the complex interactions, combinations, and spatial variability of erosion processes and components by assuming that wind erosion varies linearly with climate, soil property, and surface condition changes.

Revised WEQ (RWEQ): In 1998 (Fryrear et al., 1999), researchers improved upon the WEQ model to create the RWEQ model, which is more logically organized and more accurately captures the physical processes of wind erosion. Soil erosion at both the neighborhood and regional scales can be evaluated by combining large data sets from the field with computational models. Wind velocity, precipitation type, soil roughness, the erodible fraction of soil, crusts, and the amount of surface residue are just a few of the dynamic parameters that RWEQ uses to assess erosion. It estimates the mass transported by wind-based on the weather factor (WF), the erodible fraction of soil (EF), the soil crust factor (SCF), the soil roughness factor (K'), and the combined crop factors (COG).

Wind Erosion Prediction System (WEPS)

Due to its superiority as a process-based, continuous, daily time-stepping, computational model, WEPS was designed to succeed WEQ as the standard method of forecasting.

In contrast to WEQ, it uses physically plausible erosion processes to model wind erosion. If you're looking for reliable estimates of wind erosion, WEPS is the way to go (Visser et al., 2005).

Other wind Erosion models

(a) Wind Erosion Stochastic Simulator (WESS)

(b) Texas Tech Erosion Analysis Model (TEAM)

(c) Wind Erosion Assessment Model (WEAM)

(d) Wind Erosion and European Light Soils (WEELS)

(e) Dust Production Model (DPM)<

Computer Models of Tillage Erosion

Tillage Erosion Prediction (TEP) Model

The TEP model is a straightforward computational framework based on regression equations of soil translocation by tillage (Lindstrom et al., 2000). With the assumptions of a constant slope, homogenous soil loss or gain, and no vertical soil movement during tillage inside each slope segment, it predicts soil displacement for individual slope segments and specific tillage operations.

Water and Tillage Erosion Model (WaTEM)

Water and tillage erosion are two forms of environmental degradation that can be simulated using the WaTEM (Oost et al., 2000). Water and tillage erosion can be modeled with the help of the WaTEM. The landscape structure is a key factor in tillage erosion, which is why the WaTEM takes topography as its starting point. Two-dimensional simulations rely on topographic information acquired from aerial pictures or digital elevation models. The soil bulk density, tillage depth, tillage transport coefficient, and contributing area are all important inputs.

Soil Redistribution by Tillage (SORET)

In order to predict how soil may change due to tillage erosion, scientists have developed a three-dimensional model called SORET (Alba, 2001). It models the dynamic shifts in soil profile inversions/cuttings and transformations, soil horizon substitutions, surface characteristics, and soil-landscape interactions that occur as a result of soil erosion. The SORET can model the cumulative impacts of repeated tillage operations on soil erosion and use that information to make predictions about erosion under various tillage scenarios.

Soil Erosion by Tillage (SETi)

The SETi model (Torri & Borselli, 2002) is a process-based method for calculating the tillage transport coefficient as a result of the interplay between slope, tillage tool, and soil clod displacement.

Water and Tillage-Induced Soil Redistribution (SPEROS)

Soil redistribution due to water and tillage erosion is simulated using 137Cs data in the process-based SPEROS model (Oost et al., 2000). Using 137Cs data, SPEROS calculates water and tillage erosion rates, allowing the water and tillage erosion contributions to soil erosion to be separated. It provides approximations for both the vertical and horizontal movement of 137Cs in the soil profile due to factors such as water and tillage erosion.

Soil Conservation and Health

Crop and variety selection, fertilizer use, conservation techniques, tillage and irrigation timing and intensity, and many more are only some of the production and conservation decisions that farmers must make. Agricultural policies and readily available technologies are two examples of what influences these choices. Soil erosion, nutrient pollution, and water availability and quality are only some of the environmental impacts that can be influenced by production choices. Soil, water, and air quality are all part of the resource base that affects agricultural output, along with other elements like pollinator health and climate.

- Soil health and productivity, water quality and quantity, and nutrient and pesticide use are all sources of environmental concern associated with agriculture. Factors such as regulatory incentives for environmental protection and the availability of technological choices for inputs and field management play a role in shaping these effects.
- Cost-share incentives are available to farmers through the Environmental Quality Incentives Programme (EQIP) who voluntarily engage in soil conservation practices. Cover crops, conservation tillage/residual management, conservation crop rotation, terraces, and nutrient management have been the five most popular conservation practices sponsored by EQIP since 1998.
- By letting farmers make management decisions based on the foot rather than the field, variable rate technologies (VRT) offer the potential to lower input costs and off-field environmental consequences.

Maintaining Soil Health and Productivity

The quality of an agroecosystem, including the soil health and future productive potential of a producer, is strongly influenced by the producer's management decisions. The following are essential goals for managing soil health and productivity:

- Avoiding the undesirable accumulation of salts or other toxic substances
- Maintaining or improving soil depth by keeping soil erosion rates low enough so that soil formation replaces soil lost to erosion
- Improving soil as a rooting medium
- Using agrochemicals in amounts so low, that adverse environmental effects are avoided
- Maintaining or improving soil biodiversity
- Avoiding monocultures
- Minimizing soil compaction
- Maintaining or increasing soil organic matter (SOM) content
- Improving infiltration and percolation of stormwater, plant-available soil water, and minimizing evaporative losses.

Several methods exist for accomplishing these aims in management. In order to sustainably increase crop yields over time, these practices prioritize SOM maintenance and enhancement, erosion control, and responsible pesticide use. This section discusses fundamental management practices and factors that contribute to soil health and sustain soil production over time.

Conclusion

Soil deterioration is a cycle, and one of its effects is a thinner soil profile. Subsoil erosion is common since it is typically less organized than topsoil. Without proper mitigation, soil thinning will eventually result in bedrock exposure. Soil is essential for a wide variety of ecosystem functions, many of which would be lost without it. The future viability of our soil resources is difficult to foresee. Soil heterogeneity is a contributing factor. Understanding the potential responses of larger ecosystems and how this may affect soil functionality indirectly is essential for predicting how our soils may react to future environmental changes. While we have extensive information on soil erosion around the world, our understanding of soil formation rates is much more limited. Soil formation and erosion rates, particularly on agricultural land, are thus rarely measured in tandem. Soil erosion provides a hazard to soil sustainability, but how great that threat is can't be determined unless soil formation is also considered. Minimal adjustments to our agricultural methods are part of a toolset of established soil conservation measures. It has been proven, for instance, that switching to reduced or even zero tillage helps prevent soil erosion caused by water by keeping the soil in its natural structure. Seeding cover crops is another option for preventing soil erosion caused by keeping vast swaths of land bare. It has been proven that certain cover crops provide additional benefits, such as reducing the prevalence of weeds and fungal diseases and attracting a wider variety of wildlife. Before it's too late, we need to make the same effort to keep our soils healthy and productive for as long as possible.

References

Alba, S. de. (2001). Soil redistribution by tillage (SORET) is a computer model of 3-D simulation at a field scale. Conservation agriculture is a worldwide challenge. First world congress on conservation agriculture, 1-5 October, 2001. Volume 2: Offered contribution. 643-646. Madrid, Spain.

Balasubramanian, A. 2017. Soil erosion–causes and effects. *Centre for Advanced Studies in Earth Science, University of Mysore, Mysore.*

Bashir, S., Javed, A., Bibi, I., & Ahmad, N. 2017. Soil and water conservation. *Pakistan, University of Agriculture, Faisalabad*, 263-86.

Blanco-C. H. & Lal, R. Soil and water conservation. *Principles of soil conservation and management*, 9, (2008): 1-19.

Dumanski, J., Peiretti, R., Benetis, J., McGarry, D. & Pieri, D. The paradigm of conservation tillage. *Proc. World Assoc. Soil and Water Conserv.* P1 (2006): 58-64. Available from: http://www.unapcaem.org/publication/ConservationAgri/ParaOfCA.pdf

FAO 2001. Defining conservation agriculture. In: The economics of conservation agriculture, FAO, Rome, Italy. www.fao.org/docrep/004/Y2781E/y2781e03.htm

FAO 2005. Information page: Sustainable Land management. https://www.fao.org/land-water/land/sustainable-land-management/en/

Flanagan, D. C., Ascough II, J.C. Nearing, M. A. & Laflen, J. M. (1995). The water erosion Prediction Project (WEPP) model. 145-199. Landscape Erosion and Evolution modelling. Springer. Availabel from: https://link.springer.com/chapter/10.1007/978-1-4615-0575-4_7#citeas

Foster, G. R. & Lane, L. J. (1983). Erosion by concentrated flow in farm fields. In: D. B. Simons Symposium on Erosion and Sedimentation, Colorado State University, Fort Collins, CO, p.21.

Fryrear, D. W., Sutherland, P. L., Davis, G., Hardee, G. & Dollar, M. (1999). Wind Erosion Estimates with RWEQ and WEQ. In: 10th International Soil Conservation Organization Meeting held May 24-29, 1999 at Purdue University and the USDA-ARS National Soil Erosion Research Laboratory. 760-65.

Lal, R., & Moldenhauer, W. C. Effects of soil erosion on crop productivity. *Critical Reviews in Plant Sciences*, *5*, no. 4, (1987): 303-67.

Lindstorm, M. J., Schumacher, J.A. & Schumacher, J.A. TEP: A tillage erosion prediction model to calculate soil translocation rates from tillage. *Journal of Soil and Water Conservation.* 55, no. 1, (2000): 105-08.

Mandal, D., Dadhwal, K. S., Khola, O. P. S., & Dhyani, B. L. (2006). Adjusted T values for conservation planning in Northwest Himalayas of India. *Journal of soil and water conservation, 61*, no. 6, (2006): 391-97.

Morgan, R.P.C., Quinton, J. N., Smith, R. E., Govers, G., Poesen, J.W.A., Auerswald, K., Chisci, G., et al. (1998). The European Soil Erosion Model (EUROSEM): A dynamic approach for predicting sediment transport from fields and small catchments. *Earth Surface Processes and Landforms.* 23, no. 6, (1998): 527-44. Available from: https://doi.org/10.1002/(SICI)1096-9837(199806)23:6<527::AID-ESP868>3.0.CO;2-5.

Munyanziza, E., & Oldeman, R. A. A. Pterocarpus angolensis DC: Field survival strategies, growth, root pruning and fertilization in the nursery. *Fertilizer research, 40*, (1994): 235-42.

Oost, V.K., Grover, G. & Desmet, P. Evaluating the effects of changes in landscape structure on soil erosion by water and tillage. *Landscape Ecology.* 15, (2000): 577-589. Available from: https://link.springer.com/article/10.1023/A:1008198215674

Renard, K. G., Foster, G. R., Weesies, G. A., McCool D. K. & Yoder, D.C. Predicting soil erosion by water: A guide to conservation planning with the revised universal soil loss equation (Rusle). *United States Department of Agriculture.* Agriculture Handbook No. 703.

Schillinger, F. & Bolton, F. E. (1993). Fallow water storage in tilled vs. Untilled soils in the Pacific Northwest. *Journal of Production Agriculture.* 6, no. 2, (1993): 267-69. DOI: https://doi.org/10.2134/jpa1993.0267.

Smyth, A.J. & Dumanski, J. 1993. An international framework for evaluating the sustainable land management. FAO World Resource Report 73 and 74, FAO, Rome.

Torri, D. & Borselli, L. Clod movement and tillage tool characteristics for modelling tillage erosion. *Journal of Soil and Water Conservation.* 57, no. 1, (2002): 24-28.

Visser, S. M., Sterk, G. Karssenberg, D. Wind erosion modelling in a Sahelian environment. *Environmental Modelling & Software.* 20, no. 1, (2005): 69-84. DOI: https://doi.org/10.1016/j.envsoft.2003.12.010

Wischmeier, W.H. & Smith, D.D. 1965. Prediction Rainfall Erosion Losses from Cropland East of the Rocky Mountains: A Guide for Selection of Practices for Soil and Water Conservation. Agricultural Handbook, No. 282, 47 p.

Woodruff, N. P. & Siddoway, N. P. A Wind Erosion Equation. *Soil Science Society of America Journal.* 29, no. 5, (1965): 602-8. DOI: https://doi.org/10.2136/sssaj1965.03615995002900050035x.

15

Current Technological Aspects for On-Farm Cooling of Fresh Fruits and Vegetables

***Boris Huirem*[1] *and Pradeepta Kumar Sahoo*[2]**

[1]*Department of Horticulture and Food Science, School of Agriculture and Allied Sciences, The Neotia University, Sarisha, West Bengal, India*

[2]*Food Engineering Department, Faculty of Agricultural Engineering BCKV, University in Haringhata, West Bengal, India*

Introduction

According to research released not too long ago by the United Nations, the population of the entire planet is expected to reach around 8 billion by the end of the year 2022. As a consequence, the future demand for the production and distribution of food must be satisfied to provide sustenance for the expanding population of the world and to alleviate human suffering. However, in countries like India, food is wasted at an alarmingly high rate every year because there are insufficient facilities for cooling and storing food once it has been harvested. The majority of the losses began on the farm level since the fresh produce, which requires fast refrigeration after harvest in large quantities, was unable to be given, which led to the spoilage of the food. It is especially vital to shorten the amount of time between harvesting and first cooling to slow down the rate at which newly harvested fruits and vegetables breathe out carbon dioxide and exhale it. When keeping fresh fruits and vegetables, the two most important controlling parameters are temperature and relative humidity. These two factors work together to minimize the amount of water that is lost and to keep the product quality from deteriorating. Following their preferred ranges of temperature and relative humidity, fruits and vegetables can be divided into the following categories: (Table 15.1). For the objectives of on-farm cooling, the majority of fresh fruits and vegetables that are classified as belonging to groups 1 and 2 need to be concentrated, which necessitates maintaining an ideal temperature of less than 10°C and a relative humidity of more than 90%.

Table 15.1: Optimum temperature and relative humidity ranges of fresh fruits and vegetables for storage (7 days) (USDA Agriculture Handbook 66, 2016)

Group 1	Group 2	Group 3
• Temperature: 0–2°C	• Temperature: 7–10°C	• Temperature: 13–18°C
• Relative humidity: 85–95% (mostly fruits and melons), 90–98% (mostly green vegetables)	• Relative humidity: 85–95%	• Relative humidity: 85–95%

Evaporative Cooling

Evaporative cooling is a tried-and-true strategy for bringing down the temperature, and it is also one of the more efficient methods. The process of evaporation involves taking energy from the warm air in the environment to cause water to evaporate; as a consequence, a significant cooling impact is produced. The effectiveness of an evaporative cooler is determined by a variety of elements, including the temperature of the air, the relative humidity of the air around it, the flow of air, and the total surface area. Therefore, it is vital to find out the meteorological conditions to adequately analyze the prospective efficacy of evaporative coolers. When it comes to storing fresh fruit, most people agree that the traditional cold storage that makes use of refrigeration methods such as vapor compression systems is the most effective way. However, in order for them to function, they need to be powered by electricity, and they call for a significant expenditure of capital. This makes them an unreliable option for several farmers. Furthermore, the growing worry about the contribution to the greenhouse effect that is caused by the conventional CFCs/HCFCs refrigerant that is used in conventional refrigeration systems is a significant problem for the subsequent generations. As a result, there is an immediate requirement for an alternate cooling system that is both cost-effective and sustainable to reduce the amount of food that is wasted. Table 15.2 outlines some of the more workable cooling technologies that could be used on farms to preserve the freshness of fruits and vegetables.

Table 15.2: List of cooling technologies with their places of origin

Sl. No	Technology	Developed by	Year	Country
1	Zero Energy Cooling Chamber	S. K. Roy and R. K. Pal	1991	India
2	Charcoal cooler	Sharma and Rathu	1991	India
3	Coolbot walk-in Cooler	Ron and Kate Khosla	2006	USA
4	Ecofrost	Devendra Gupta, Prateek Singhal, and Vivek Pandey	2010	India
5	Evaptainers	Spencer Taylor, Quang Truong, and Jeremy Fryer-Biggs	2014	USA
6	SAVARS	Theakston, FH	1982	USA

Different Methods of On-Farm Storage

Zero Energy Cooling Chambers (ZECC)

The Zero Energy Cooling Chamber, also known as the Pusa Cooling Chamber, is a straightforward cooling storage structure that is also environmentally friendly, affordable, and efficient. Farmers can adopt this structure for the short-term or on-farm storage of fresh fruits and vegetables before transporting the bulk of their produce to the market. It utilizes the basic concept of evaporative cooling, in which the heat from the warm air is absorbed by the water, which then causes the water to change its phase from liquid to gas, so producing a cooling effect. In ZECC, it is possible to produce a temperature drop of 10–15°C and a rise in relative humidity of 90–95% compared to the surrounding ambient condition, respectively. The fact that this storage structure may be built out of readily available raw materials in the area, such as bricks, sand, bamboo, straw, gunny bags, and so on, is one of the most significant advantages it offers. In addition, the method of building is rather easy, and even a person with no prior experience should have no trouble putting it in place.

The diagram of a zero-energy cooking chamber is shown in Figure 15.1. It is built with double brick walls (outer and inner walls) that are divided by a tiny cavity measuring around 7–8 cm on all sides of the walls. This cavity is filled with coarse sand material that functions as an evaporative medium. The chilling chamber contains perforated plastic crates in which fresh meals are stored. These crates can have the tops covered with transparent polyethylene sheets to further limit the amount of oxygen that is lost through respiration and transpiration. There is a very slight gap between the walls and the boxes to ensure that there is adequate air circulation within the chamber. During construction, the inner brick wall is designed to be a little bit higher than the outside brick wall by making it one brick higher. This is done to give cover inside the cooling chamber. The sand hole and the brick walls need to be maintained wet there for adequate cooling, and this can be accomplished by utilizing a water bucket to apply water three to four times a day. Alternatively, one can construct a drip system made of plastic pipes to send a continuous supply of water into the sand chamber from an overhead water tank. This can be done in several different ways. A thatch roof is typically supplied to shield the building from the intense heat of the sun as well as precipitation.

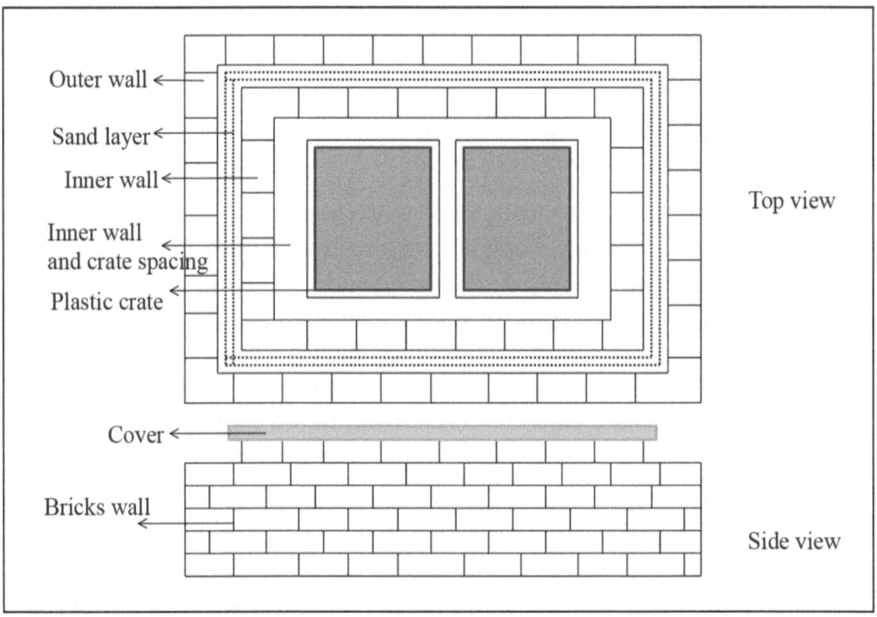

Figure 15.1: Schematic diagram of a typical zero energy cooling chamber.

Charcoal Cooler

A charcoal cooler is a simple and efficient structure for the short-term preservation of fruits and vegetables that makes use of raw materials that are readily available in the surrounding area, such as bits of charcoal. It is an inexpensive structure that even a farmer with no construction experience may erect without much trouble. An evaporative cooling process is used, which means that it cools the air by causing the water that is on the porous structure of charcoal to evaporate when it is exposed to heated air. This results in the air getting cooler. It is possible to cut post-harvest losses by as much as 80%, and a conventional charcoal cooler with these dimensions may hold green leafy vegetables for up to 6 days.

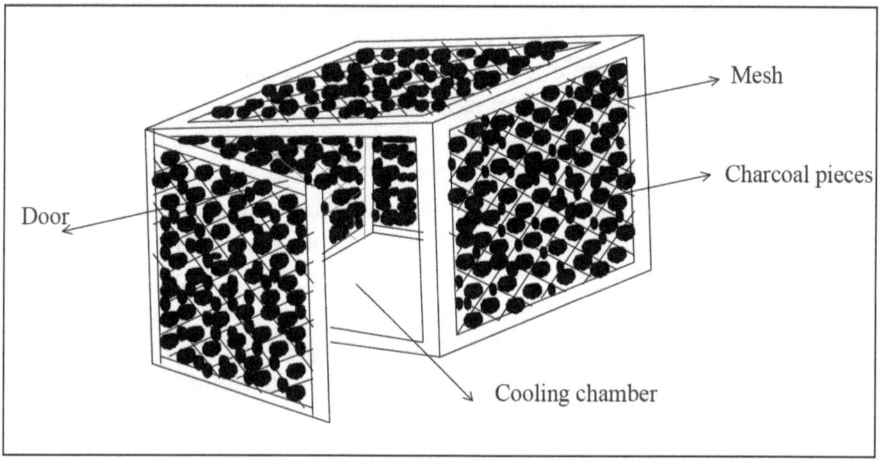

Figure 15.2: Schematic diagram of a typical charcoal cooler.

A diagrammatic representation of a charcoal cooler is shown in Figure 15.2. The cooler is made up of charcoal-filled walls that are completely enclosed within a wire mesh roughly 10 cm apart and are supported by a hardwood framework. The roof has a gap that is approximately 15–20 cm wide so that air can circulate freely. Inside the chilling chamber, perforated plastic crates or shelves are used to store fresh produce such as fruit and vegetables. The charcoal is kept moist utilizing an overhead water tank and drip pipe system, which works to distribute water evenly on the pieces of porous charcoal for continual cooling action. This keeps the charcoal from becoming dry and brittle. In most cases, a shade is supplied to cut down on the amount of water used and achieve the desired level of cooling in a shorter amount of time.

Evaptainers

Evaptainers are portable refrigeration machines simply powered by liquid water. They are very inexpensive and do not require any electricity to function. Evaporative cooling is utilized, and modern design and materials are incorporated, to produce a cooling system that is not only effective but also lightweight, long-lasting, simple, and user-friendly. This structure has been shown to be effective for agricultural products at the site of production, extending the shelf life of those products by three to four times longer than it would be under normal ambient conditions. A temperature difference of 10–15°C might be achieved between the outside environment and the interior chamber of an evaptainer, much like it is possible with other types of evaporative coolers. A diagrammatic representation of an evaptainer is shown in Figure 15.3. The device features a clamshell design that is foldable and collapsible, making it simple to both operate and move. The device consists of extremely effective evaporation side walls made from a membrane that is highly evaporative on the outside and a waterproof membrane on the inside to form a sandwich structure. The interior, which is encircled by water in the interstitial space, is used for food storage. The amount of water used is extremely minimal, and the water that is trapped between the layers turns into vapor by extracting the heat that is absorbed from the hot and dry air. This results in a cooling effect being achieved within the chamber. Additionally, it prevents unsightly puddles of water from forming below the unit as a result of the drips. The outer layer features expanded polypropylene materials that provide excellent insulation.

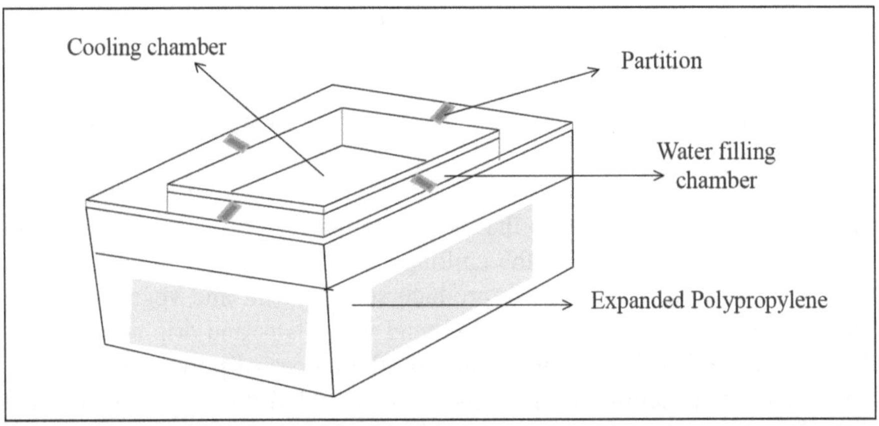

Figure 15.3: Schematic diagram of an evaptainer.

Coolbot Walk-In Cooler

Coolbot walk-in cooler uses a typical air conditioner (of the window and wall-mounted mini split air conditioner type) and a coolbot temperature controller unit and it is an innovative cooling technology that can turn any room adequately insulated into a cooling chamber, that is, the room is converted into a walk-in cooler. When compared to the standard walk-in configuration, it uses less energy to operate, making it more cost-effective. Additionally, it is simple to install and does not call for the use of any special tools or prior electrical expertise. A sample example of a coolbot walk-in cooler's schematic diagram is presented in Figure 15.4. It is constructed with a cooling chamber as its core component, with the air conditioning unit and the coolbot temperature controller unit attached to the walls of the cooling chamber. Coolbot is equipped with several sensors, a heating element, and a microcontroller unit programmed with instructions. These instructions tell the air conditioner's compressor unit to cool the room to approximately 34–36°F (1.1–2.2°C) without causing the air conditioner unit to freeze. This cooling system has a wide range of applications and can be utilized by a variety of industries, including agriculture, floristry, brewing, and dining.

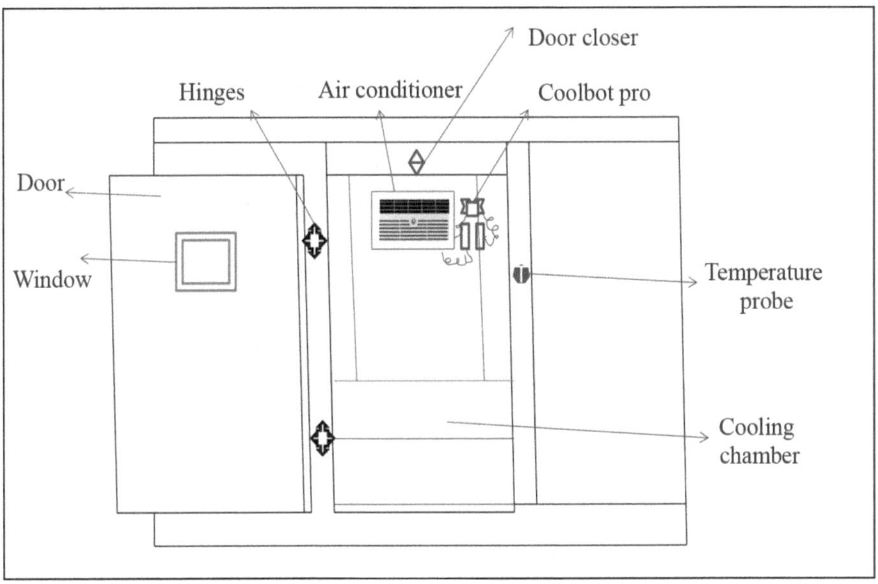

Figure 15.4: Schematic diagram of a typical Coolbot walk-in cooler.

Solar Powered Cold Room (Ecofrost)

On-farm cooling of fruits and vegetables is made possible with the help of Ecofrost, a portable micro cold storage device that uses solar energy to power its refrigeration system. Smallholder farmers in rural areas are the focus of this initiative, with the end goal of enhancing food security by lowering the rate of spoilage among perishable food products. The system is able to maintain relative humidity between 80% and 95% and pre-cool items to between 10% and 15% in a matter of hours (about 6 hours). The diagram shown in Figure 15.5 is an example of a standard schematic for a solar-powered cold room. A separate refrigeration unit that offers a temperature range of 4–10°Cand a humidity controller unit that offers a humidity range of 65%–95% are both included in this system. Foods are stored in the cooling chamber, which is effectively insulated with materials made of polyurethane foam (PUF). The foods are maintained on shelves or in perforated plastic crates. Several flat-plate solar collectors are mounted on the roof of the storage unit. Together, these collectors have the capacity to generate 5 kW of solar power on their own. It is powered by electricity from the grid and only requires 5 to 6 hours to fully recharge itself. There is a provision for an alternate power supply unit that comes from a generator, and if the weather is cloudy, the switch will take place automatically. This cooling system's capacity for remote monitoring is one of its most valuable characteristics; also, it enables farmers to run the system with relative ease using an Android application.

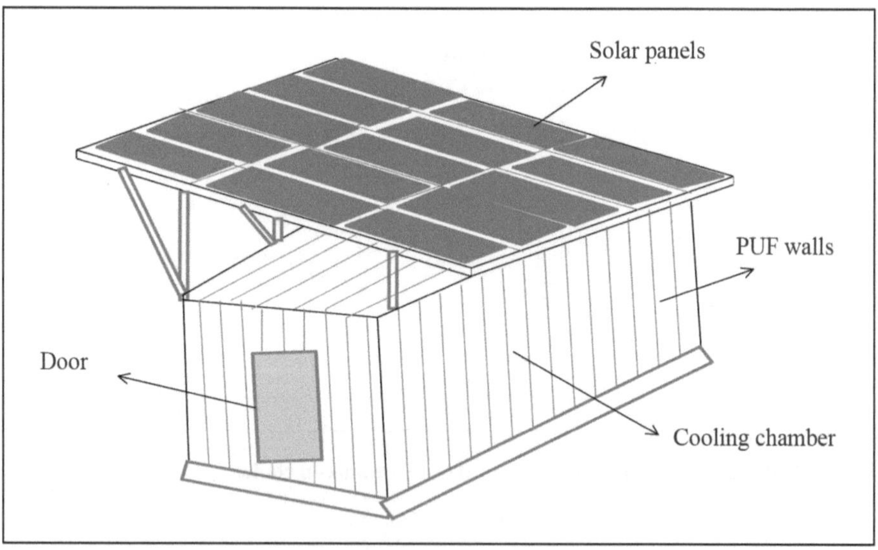

Figure 15.5: Typical schematic diagram of a solar power cold room.

Solar Assisted Vapor Absorption Refrigeration System (SAVARS)

An alternative to a vapor compression refrigeration system is a vapor absorption refrigeration system. This type of system makes use of the low-grade heat that is produced by a normal solar collector, which is typically a less expensive option. A properly functioning vapor absorption refrigeration system can make on-farm or transit cooling of fruits and vegetables possible. This is due to the fact that demands for complex operation can easily be substituted by simpler devices such as pumps, as opposed to reciprocating or rotary compressors. Both evaporation and condensation processes take place in enclosures that are physically separated from the food items in a refrigerator chilling system that uses vapor absorption. As this prevents the formation of condensed water on the surface of the fruit, almost none of the food ingredients are lost in the process.

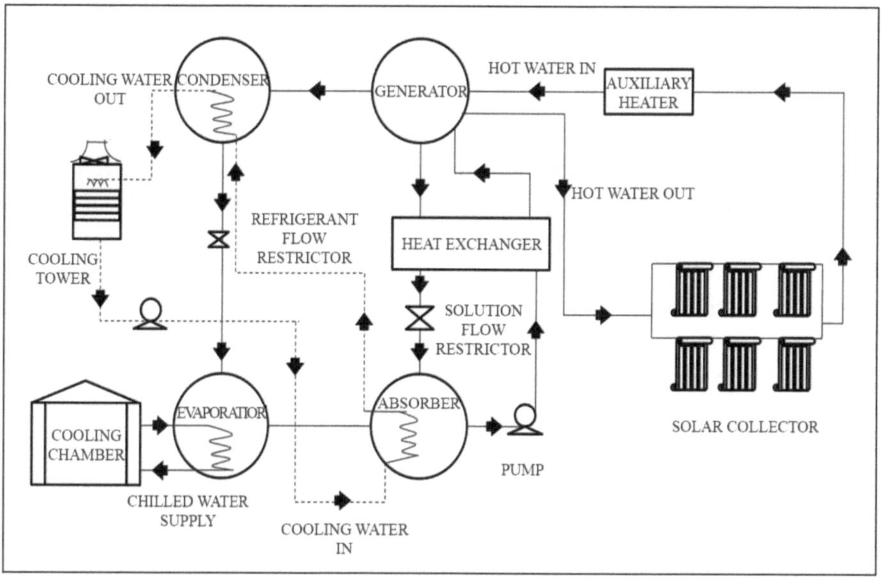

Figure 15.6: Schematic diagram of a solar-assisted vapor absorption refrigeration system (SAVARS).

The schematic diagram of a solar-aided vapor absorption refrigeration system is shown in Figure 15.6. This system is based on a LiBr-water binary fluid, with LiBr acting as the absorbent and water serving as the refrigerant in the system. It consists of (a) a vapor absorption machine that includes an evaporator, generator, condenser, absorber, solution heat exchanger, pumps, pipings, and other components; (b) a solar collector network; (c) a cooling tower; and (d) a cooling chamber cum air handling unit are the components that make up a solar-assisted single-effect vapor absorption refrigeration system. A continuous supply of hot water is necessary in the system in order to operate the generator of the vapor absorption machine. This hot water supply is being given by the network of solar collectors in the system. Because the absorber and generator components of the vapor absorption refrigeration system are responsible for the processes of absorbing and desorbing the refrigerant, respectively, heat is produced in these parts of the system. Through the use of a cooling tower, this heat is dissipated into the surrounding environment. It is accomplished through the circulation of cooling water between the system and the cooling tower, which allows for the simultaneous transfer of heat and mass between the ambient air and the hot water. It is a relatively low-cost technology that has the ability to save water and is reliable for eliminating low-grade heat from cooling water.

This is accomplished by enhancing the water use efficiency, which is important in light of today's growing water costs and diminishing water resources. When using a vapor absorption machine, chilled water is produced in the evaporator as a byproduct of evaporative cooling. This water is then circulated throughout the machine to give the required amount of cooling, and the machine is attached to a cooling chamber. An air handling unit (AHU) installed inside the cooling chamber makes the recirculation of the chilled water possible. The AHU is made up of several cooling coils that function as a heat exchanger between the chilled water and the warm air that is produced by the respiration of food products stored inside the chamber. This warm air is generated by the food products themselves as they are kept inside the chamber. In the cooling chamber of the vapor absorption refrigeration system, which is suitable for the on-farm cooling of food produce, the cooling air temperature can be maintained within the range of 6–10°C and the relative humidity can be maintained within the range of 90%–95%.

Selection of Cooling Technology Consideration

Deciding on the type of cooling or refrigeration system to be installed is not a simple task. Before making a decision, it is vital to carefully examine each technology, taking into consideration the requirements for cooling, whether or not it is feasible, and the various aspects that are involved. The following questionnaire could be helpful in determining the best course of action to take:

i. **How much cooling load is required?**

   When it comes to cooling, various types of food products demand different temperatures. In addition to this, it is essential to have a working knowledge of the compatibility group when it comes to the preservation of fruits and vegetables.

ii. **What is the average temperature and relative humidity of the area where cooling is needed?**

Evaporative cooling requires a low relative humidity in order to be effective, and this must be maintained throughout the cooling process. Evaporative cooling could not be effective if there is a high relative humidity in the area where it is being used.

iii. **How windy is the area? Is the water source close to the area?**

Evaporative cooling is greatly dependent on the presence of wind. It is necessary for effective cooling to ensure that warm air is properly circulated with the cool air that is all around it. In addition to this, evaporative cooling needs a constant supply of water in order to function properly; hence the water supply should be situated close to the area that needs to be cooled.

iv. **Are the raw materials and skilled personnel easily available?**

During the construction process, one of the most crucial considerations to make is how to obtain the raw materials needed. Additionally, the availability of skilled employees is a must for this endeavor.

v. **Are mechanical cooling or refrigeration systems available in the area? Are they costly?**

If a commercial refrigeration system is readily available and does not require an excessive investment of money, then it is possible that this would be a preferable choice.

Conclusion

This chapter highlights several different technologies used to cool fresh fruits and vegetables, particularly for cooling in a short amount of time or on the farm itself. The majority of these technologies, such as zero energy cooling chambers, charcoal coolers, and evaptainers, do not need any form of electricity in order to function. On the other hand, the coolbot walk-in coolers, ecofrost, and solar-assisted vapor absorption refrigeration systems require either a small amount of electric power or a combination of solar power and conventional electricity in order to function. However, in comparison to the typical vapor compression refrigeration system, which is wholly reliant on the grid system, the electricity usage of this system is significantly lower. Therefore, farmers who live in rural areas or in locations with limited access to electricity can take advantage of these technological advancements to preserve and extend the shelf life of their freshly harvested product while it is still in the on-farm condition. Farmers can easily store their harvested produce for a few days before sending it to the market using these technologies, thereby avoiding the distress sale of the produce and simultaneously enhancing their source of income. This allows farmers to avoid the need to sell their produce at a discount.

References

Ashitha, G., Sonaraj, P. R., Sooraj Krishna, P. M., Sreehari, T., & Shilpa, P. (2021). Zero energy cooling chamber and zero energy cooling system in building wall. *International Journal of Engineering Research & Technology, Special Is,* 6, (2021): 100-6.

Chinenye, N. M., & Manuwa, S. I. Review of research and application of evaporative cooling in preservation of fresh agricultural produce. *International Journal of Agricultural and Biological Engineering*, 7, no. 5, (2014): 85-102.

Defraeye, T., Schudel, S., Shrivastava, C., Motmans, T., Umani, K., Crenna, E., Onwude, D. The charcoal cooling blanket: A scalable, simple, self-supporting evaporative cooling device for preserving fresh foods. *EngrXiv*, (2022): 1-24.

Ecozen Solution. 2009. Ecofrost Solar Cold Room. Retrieved from: https://www.ecozensolutions.com/wp-content/uploads/2019/11/Ecofrost-brochure.pdf

Evaptainers. Available from: http://www.evaptainers.com/ev8/. (Assessed on 15-10-2022).

Gross, K., Wang, C., & Saltveit, M. The Commercial Storage of Fruits, Vegetables, and Florist and Nursery Stocks, Agriculture Handbook No. 66, 5thedn. (U.S. Department of Agriculture, Agricultural Research Service, Washington, DC, 2016), pp. 54–6.

Huirem, B., & Sahoo, P. K. 2020. Thermodynamic modeling and performance optimization of a solar-assisted vapor absorption refrigeration system (SAVARS). *International Journal of Air-Conditioning and Refrigeration*, 28, no.1, (2020): 2050006-1-18. DOI: https://doi.org/10.1142/S2010132520500066

Obura, J. M., Banadda, N., Wanyama, J., & Kiggundu, N. A critical review of selected appropriate traditional evaporative cooling as postharvest technologies in Eastern Africa. *Agricultural Engineering International: CIGR Journal*, 17, no. 4, (2015): 345–54.

Odesola, I. F., & Onyebuchi, O. *A Review of Porous Evaporative Cooling for the Preservation of Fruits and Vegetables. The Pacific Journal of Science and Technology*, 10, (2010): 935-941. Retrieved from: http://www.akamaiuniversity.us/PJST.htm

Revolutionary Walk-in Coolers. Available from: https://www.storeitcool.in/. (Assessed on 15-10-2022).

16

Sustainable Urban Landscaping

C. Kanimozhi[1], R. Angelin Silviya[2], S. Anandhi[3], and B. Gopu[4]

[1–4]SRM College of Agricultural Sciences, SRM Institute of Science and Technology, Chennai, Tamil Nadu, India

Introduction

Landscaping is an expanse of scenery, a great way to add beauty, and curbs appeal to the place. The urban landscaping designs the outdoor and indoor places marveling the beauty of nature, i.e., the warmth, breeze, sunrays, shade, and the cool climate. Sustainable Landscaping can be defined as the designing of a garden considering the environmental issues at hand and landscaping in such a way that the designs, the planting materials, and the management are in balance with the locality with minimal resources. According to Loehrein (2013), organic gardening practices are integral to the design, construction, and maintenance of sustainable residential and commercial gardens.

Precision sustainable gardening, which involves applying specific amounts of care and resources to individual plants at precisely the right time and in the right place, is a form of precision farming. The local community, the environment, and the health of birds, bees, and other wildlife can all benefit immediately from a sustainably landscaped yard. Xeriscaping, grasscycling, erosion control, water conservation, carbon sequestration, habitat creation, mulching, and composting are all practices that contribute to more sustainable and climate-appropriate landscapes (Kamakshi, 2022).

Figure 16.1: The Era of Anthropocenes.
Source: https://www.anthropocene.info/

Why Sustainable Urban Landscaping?

The recent climate change and the pandemic forecast humans what the future is and how fast resilience could save us from threats. Mother Earth is not only for the present generation to live upon and can't become a garbage space of the universe like paper thrown into the garbage after the sweets have gone. The current Anthropocene epoch is the unofficial unit of geological time which describes the impact of human intervention over the earth's history affecting the planet's climate and the ecosystem. The future generations should not superscribe the present generation as the **Anthropocenes** who have altered and damaged the earth comprehensively. Though humans are tagged as anthropocenes the days are at hand to change the scenario by paying individual attention to the earth (Figure 16.1).

Urban Heat Island (UHI)

The urban haze, the closer constructions, the skyscrapers, the paved roads, the concretes, and the asphalts cause pollution over the cities due to the greenhouse effect and cause urban heat island effect. The Urban Heat Island (UHIs) in the densely populated cities are 2 degrees higher than the peri-urban and the rural areas and this urban-rural difference happens usually 3–5 hours after sunset.

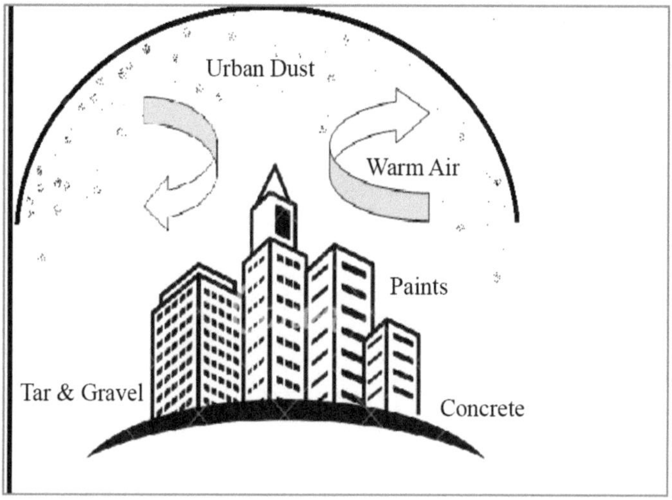

Figure 16.2: The era of urban dust domes.
Source: Sarkar & Aldag, 2018

Li-Yang (2016) reported that the influence of Urban Heat Island (UHI) increases the land surface temperature and that impacted directly on energy flow, material flow, ecology, environment, urban hydrologic situation, soil properties, atmospheric conditions, habits, people's health, and metabolism.

Thus, the urban population is surrounded by huge storms of dust over their dwellings causing a huge impact on the health of the residents and causing ill effects on the environment. These dust domes (Figure 16.2) can add fuel to wildfires, prevent cloud formation, cause heat strokes for older people, and increase energy demands like using air coolers. The solutions for the dust domes are planning the city with perfect landscaping with social concerns, for example the afforestation measures like green belt management for the sweaty city, and introduction of park cool islands which can cool their neighborhood for a few degrees during the summer in the city are the measures which can be planned by the city municipal corporation

Threats for Sustainable Urban Landscaping

The challenges (Figure 16.3) stand in the path of implementing sustainable urban landscaping include various parameters viz., the uncertainty in climate conditions, adopting exotic and invasive species, disturbing the landscapes, loss of different habitats, cultural landscapes, and the heritage and old monuments (Ivo, 2022, Machar *et al.,* 2017, Oprasal *et al.,* 2018 and Simon *et al.,* 2014)

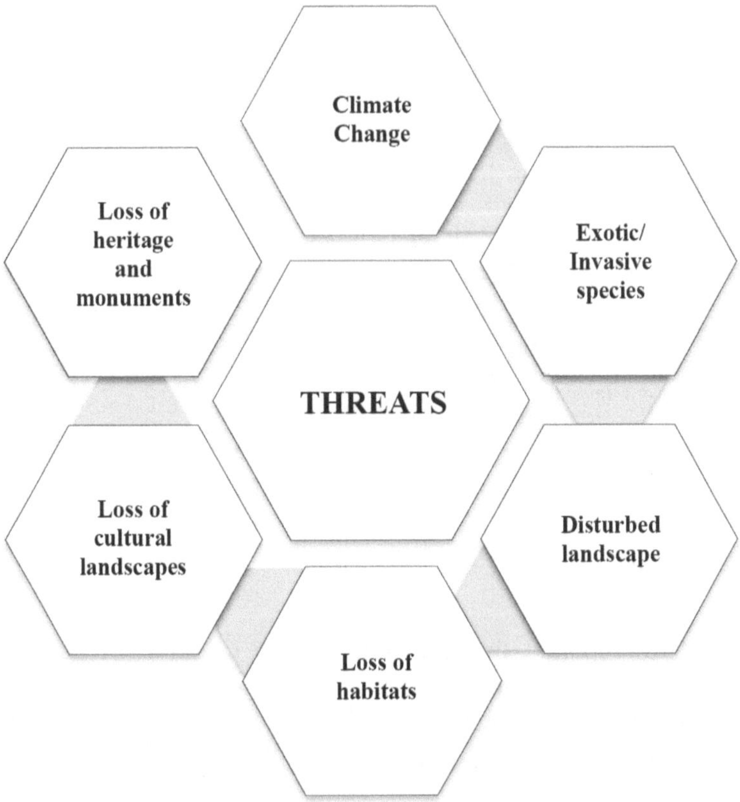

Figure 16.3: Threats for imposing sustainable urban landscaping.
Source: Machar *et al.,* 2017

Approaches for Sustainable Urban Landscaping

The wealth of a country is decided by the potential flora and the fauna as inhabitants. Big revolutionary contributions must be permeated immediately by the government with exemplary design considering climate change and other environmental issues. Also, sustainable landscapes should start from the individuals who are planning for the landscaping because we live in a world of limited resources. Whether, being a landscape designer, individual homeowner, community gardener, the government, or nurseryman sustainability and environmental concerns are to be considered before the implementation of the new landscaping (Li and Zhang 2020 and Machar *et al.,* 2017).

The advantage of a sustainable landscape is that it protects Mother Earth by the conservation of energy and resources while planning a long-lasting garden

(Figure 16.4) with efficient resource management and water management methods such as Xeriscaping, labor Management through artificial Intelligence and smart sensors, soil health management through mulching, and plant protection methods by usage of least toxic chemicals and by the adaptation of carbon neutrality methods like burlapping.

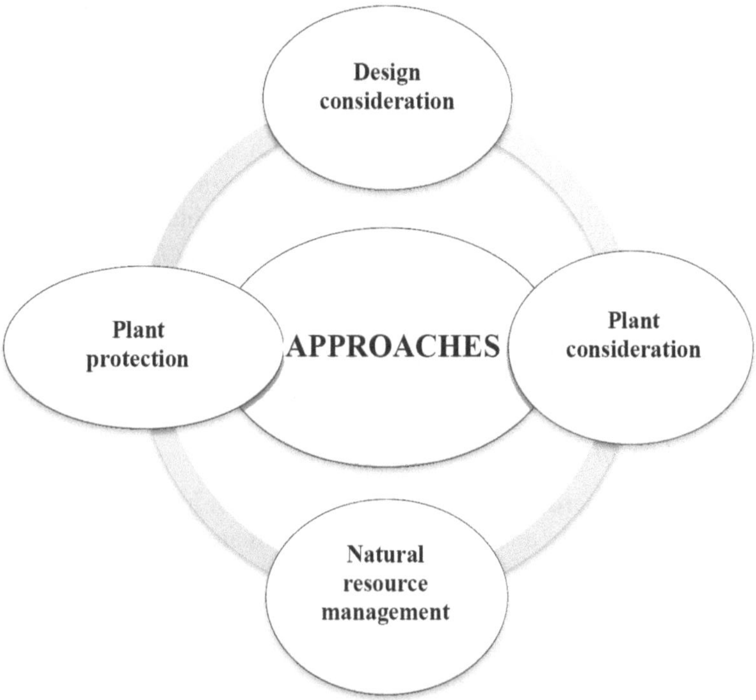

Figure 16.4: Different approaches for attaining sustainable urban landscaping.
Source: Li and Zhang, 2020

Design Considerations

For holistic sustainable landscaping, the designing of a garden should involve a horticulturist, an architect, and an environmentalist for the large-scale implementation of designs. And as an individual home grower, the pros and cons from each end of the scientist should be taken into consideration. The designs should be long lasting, expandable, water saving, energy efficient, and adaptive besides being carbon neutral. The designing of a landscape garden is completely different from the field establishment of the agricultural and horticultural crops because there are no standardized procedures to be followed on, as it does not depend upon the yield attributes. Rather the designing is just one's imaginary skills or the aesthetic sense, which is directly dependable on the qualitative attributes in the form of appreciation from the viewer. The attention seeking, applause, the wow factor, the calmness, the serendipity, and the therapy it brings to the mind are the factors to be considered as the yield of the garden.

The state of mind of the individual is the state of art over the ground; hence the mind must enhance itself that while designing the new garden the environmental aspects should also be toned up in the mind so that the art over the soil looks beautiful, healthy, and sustainable. The site analysis is the first step in gardening considering the contour, soil health, and soil wealth. The design should go with the contour available, instead of changing the topography the design should quench with the available contour forms. The plan should navigate with the existing place wealth, for example, the coastal area design should not thought to be fully reclaimed with new red soils, instead, the plan should bring in designs with localized plants acclimatized to the soil condition. Choosing the right plant materials that are drought tolerant, pest and disease-resistant while adapting to low fertilization is a prerequisite in sustainable landscaping.

Hardscape Element

The hardscape of a garden improves the aesthetics of a garden, but it should be environment-friendly and sustainable. These materials include the use of wood, concrete, and stone, among other options. If unsustainable materials are chosen, the hardscape can lead to negative impacts. In the same way, too much hardscape in a landscape, or poorly chosen elements, can also detract from the environment. The installation of rainwater catchments aesthetically will manage the water needs of the plants in the year. As an alternative to traditional hardscapes, permeable hardscapes can filter and store stormwater (surface water from heavy rains or snow). Instead of collecting in low areas or running off into storm drains or nearby streams, stormwater can seep into the ground (permeate) through permeable hardscape materials. Driveways, walkways, patios, and any other areas where a hard surface is preferred can all benefit from permeable hardscapes. To lessen the severity of floods, replenish groundwater supplies, and lessen pollution in local waterways, permeable hardscapes are a great option. Porous asphalt, pervious concrete, and permeable interlocking concrete pavers are the three most common types of permeable hardscapes.

Biodiversity

The success of a garden lies in welcoming wildlife into the garden in the form of chirping birds, soft butterflies, colorful moths, buzzing bees queuing ants, and other wildlife to the garden. To welcome this wildlife, the garden should consider plants with attractive colors, fragrant flowers, shattered seeds, vibrant berries, and rippling water to provide food, water, and shelter. Avoid pesticides in the garden which may be poisonous to the useful predators in our garden. Also, the pets that are allowed to threaten the wildlife will have negative effects on welcoming the wildlife says Aubrey (2014).

Green Belt Management

Imposing green belts in urban areas will improve the ecological benefits. These green belts improve the atmospheric air quality and help in lowering the atmospheric temperature and humidity. Zhu *et al.* (2017) reported the benefits of green belt management that may benefit the urban planners/designers to achieve the goal of utilizing urban areas with greenery which may optimize the ecological effects and render cultural benefits. A Pan-African movement was initiated in the Sahel—Saharan desert of Africa during 2007 to construct a huge tree wall, the green belt action over 15 km wide and 7775 km long from Senegal in West Africa to Djibouti in East by covering 40,000 hectares as an act against the desertification. Similarly in India, a green belt planning has been initiated across Porbandar to Panipat covering the entire Aravalli range and beyond it. This green wall will act as a barrier by protecting against dust from the west and checking eastward of the Thar desert. It may help against desertification by restoring the degraded lands by massive afforestation in the Thar desert region of India (Figure 16.5).

Xeriscaping

Although Earth has been covered with water, fresh water consists of only 2.5% so the problem of water scarcity is growing higher. This scenario alarms us in meeting the water needs of plants in the garden. This can be solved by adapting Xeriscaping, in which the garden is specifically designed with the usage of less water without compromising the aesthetic appeal of the garden. The term "xeriscaping" refers to a style of landscaping that relies less on a constant supply of water. Using minimal irrigation effectively, strategically placing drought-tolerant plants, and covering the ground are all part of this method. In this case, Xerophytic plants, such as cacti and succulents, and other plants adapted to dry conditions are used (e.g., small leaves, succulent leaves, white or silvery reflective foliage).

Building Material Considerations

The current generation is dwelling in multistoried apartments where land is entitled to them in minimal spaces. The living space is their living hall with their vision thronged to the four walls. Their feet would puddle into the soil only when they happen to meet their grandparents in the villages once in a year or a lifetime. The lives in these concrete halls make no sense of greenery unless it's been replaced with indoor plants and vertical walls. The spaces available for the plants in the urban dwellings are the walls, the rooftops, the front yard, and the backyards of individual homes and while some of the bigger residential apartments are planning to have a community garden. The handling of minimal space is a real constraint. So, making the spaces more energy efficient is in the hands of the engineers.

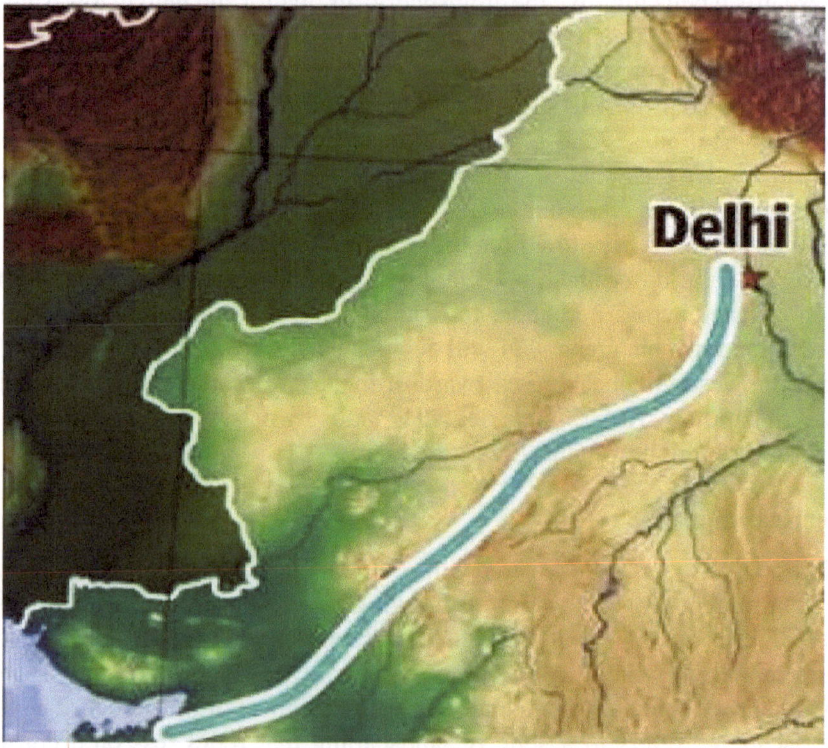

Figure 16.5: Rough contours of proposed 1400 km green wall in India.
Source: http://myforests.eu/india/

Planting Considerations

In general, what does a plant require to have successful growth? The light requirement, the humidity, the temperature, the media where it grows, the nutrients for its growth, all this put together is the basic requirement of any type of plant. When we provide this sophisticated environment, the plants will thrive whereas when the conditions are not optimized the growth will fall. Being ornamental plants for landscaping these plants in general can be categorized as tropical, subtropical, and temperate plants where the plants flourish to their maximum in their specific zone. When these plants are being altered to grow in between these zones, naturally the environment must be provided to the plant according to the parent zone.

Shifting its place of origin and expecting to shine in an uncomfortable zone is the expectation of the human leading to the failure of the crop. For example, the growth of hydrangeas excels in temperate zones having acidic soil, by appreciating its bloom and enhanced by its beauty planning to plant the hydrangeas in the backyards of the tropical plains will lose its aesthetic sense by means of disappointment to the gardener and limited growth attributes to the plants. Hence sustainable landscaping encourages the gardeners and the landscapist to go for plants suitable for the locality and the Act like EPA (Environmental Protection Act) suggests having native plants. The native plants will not only flourish well in the environment it also takes care of the soil health too. The season should also be taken into consideration for the annuals.

Utilization of Forms of the Tree

The forms of the ornamental plants should be selected before planting and the expansion of the perennial plants after certain years should be taken into consideration to avoid damage to the building site. If the form of the plant is poorly exhibited, then its adaption to the limited spaces affects the aesthetics. Poor forms will lead to the removal of the plants. When there is a situation to remove the plants, determine whether the plants truly need to be removed or relocated to another area of the garden.

Choosing right plant: In landscaping the plants are generally chosen based on the aesthetics, their texture, their form, their flower colors, and their blooming season. Fancied by the shape of the Christmas trees many chose plants which are not appropriate for the location. The plant grown in the pots has a contained growth because of the root coiling around the containers, but once they are being planted in the soil it excels and puts forth its maximum potential. Admired by their form and beauty many ought to buy the plants from the nursery based on the size they view, but once planted they occupy too much of space in the hardscapes, compelling the owner to cut off the branches in uneven form or in worst case the complete removal of the plants on his own expenses. The right plant selection should be planned during the layout of the plan itself and should be bought considering the growth stages it will put forth in the years. One such example of the poor growth form maintained by the gardeners is the Almond trees, the *Terminalia* species has an excellent branching lateral extending on both sides widespread around 4–5 m horizontally. But most of the species of *Terminalia* planted as avenue planting through some social activities do not consider the tree form rather, they concentrate on the number of tree seeds sown that day. The social workers and other volunteers who take steps for the tree sowing activities like seed ball making and palm seed sowing or any seed sowing activities especially in the avenues should consider the landscaping principles and basic elements before sowing.

Utilization of planting material resources: When seeds are sourced from the traditional means they can be used again and again, whereas the hybrid seed companies do not come forward to disturb their cash flow. Another important sustainable management practice is the rooting of cuttings of some of the species can be done at home easily. For example, the softwood cuttings can be just clipped and used for the next successive generation (example, the plants like Coleus, Pedilanthus, Crotons, etc.). Likewise, the seeds from species like Gomphrena, Marigold, Celosia, and a few of the Amaranthus can be restored to allow the annuals to full maturity and the seeds can be collected dried and stored for the next generation before senescence of the plant.

Natural Resource Management

Placing the plant in the correct place is necessary with a basic understanding of the low rays of the Sun in the early morning and late evening. The trees should be placed facing southeast and southwest of the building. When the spaces are limited, choose the climbers that take the vertical place rather than horizontal growth. Poor sunlight leads to stunted growth due to the absence of photosynthesis, and each plant has a specific need for the light, for some plants keeping in bright sunshine for two days is enough, whereas some plants require rotating of the pots and some plants will exhibit only in full sunshine. The artificial lighting is cumbersome in the long run hence place the plants according to their light requirements. The relative humidity or the moisture in the air is important for the plants. To maintain humidity, the plants are to be protected from dry winds by misting on the shoots, keeping moist soil around the pots, adding charcoal, coco peat, and brick pieces, and by mulching the root zone. In places of dry climate, the cacti and succulents can be placed which require less water. Though the earthen pots are organic packages they are not good at retaining water than the plastic and other containers. Do not overwater and overfeed the plants, make the plants stress tolerant.

a. **Irrigation management:** Lawns, gardens, and individualized landscapes all benefit from landscape irrigation systems. Its primary use is to guarantee a constant and uniform distribution of water across a landscape. These techniques can be used to increase the robustness of gardens and landscapes.
    - **Reduce the need for watering in the first place.** This can be achieved through the use of drought-resistant plant species, careful runoff management, and rainwater harvesting and storage facilities.
    - **When irrigation is needed, use practices to minimize water waste.** Keeping an eye on the moisture content of the soil, mulching, watering at the right time and only the most important plants, and using the least amount of water possible from your irrigation system are all important parts of this strategy.
    - **Micro irrigation system:** Soaker hoses or a drip irrigation system are better than sprinklers for watering your garden. These run on low pressure to avoid wasting water through runoff and evaporation and to provide steady, targeted irrigation to the plant's root zone.

b. **Drought management:** This can be achieved by adapting the xeriscaping and choosing the native plants that fit well in a drought-tolerant garden.

c. **Soil management:** Application of organic matter regularly increases soil's water-holding capacity. Soils with greater organic matter and less compaction allow rainwater and irrigation water to infuse in more gradually rather than running off to drainage systems. For an established lawn and gardens, organic matter can be added in the form of grass clippings, chopped leaves, or compost. The use of mulch to retain soil moisture can be achieved by covering bare soil with mulch helps to moderate soil temperature and slow evaporation from the soil so it will stay moist longer. The rough texture of mulch traps water and allows it to percolate slowly into the soil layer rather than running off the surface. Organic mulch such as hardwood chips, straw, leaves, pine needles, or grass clippings will help improve the condition of soil, by adding nutrients as it decomposes. Inorganic mulches like rocks, pebbles, or gravel may help to eliminate weeds. The long-term success of a garden depends upon the growing medium, especially in the urban environment the space is limited and the dumping of heavy soil in the building structures affects the building. When the space is confined to the building walls, roof, and balcony, alternate techniques like vertical garden, hydroponics, aquaponics, and growing in Low-density polyethylene (LDPE) & High-density polyethylene (HDPE) bags, jute bags, and miniature gardens are considered. The medium selection for the urban apartments encourages sustainable materials like sand, clay, pumice, perlite, gravels, coir, and water which are the natural mineral components. The aquaponics in the urban places where the Recirculating (RAS) techniques and Nutrient Film Technique (NFT) technology together encourage sustainable and alternate methods of growing. Maintaining a suitable habitat for soil food web's primary consumers and higher-level consumer interactions is key to fertile soil. Growing plants in the soil as often as possible; providing a continuous and variety of food sources by growing as many different species of plants as practical; cover the soil at all times to minimize variation in the soil environment including temperature and erosion; and limit equipment or foot traffic in the garden especially when the soil is wet to minimize soil disturbance and compaction. Low or no tilling is one way to build up soil organic matter (carbon storage). Tilling the soil accelerates microbial breakdown of organic matter and the release of carbon dioxide (CO_2) into the atmosphere. Earthworms have also been noted to till the soil as well as leave behind vermicasts which attracts decomposers like fungi and bacteria. Cover cropping or integrating legumes in gardens or lawns is also a way to minimize dependency on supplemental nitrogen. Cover crops also aid in conserving soil moisture and preventing erosion. Climate-friendly sustainable solutions promote soil health and supplement nutrients that meet specific plant needs. Organic sources, such as manures and compost, should also be used strategically and sparingly. In addition to macronutrients, plants also require micronutrients. A plant's needs and soil test results should then serve as the basis for deciding if supplemental nutrients are needed. If supplemental nutrients are needed, never over-apply (this includes too much compost). Over-fertilizing can lead to excess plant growth, which can lead to greater susceptibility to diseases. Another important factor is the pH of the soil, the Acidophilic and basilophilic plants will never grow together in the same soil.

d. **Water management:** Rainwater barrels are a component of sustainable urban landscaping, which involves routing rainwater from a home's downspout into one or more rain barrels or a cistern in order to collect and store it for later use in the garden. Rainwater barrels are also referred to as cisterns. In addition, the downspouts can be redirected so that rainwater flows into a garden or other part of the yard where it can naturally soak in and recharge groundwater. The two most efficient strategies for reducing the amount of runoff that occurs at the surface are bioswales and rain gardens when water from rain begins to move in accordance with the principles of gravity. Runoff is the residual water that travels downhill after some of it has soaked into the ground and helped to refill the groundwater reserve. Rain gardens and bio-swales are meant to use plants, soils, mulch, and microbes in order to slow down and treat the runoff of stormwater. These man-made features are fashioned after natural ecosystems. They lessen the likelihood of floods further downstream and add to the amount of groundwater that is available. Rain gardens can be any size or shape and are purposefully placed in low-lying locations to collect rainwater. The surface runoff from an area can be directed to rain gardens or other areas where it can be absorbed via the use of bio-swales, which are shallow channels that are planted with plants.

e. **Carbon management:** The use of organic farming methods results in a smaller carbon footprint. Organic farming practices emphasize the enhancement of soil quality and the use of an integrated pest control approach rather than the application of synthetic fertilizers or pesticides derived from petroleum. When there is a greater quantity of humus in the soil, carbon emissions are prevented from being released into the atmosphere. It is critical to ensure that gardens are filled with plants that are actively growing; in particular, perennial gardens are the best option. By reducing the need to acquire products based on petroleum, organic gardening lowers the amount of pollution caused by fossil fuels during the manufacture and transportation phases of those products. Black, crumbly organic manure is produced as a byproduct of the breakdown of organic materials. This manure helps plants meet their nutrient requirements and also contributes to the improvement of the surrounding environment.

Plant Protection

The element of plant protection encompasses weed management, insects management, nematodes management, and disease management in garden plants. Garden plants face competition from weeds for water and nutrients. Gardens having an integrated weed management plan ensures appropriate management of both water and nutrients. Early detection of pests can reduce the spread of the pest, thereby reducing the amount of pesticides used, and the cost of control. It may also allow for the use of natural enemies or other slow-acting, less toxic pesticides when there is no immediate threat of damage that may be followed. Early detection of pests can also reduce the amount of pesticides used. Pests such as beetles, eggs, and snails can sometimes be removed by hand. Neem oil can be used as a deterrent in home gardens, mechanical control of pests and diseases, encouraging the population of predators and parasitoids, pruning diseased branches, pruning plants to get uniform sunlight, mechanical control of pests and diseases.

Conclusion

Introducing the garden in the available spaces to improve human health became necessary as the number of large buildings, high-density populations, and pollution levels seemed to increase at an exponential rate in the urban areas. As a result of this approach, the potential for attracting biodiversity by opening the doors to welcome it increases, and the ecological balance in the urban areas also improves. As a result, sustainable landscaping promotes resource-efficient plants, elements of hardscape that are friendly to the environment, and a toxic-free environment for both plants and humans. This is accomplished by preserving nature and adopting an environmentally responsible way of life in urban areas.

References

Ivo Machar. Sustainable Landscape Management and Planning. *Sustainability*, 12, (2020): 1–4.

Kamakshi Kanchana, M. Low-cost sustainable composting for sustainable landscaping. *International Journal of Engineering, Research and Technology*. 11, no. 10, (2022)): 149-55.

Li, J. and W Zhang. Fiscal incentives and sustainable urbanization: Evidence from China. *Sustainability*. 12, (2020): 103.

Loherlein Marietta. 2013. Sustainable Landscaping: Principles and practices. *CRC Press*. ISBN: 9781466593206.

Liyang, Feng Oianab, De Xuan Songa and KeJiaZhenga. Research on Urban Heat Island Effect. *Procedia Engineering*. 169 no. (2016): 11-18. Available from: https://doi.org/10.1016/j.proeng.2016.10.002.

Machar, I., V. Vozeenilek, K. Kirchner, V. Vlckova and A. Bucek. A biogeographic model of climate conditions for vegetation zones in Czech Republic. *Nat. Conserv. Bulg*. 24, (2017): 21-41.

Machar, I., V. Vozenilek, J. Simon., V. Pechanec. J. Brus. P. Fulnecek and T. Vitek. Joining the historical research and future predictions as a support tool for the assessment of management strategy for European beech dominated forests in protected areas. *Nat. Conserv. Bulg*, 22, (2017): 51-78.

Oprasal Z., J. Harmacek, P. Pavik and I Machar. What factors can influence the expansion of protected areas around the world in the context of international environmental and development goals? *Probl. Ekorozw.* 13, (2018): 145-57.

Sakar, E. and Aldag, M. C. (2018). Vertical Green Areas for the Protection of Ecological Balance. *4th International Scientific Conference, Sustainable Business and Ecological integration and collaboration.*

Simon J., I. Machar and A. Bucek. Linking the historical research with the growth simulation model of hardwood floodplain forests. *Pol. J. Ecol.* 62 (2014): 273-88.

Zhu, C., P. Ji and S. Li. Effects of Urban green belts on the air temperature, humidity, and air quality. *Journal of Environmental Engineering and Landscape Management.* 25, no. 1, no. 2017: 39-55.

Web References

Aubrey Bray. 2014. Available from: https://ucanr.edu/blogs/blogcore/postdetail.cfm?postnum =13669.

https://calrecycle.ca.gov/organics/landscaping/#:~:text=How%20to%20Landscape%20 Sustainably,habitat%2C%20mulching%2C%20and%20composting.

https://www.mastergardenproducts.com/sustainablelandscape/

https://education.nationalgeographic.org/resource/anthropocene

https://www.researchgate.net/publication/346514588_Xeriscapping_-Water_Efficient_ Gardening

India – MyForests

Senegal has championed the effort to build the Great Green Wall to fight back against desertification in west and north Africa (qz.com).

17

Organic Seed Production Implication for Regenerative Agriculture

Richa Bharti

Doon Business School, Dehradun, Uttarakhand, India

Introduction

Seed technology improves seed genetics and physiology. Variety creation, release, seed production, processing, storage, and certification are included. (Feistritzer, 1975). Organic farming began in the early 20th century due to constantly changing farming practices. Over 50 million hectares of certified organic agriculture are in Australia. "That discipline of study having to do with seed production, maintenance, quality and preservation" is seed technology. Organic seeds are fully produced with organic matter practices made with proven organic function and natural seed production, including seeding plants, with a set of guidelines prohibiting the use of manufactured products or chemicals to be sold as such. They must be produced by an accredited institution by an authorized person government agency. Organic seed is created by a planted crop and organically raised at least one generation for annual crops and two generations for annual and perennial plants. Organic farming is sustainable and eco-friendly. Farming needs better norms and standards to achieve biological goals. It is usually cold and people need lots of vegetables but have few organic, clean supplies. Organic farming in India needs a certain space and high-yield vegetable crops in national marketplaces to produce high-quality products. Indian agriculture uses chemical fertilizers, which can be minimized. Experiments have shown that it works well in conjunction. Inorganic and biological fertilizers boost yields, soil quality, and productivity.

Role of Seed Technology in Seed Production

Feistritzer (1975) outlined the following roles of improved seed:

- **Improved seed** – A carrier of new technologies. The introduction of quality seeds of new varieties wisely combined with other inputs significantly increases yield levels. In India, the cultivation of high-yielding varieties has helped to increase food production from 52 million tons to nearly 180 million tons over 40 years.
- **Improved seed** – A basic tool for secured food supply. The successful implementation of the high-yielding varieties program in India has led to a remarkable increase in production and food imports from other countries have been brought down despite a rapid increase in population.
- **Improved seed** – The principal means to secure crop yields in less favorable areas of Production. The supply of good quality seeds of improved varieties suitable to these areas is one of the important contributions to securing higher crop yields.
- **Improved seed** – A medium for rapid rehabilitation of agriculture in cases of natural disaster. In case of floods and drought-affected areas, the Government will provide the improved seeds from national seed stocks to rehabilitate the agricultural production of food grains in the country.

Deterioration Causes of Crop Varieties and Their Control

Variety: It is a group of plants having clear distinguished characters which when reproduced either sexually or asexually retains these characters. The main aim of seed production is to produce genetically pure and good-quality seed. But why or how the genetic purity of a variety is lost or deteriorated during seed multiplication? The several factors that are responsible for loss of genetic purity during seed production as listed by (Kadam, 1942) are:

1. Developmental Variation
2. Mechanical Mixtures
3. Mutations
4. Natural Crossing
5. Genetic Drift
6. Minor Genetic Variation

7. Selective influence of Diseases
8. Techniques of the Breeder
9. Breakdown of Male Sterility
10. Improper / Defective Seed Certification System

Seed: Seed is a matured ovule that consists of an embryonic plant together with a store of food, all surrounded by a protective coat. The formation of a seed completes the process of reproduction in a seed plant. At present, most optimistic estimates show that about 25–30% of the nutrient needs of Indian agriculture can be met by various organic sources. Supplementation of the entire Nitrogen (N) through Farm Yard Manure (FYM) sustains crop productivity at more than the use of conventional N fertilizers. Since the estimates of Nitrogen Phosphorus and Potassium (NPK) availability from organic sources are based on total nutrient content, the efficiency of these sources to meet the nutrient requirement of crops is not as assured as mineral fertilizers, but the joint use of chemical fertilizers along with various organic sources is capable of sustaining higher crop productivity, improving soil quality, and productivity on a long-term basis. These organic sources besides supplying Nitrogen (N), Phosphorus (P), and Potassium (K) also make unavailable sources of elemental nitrogen, bound phosphates, micronutrients, and decomposed plant residues into an available form to facilitate the plants to absorb the nutrients. Application of organic sources encourages the growth and activity of mycorrhizae and other beneficial organisms in the soil and is also helpful in alleviating the increasing incidence or deficiency of secondary and micronutrients and is capable of sustaining high crop productivity and soil health. (Nambiar, et al., 1992). Nutrient content in Farm Yard Manure (FYM) is dependent upon the product i.e., the manure and manure should be in good condition and not so fresh. It is reported that on average, well-rotted Farm Yard Manure (FYM) contains 0.5% N, 0.2% P_2O_5, and 0.5% K_2O (Gaur,1992). This system makes use of organic wastes and crops are raised in such a manner that it keeps the soil healthy and alive. Microbes are used as bio-fertilizers to increase production without polluting the environment (Bagchi,et al., 2015).

Suggested Organic Farming Techniques

i. **Green manuring:** The present decline trend in crop production or its stagnation of crop yields in India is the accumulative after-effect of abounding soil accompanying constraints. The documents of low crop yields are acceptance of inappropriate methods of soil management, across the board of the afire of crop residues, nonjudicious adoption of agronomic practices, and asymmetric nutrients management. Further, depletion of soil organic matter, scant use of bio- and organic fertilizer, and poor management are also amenable to lowering crop yield per unit area of land. Trees, shrubs, crops, grain legumes, grasses, weeds, ferns, and algae provide green manure an inexpensive source of organic fertilizer to build up or maintain soil organic matter and fertility. The cumulative effects of continued use of green manures are important not only in terms of nitrogen supply but also with regards to soil organic matter, phosphate, and micronutrients which are mobilized, concentrated in the topsoil, and made available for the plant.

ii. **Organic manure:** FYM, sheep manure, crop residues, organic manure: Poultry manure, oil cakes, composts, and other farm wastage. The indigenous and biodynamic preparation such as compost preparation can be used in organic nutrition management.

iii. **Vermicompost:** It is the compost prepared by earthworms, which is biologically degradable and decomposable organic wastes and is used as manures and earthworm feed.

iv. **Biofertilizer:** Microorganisms fix atmospheric nitrogen, to release and mobilize phosphorous and other nutrients. Natural fertilizers containing carrier-based micro-organisms viz., *Rhizobium*, *Azotobacter*, *Azospirillum*, Blue Green Algae, *Mycorrhizae,* and Phosphobacteria.

v. **Plant products (botanicals):** Seed hardening done with garlic extract, leaf extracts of *Prosopis*, *Pungam*, *Acacia*, *Calotropis*, etc. The decayed plant extracts can be used as liquid manure for Promoting plant growth.

vi. **Mulching:** It is an important technique for improving soil microclimate, enhancing soil life, structure and fertility, conserving soil moisture and energy, reducing weed growth, preventing damage by impact from solar radiation and rainfall (erosion control) and reducing the need for tillage. Widely used traditional mulches include layers of dry grass, crop residues (straw, leaves, etc.), fresh organic material from trees, bushes, grasses, and household refuse green manures.

vii. **Panchmukhi**: Five factors of agricultural treatments adopted by natural farmers as seed treatment, soil treatment, water treatment, environmental treatment, and crop treatment are cumulatively known as panchmukhi farming process for boosting agricultural yields.

viii. **Panchagavya:** It is a foliar spray prepared by organic growers using the following ingredients and methodology: Biogas slurry/cow dung 5 kg, cow urine 3 l, cow milk 2 l, curd 2 l, clarified butter/ghee 1 l, sugarcane juice 3 l, palm sugar 1 kg, tender coconut water 3 l, and banana. The ingredients are mixed in a mud pot after stirring them well. Then, it is kept in a shady place for one week for fermentation. Then 3 l of panchagavya are diluted in 100 l of water. This mixture is sufficient for spraying four acres. The diluted mixture has to be thoroughly stirred for 20 minutes before spraying. It can be stored for 1 month. It reduces vegetative growth, enhances quick flowering, and also gives resistance against pests and diseases.

Organic Seed Production Practices

i. **Land selection:** Land should be organically managed. Avoid the low-lying areas to restrict the run-off water contamination from conventional farming systems. To avoid contamination from wind, the organic farm shall be separated from the conventional farm by a live fence or manmade organically managed crop can be maintained as a buffer zone. A buffer zone of at least 3 m shall be maintained between conventional and organic management land. The equipment or implements used for organic management shall be cleaned before use. Crops should be rotated to reduce pest problems and any potential for seed contamination by open pollination with similar species types. The seed production field should not have known weed problems that are too difficult to control through organic means.

ii. **Land preparation:** Soil should be tilled to ensure a fine seed bed, which is critical for germination, particularly with small-seeded crops. The soil should have good water-holding capacity to allow for uniform germination and continued vegetative growth. The beds should be raised and shaped depending on rainfall. Uniformity of seedbeds is especially important because the seeds are often precision planted and uniform emergence and seedling developments are required for optimum management.

iii. **Soil fertilization:** The fertility of the soil must be improved when producing organically since chemical fertilizers cannot be used. To ensure good soil fertility and fewer soil-borne diseases, crop rotation, use of a cover crop, green manure crops, mulch, animal compost, and plant material compost can be used. There are many commercial organic fertilizers available for organic crops; the major ones are listed below:

a. Composted manure to increase nitrogen content
b. Crop residue and green manure
c. Straw and other mulches
d. Biofertilizers
e. Wood ashes to increase potassium
f. Vermiculite and peat
g. Plant preparation and botanical extract
h. Rock phosphate
i. Inoculated beneficial fungi, which help plants roots to fix nitrogen from the atmosphere

iv. **Choice of crop and varieties:** Any crop of variety or hybrid except genetically modified organisms or crops which suits the location shall be used or grown. Pest and disease-resistant varieties are mostly preferred.

v. **Seeds and planting material:** Seeds or planting materials shall be used from organically certified source. In case of unavailability of organic seeds, untreated seeds from conventional farms shall be used for the first year, and for subsequent years organic seeds shall be used. If the crop varieties are not grown in the first year, chemically untreated conventional materials shall be used. Genetically engineered seeds, pollen, transgenic plants, or plant materials shall not be allowed in organic Seed Production.

vi. **Planting techniques:** Seeds are generally planted directly by drilling in the field or transplanting from a greenhouse-grown seedling. These seed transplants should be organically produced. The seeds must be planted in such a way that proper vegetative development occurs that will support Fruit and seed development with proper spacing and depth in the bed. Row spacing and plant density must allow for maximum plant development of the flower and unrestricted access to inflorescences for pollinators to ensure proper seed set. Proper spacing will also allow for improved air movement, reducing pathogens and providing space for harvest operations at the end of the season.

vii. **Rouging:** Rouging at periodical intervals to remove the off types from both male and female lines during hybrid seed production.

viii. **Weed, pest, and disease management:** Management of weeds and pests is critical to ensure that organically produced seeds have high yield and quality. Weed can be managed through mulching with plant residues and other fully biodegradable materials, livestock grazing, and hand weeding coupled with mechanical cultivation. The seed crop is in the field for a long period, there are many opportunities for multiple pathogens to interact with a single crop. To control these pathogens organically is complex and requires proper growing conditions.

Biological Pest Control

This practice is highly compatible with organic seed production. Biological control utilizes three sources of natural enemies that can be used to control harmful pests and reduce the use of organic pesticides. The first group includes parasitoid insects, which lay their eggs inside another insect. An example is the wasp, *Aphidius colemani* which lays its eggs in aphid adults. The second group of beneficial insects is predators, which eat other insects. A common predator is the lady beetle, which prey on insects and mites. The third group includes the weed feeders. Insects such as a weevil, and *Hylobius transverse vittatus* feed on certain weeds, including purple loosestrife. Other bio-control agents and practices that can be employed are:

- Viral, Fungal Bacterial and Protozoa
- Introduction of predators or parasites of the pest
- Natural enemies like spiders, insects, mites, nematodes and birds
- Nonsynthetics control such as lures, traps, and repellant
- Mulches and nets
- Sanitation to remove disease vectors, weed seeds, and habitat for pest organism
- Development of habit for the natural enemy of the pest
- Botanical pesticides
- Crop rotation, trap crops, and alternate host crop
- Insect trap pheromones

Physical method: In this method, there is the use of human effort to control insect pests and diseases, which includes the following:

- Regulation of temperature: Applicable for stored place
- Regulation of light: Applicable for field crop
- Regulation of moisture: Use for stored insect pest control
- Use of sound waves

Mechanical method: The mechanical method of insect pest management includes:

- Hand picking
- Sieving and winnowing
- Shaking and beating
- Netting
- Wrapping
- Painting

Seed Production through Indigenous Traditional Knowledge

Local and indigenous knowledge refers to the understandings, skills, and philosophies developed by societies with long histories of interaction with their natural surroundings. For rural and indigenous peoples, local knowledge informs decision-making about fundamental aspects of day-to-day life. Indigenous Technical Knowledge (ITK) refers to the unique traditional local knowledge existing within and developed around the specific conditions by women and men indigenous to a particular geographic area. This ITK that people in a given community have developed over time and continue to develop, is based on human experiences on a mass scale, dynamic and changing, tested in most cases over centuries of use, endowed with the highest possible adaptability to local culture and environment and put greater weightage on minimizing risks rather than maximizing profit. The ITK covers a wide range of subjects, viz. crop production, livestock rearing, natural resource management, food preparation, healthcare, insect pest management, and many others.

ITK and Plant Disease Control

Traditional systems of crop protection that farmers have learned from their long association with the land, its flora and fauna were based on eco-friendly systems of suitable cultural practices, judicious rotation of crops, and knowledge of pests and their life-cycle.

- In case of bunchy top disease in chilies, dusting of ash and spray of liquid waste of tanned leather in tribal areas.
- Some farmers used to apply fresh cow dung near the collar region of chilli plant to control fungal diseases, viz. damping off and dieback.
- In case of soil-borne disease, root rot and collar rot used castor cake, karanj cake and neem cake as control measures.
- 20 kg of *Casuarina equisetifolia* leaves are boiled in water for 20 minutes. After cooling, the solution should be filtered. Then the extract is diluted with water and can be given to control some bacterial and fungal disease.
- Prepare solution from 2 kg fresh leaves of papaya in 3–4 l of water and keep it overnight. After filtration, this is diluted with 50–60 l of water and 250 ml soap solution added to it, which is effective in controlling brown spot disease of rice.
- Cultivation of marigold followed by solanaceous vegetable crops is effective to control bacterial diseases.
- Leaves of khair (*Acacia catechu*) can be put into water channel to control brown spot disease of rice.

ITK and Insect-Pest Control

The application of chemicals for pest control leads to food poisoning, soil, water, and environmental pollution. These chemicals create ecological imbalance and allow insect pests to develop resistance. At these moments, an indigenous knowledge related to plant protection in agriculture would play a vital role.

i. **Management of insect pests through ash**

   Sprinkling of ash over and around the vegetable crops like onion, okra, brinjal, tomato, and cucumber and in fields is effective against insect pests viz, beetles (pumpkin beetle, hadda beetle, etc.), leaf defoliating insects, leaf miners, thrips and aphids. It is the cheapest practice for small farmers. Ash acts as a detergent where chewing and sucking type of insects, find it difficult to chew plant parts due to the deposition of ash.

ii. Management of insect pests of mustard through *Aloe barbadensis* (Gwarpatha) + *Nicotiana tabacum* (tobacco) + *Azadirachta indica* (neem) + *Sapindus trifoliatus Linn.* (Aritha)

The mixture is effectively used against the insect pests of mustard crops. Leaf decoction (1 kg) of gwarpatha and tobacco powder extract (200 gm) is prepared in 5 l of water for 3–4 h to make a 2 l solution. Neem leaf extract (200 ml) is added after evaporation process and decoction of 50 gm aritha powder is added to the above solution and mixed thoroughly. This is sprayed on the mustard crop at intervals of 2–3 weeks. It is practiced throughout the hilly areas of Mandi district of Himachal Pradesh.

iii. Management of insect pests through cow urine + *Vitex negundo* (Nirgundi) + *Ferula asafoetida* (Hing)

These mixtures are very effective, eco-friendly insecticidal treatments against insect pests of wheat crops. It repels the insect pests. Leaf decoction of nirgundi (about 30–40 leaves in 10 l of water) is prepared till it is left 1 l. This mixture is filtered properly. About 10 gm hing is mixed in 1 l water and then the above ingredients are mixed in about 5 l of cow urine and sprayed over affected crops. It is effective for all sowing seasons (early; normal or late sowing seasons) of wheat and paddy crops. It is practiced most in hilly mountain villages of Shimla, Himachal Pradesh.

iv. Management of pod borers in gram crop through whey (lassi) + *Aloe barbadensis* (Gwarpatha) + *Nicotiana tabacum* (tobacco)

This method is very effective against the pod borers infesting the gram crop. Tobacco powder (200 gm), lassi (2 l), and Gwarpatha (2 leaves) are dissolved in 15 l of water. This solution is left undisturbed for 15 days. It is then filtered with a muslin cloth and the filtrate is sprayed over the infested crop at an interval of 2–3 weeks.

v. Management of paddy insect pests through *Vitex negundo* (Nirgundi)

- Farmers sweep the paddy field with brooms made up of branches of *Vitex negundo*, which are known to act as an insecticide and enhance the growth of paddy. It is practiced throughout the hilly areas of Himachal Pradesh.
- Farmers sweep the paddy field with brooms made up of branches of *Vitex negundo*, which are known to act as an insecticide and enhance

the growth of paddy. It is practiced throughout the hilly areas of Himachal Pradesh.

- 4 kgs of tobacco leaves and twigs are boiled in 40 l of water for 40 minutes. After cooling, 1 kg soap powder is mixed and the solution is diluted 7–8 times and sprayed to control aphids and white flies in citrus plants.
- Rice seedlings raised from seed treated with extract of neem kernel are resistant to leaf hopper.
- For prevention of infestation of shoot borer in mango trees, common salt is mixed with soil near the collar region of tree.
- In the case of insect holes made by shoot borer and bark eaters in mango, jaggery is placed in the holes to attract other predators to feed on the insects present in the holes.
- Similarly, the practices of pouring kerosene or petrol in holes and blocking holes with mud or cow dung are also common in citrus plants.
- 1 extract of an equal quantity of crushed green chilies and garlic mixed with 200 l of water is effective in controlling aphids and jassids.
- Filtrate prepared from a solution of 1 kg *Calotropis* leaves and 5 l of water is effective to control leaf-sucking pests.
- An extract of tobacco waste with 250 g of soap solution in 200 l of water as spray control stem borer.
- A solution prepared from neem leaf paste in water (10 kg: 2L) is effective in controlling leaf folder in rice.
- A solution prepared from 100 g tulsi (*Oscimum* sp.) leaves in 1 litre of water with 1 mL soap solution can be used for effective control of aphid, army worm, red cotton bug, mosquito, etc., is used to control cotton semilooper, mites, green leaf hopper, aphid, etc.

vi. **Control of stored grain pests**

- Garlic leaves are used for safe storage of food grains.
- Leaves of *Vitex negundo* (Nirgundi) are incorporated in any pulse seeds for long time preservation

- There is a common practice of storing food grains like wheat and rice use of neem leaves to prevent storage pest damage.
- Milled chickpea, green gram, and other pulses are stored after being thoroughly treated with mustard oil.
- Green gram can be kept free from any pest infestation by treating it with 1% neem leaf powder.
- Seed mixed with *Acorus calamus* (baje) powder in the ratio of 10 kg:1 kg, would help in preserving the seed free from stored pests for a long time.

Plant Production through Indigenous Traditional Knowledge

Agroforestry

Agroforestry involves the deliberate maintenance and planting of trees to develop a microclimate that protects crops against extremes. Blending agriculture with forestry techniques, this farming system helps to control temperature, sunlight exposure, and susceptibility to wind, hail, and rain. This system provides a diversified range of products such as food, fodder, firewood, timber, and medicine while improving soil quality, reducing erosion, and storing carbon.

Crop Rotations

The principles of crop rotation have been successfully used for thousands of years in agriculture and are still used today. Crop rotation is the practice of growing different crops on the same land so that no bed or plot sees the same crop in successive seasons. It is a practice designed to preserve the productive capacity of the soil, minimize pests and diseases, reduce chemical use, and manage nutrient requirements, all of which help to maximize yield. The practice of crop rotation builds better soil structure and increases the ability to store carbon on farms.

Mixed-/Inter-Cropping

Mixed cropping, also known as intercropping, is a system of cropping in which farmers sow more than two crops at the same time. By planting multiple crops, farmers can maximize land use while reducing the risks associated with single-crop failure. Intercropping creates biodiversity, which attracts a variety of beneficial and predatory insects to minimize pests and can also increase soil organic matter, fumigate the soil, and suppress weed growth.

Polyculture

Polyculture systems involve growing many plants of different species in the same area, often in a way that imitates nature. By increasing plant biodiversity, polyculture systems promote diet diversity in local communities, are more adaptable to climate variability and extreme weather events, and are more resilient to pests and diseases. Polycultures are integral to permaculture systems and design and provide many advantages such as better soil quality, less soil erosion, and more stable yields when compared to monoculture systems.

Water Harvesting

Water harvesting is defined as the redirection and productive use of rainfall, involving a variety of methods to collect as much water as possible out of each rainfall. Many water harvesting structures and systems are specific to the eco-regions and cultures in which it has been developed. This may involve collecting water from rooftops, swollen streams and rivers during monsoon season, or from artificially constructed catchments. This ensures that farmers have a substantial amount of water stored up in the case of drought or limited rainfall.

Harvesting, Threshing, and Drying

The method of harvesting depends on the type of seed being produced. When harvesting dry-seeded crops, Seed shattering must be prevented because seed harvest generally occurs after the crop reaches physiological maturity. To reduce shattering, the stalks of the plant need to be cut while still green and field-dried, allowing for uniform seed maturation.

Cleaning and Storage

Once seeds are harvested, then cleaning and storage are done. They should be evaluated to determine their physical purity. All seeds should be single units and all should be stored according to their individual temperature/humidity requirements. Generally moisture content should be below 12% for seed storage.

Seed Treatment

The organic seeds are treated normally with materials from organic sources such as

- **Botanicals**: Neem leaf extract, mint leaf extract, sarani leaf extract, *Prosopis* leaf extract, Arappu leaf extract, etc.
- **Biofertilizer**: *Rhizobium, Azotobacter, Azospirilium, Phosphobacteria,* etc.

- **Cows' products**: Panchagavya, cow milk, curd, curd urine, cow dung, etc.
- **Bio-control agent**: *Pseudomonas* spp, *Trichoderma* spp.
- **Others**: Coconut oil, tender coconut, vermicompost, vermiwash, etc.

As it is often not possible to produce disease-free seeds and as conventional effective seed treatment with synthetic compounds is not possible, a lot of work on alternative seed treatments has been done and is still going on. The various tested treatments can be classified into several categories:

a) **Thermal treatment**

Hot water seed treatments are efficient on several crops but have to be applied with caution to avoid killing the seed. The limitation is that seed must be dried rapidly after the treatment and this is difficult at an industrial scale. To avoid that difficulty, an aerated steam method has been proposed. Because the seed is not immersed in water but exposed to hot moist air, drying is no longer a problem. The choice of temperature and its control is critical.

b) **Use of antagonists**

Several antagonists have been tested and the list is long. Some nonexhaustive results are as follows:

- *Trichoderma* spp against collar rot disease (*Aspergillus niger*) on groundnut
- *Pseudomonas chlororaphis*, *Bacillus subtilis, Fusarium oxyporum, Streptomyces spp* against *Alternaria spp* on *Brassica* seed
- *Bacillus subtilis* against *Tilletia* caries on wheat
- *Trichoderma viride* against *Fusarium spp* and *Bipolaris sorokiniana* on wheat and barley
- Several antagonists against *Rhizoctonia solani*

c) **Natural compounds**

Essential oils, sometimes with chelator and natural detergent have been tested. Thyme and oregano oils are reported to give good results against *Xanthomonas campestris pv. campestris, Clavibacter michiganensispv, C. michiganensis, Botritis aclada,* and *Alternaria dauci*. A yellow mustard flour-based product, Tellecur, is reported to give good results against various pathogens, in particular Telletia caries in wheat. Chitosan is reported to give good results against *Fusarium* spp. *and Bipolaris sorokiniana* on wheat and barley. A complex product, Biokal (57% of medicinal herb extracts, 38% bio-humus extracts, 5%

volatile oil, and metal and trace elements) is reported to give some good results against *Ascochyta pision* pea seed.

d) **Other Products**

In recent years, organic acids (lactic, acetic, citric, propionic, and ascorbic) and antiseptic products such as $KMnO_4$ and $CuSO_4$ are also under test and utilization.<

Benefits of Using Organic Seeds

Organic Seed Is Better for the Environment

The importance of organic seed is often overshadowed by the importance of organic food; since food goes directly into our bodies, the methods by which it was produced feel more pertinent than the methods by which seed is produced. However, whether or not seed crops are grown organically does have a significant impact on the environment and health of the surrounding communities. As the organic seed growers and trade association asks: "Is it fair for those of us in organic agriculture to want our own farms and environments to be as free of toxins as possible, but expect seed production communities to carry a heavy toxic load so that we can plant cheap conventional seed?"

Organic Seed Is Better Suited to Organic Growing Conditions

In the last 50 years, the majority of breeding work and development of new varieties has been done with conventional farming practices in mind. For example, in a conventional system, plants receive infusions of fertilizer delivered directly to the roots and are sprayed with a range of chemicals to protect from pests and diseases. While some conventional breeding programs do consider flavor, more often selections are based on traits like how fast it grows and how well it ships. In organic farming systems, on the other hand, nutrients come from complex fertilizers such as compost, so plants need vigorous roots to seek out dispersed nutrients in the soil. Organic growers use less and milder applications for pest and disease protection, so plants are required to provide more of their own defense. In organic variety selection programs, flavor and appearance consistently top the list of traits to evaluate. In recent years, more research has gone toward developing and evaluating varieties for organic farmers, high quality organic seed comes from varieties that have proven themselves under organic conditions.

Organic Certification

It is a certification process for producers of organic food and other organic agricultural products. In general, any business directly involved in food production can be certified, including seed suppliers, farmers, food processors, retailers, and restaurants. Requirements vary from country to country, and generally involve a set of production standards for growing,

Maintaining Strict Physical Separation of Organic Products from Noncertified Products

Undergoing periodic on-site inspections, storage, processing, packaging, and shipping that include: Avoidance of synthetic chemical Inputs (e.g., fertilizer, pesticides, Antibiotics, food additives, etc) and genetically modified organisms; Use of farmland that has been free from chemicals for several years (often, three or more); keeping detailed written production and sales record (audit trail). In some countries, certification is governed by the government, and commercial use of the term organic is legally restricted. Certified organic producers are also subject to the same agricultural, food safety, and other government regulations that apply to the noncertified producers.

Purpose of Seed Certification

Organic certification addresses a growing worldwide demand for organic food. It is intended to assure quality and prevent fraud. For organic producers, certification identifies suppliers of products approved for use in certified operations. For consumers, "certified organic" serves as a product assurance, similar to "low fat", "100% Whole wheat", or "no artificial preservatives". Certification is essentially aimed at regulating and facilitating the sale of organic products to consumers. Individual certification bodies have their own service marks, which can act as branding to consumers—a certifier may promote the high consumer recognition value of its logo as a marketing advantage to producers. Most certification bodies operate organic standards that meet the national government's minimum requirements. In simplified terms, the National Organic Program Standards (NOPS) required for crop farms are as follows:

- 3 years (36 months prior to harvest) with no application of prohibited materials (no synthetic fertilizers, pesticides, or GMOs) prior to certification
- Distinct, defined boundaries for the operation

- Proactive steps to prevent contamination from adjoining land uses
- Implementation of an organic system plan, with proactive fertility management systems; conservation measures; and environmentally sound manure, weed, disease, and pest management practices
- Monitoring of the operation's management practices to ensure compliance
- Use of natural inputs and/or approved synthetic substances on the national list, provided that proactive management practices are implemented prior to the use of approved inputs
- No use of prohibited substances
- No use of genetically engineered organisms (GMOs), defined in the rule as "excluded methods"
- No use of sewage sludge or irradiation.
- Use of organic seeds, when commercially available (must not use seeds treated with prohibited synthetic materials, such as fungicides)
- Use of organic seedlings for annual crops
- Restrictions on the use of raw manure and compost
- Must maintain or improve the physical, chemical, and biological condition of the soil, minimize soil erosion, and implement soil-building crop rotations
- Fertility management must not contaminate crops, soil, or water with plant nutrients, pathogens, heavy metals, or prohibited substances
- Maintenance of buffer zones depending on risk of contamination
- Prevent commingling on split operations (the entire farm does not have to be converted to organic production, provided that sufficient measures are in place to segregate organic from nonorganic crops and production inputs)
- No field burning to dispose of crop residues (may only burn to suppress disease or stimulate seed germination flame weeding is allowed)
- No residues of prohibited substances exceeding 5% of the EPA tolerance (certifier may require residue analysis if there is reason to believe that a crop has come in contact with prohibited substances or was produced using GMOs).

Conclusion

Organic agriculture with organic seed includes the growing of crops by a set of guidelines that prohibit the use of synthetic products or chemicals such as fertilizer, pesticides, and herbicides. Therefore, soil fertility and pest management are achieved through cropping patterns (rotations, intercrops or mix crops, pest, and disease-resistant genotypes), manure (green manure, organic manures, and compost), biofertilizers, cultural practices (weeding, planting, conventional tillage) and biopesticides, including plant-derived products. At present, this system seems to be an ideal and valid solution to produce seeds aside from agricultural production. The overuse of plant growth regulator, pesticides and fertilizers for faster growth of agricultural produce is detrimental to human health and the environment as a whole. Further, consumers are becoming conscious and critical about the quality of food and by-products that affect their health though the toxicity depends to some extent on the type of food consumed.

References

Bagchi, A. Organic Farming Practice for Quality Improvement of Tea and Its Anti Parkinsonism Effect on Health Defense. *J Phys Chem Biophys.* 5, (2015):178.

Feistritzer, W.P. (1975). The role of quality seed use for world crop production with special reference to the developing countries. *Outlook on Agriculture*, 8, no. 5, (1975):267-70. DOI: https://doi.org/10.1177/003072707500800503

Gaur, A.C. 1992. "Bulky organic manures and crop residues," in Fertilizer Organic Manures Recyclable Wastes and Bio-Fertilizers. In H. L. S. Tandon, (ed.). Fertiliser Development and Consultation Organisation, New Delhi, India.

Kadam, B.S. Deterioration of varieties of crops and the task of the plant breeder. *Indian J Genet Plant Breed.* 2, (1942):159-72.

Nambiar, K.K.M., Soni, P.N., Vats, M.R., Sehgal, D.K., Mehta, D.K. 1992. "AICRP on long term fertilizer experiments," Annual Reports 1987-88 and 1988-89, New Delhi, India: IARI.

18

Information and Communication Technology (ICT) for Smart Agriculture

Deepa Thangjam[1], Sashilila Ao[2], Mary N Odyuo[3] and Haobijam James Watt[4]

[1]*Department of Rural Development and Planning, School of Agricultural Sciences, Nagaland University, Dimapur, Nagaland*

[2,3]*Department of Agricultural Extension, School of Agricultural Sciences Nagaland University, Dimapur, Nagaland*

[4]*Department of Agricultural Economics and Extension, School of Agriculture Lovely Professional University, Phagwara, Punjab*

Introduction

The current science and technology is highly developed and everything is at our fingertips. The drudgery that was involved earlier while doing any kind of activity is so much reduced at present. The use of technology is so important nowadays that it has become unavoidable in all fields. With the tools and techniques, most of the problems can be solved and so technology has become a part of our lives thereby making our lives better. Some of the trending technologies are Artificial Intelligence (AI), robotic process automation, 5G, Information and Communication Technology (ICT), Machine Learning (ML), Virtual Reality (VR) etc. With the application of all these technologies, every discipline whether it is infrastructure, medical science, transport, communication, etc. has become stronger than ever. Even the field of agriculture has become so advanced with these technologies. The enhancement of agricultural activities has become so important with the depletion of natural resources, less human labor, and the changing climatic conditions. All over the world agricultural technologies are adopted, disseminated, and practiced to increase production and productivity, improve livelihood and alleviate poverty (Kilima et al., 2013). Similarly, the main reason for the growing demand for ICT in agriculture is for the farmers to get up-to-date knowledge and apply the tools and techniques to improve their livelihood. However, the ultimate aim is to adopt sustainability in agriculture. Through ICT, farmers can get updated information on the market prices, weather forecast, emergence of pest and diseases and their management practices, planting materials, varieties, etc. This knowledge not only gives the farmer an efficient method of farming for more production but also empowers them with modern agricultural ways. The use of ICT in combination with other techniques enables the farmer to save time, use resources efficiently, and produce better quality products.

What Is Smart Agriculture?

Smart agriculture is becoming increasingly important in the global context as the world population continues to grow and the demand for food increases. The United Nations projects that the world population will reach nearly 10 billion by 2050, and food production will need to increase by 70% to meet this demand. Smart agriculture can play a crucial role in meeting this challenge by increasing crop yields, reducing waste, and improving the efficiency of farming operations. In developed countries, smart agriculture is often used to optimize production on large commercial farms. In developing countries, it can help to improve food security and increase the incomes of smallholder farmers. In these countries, smart agriculture can help to bridge the digital divide and provide farmers with access to the same technologies and information as their counterparts in developed countries. Smart agriculture is also seen as a key tool in addressing the challenges of climate change, by reducing the environmental impact of farming and increasing the resilience of crops to changing weather patterns. In summary, smart agriculture is seen as a crucial tool in addressing global food security and environmental challenges, by increasing crop yields, reducing waste, and improving the efficiency of farming operations. It can also help to bridge the digital divide, especially for farmers in developing countries, and increase their income and resilience to the changing climate.

In India, this includes the use of precision farming techniques, such as precision irrigation and precision planting, as well as the use of drones and other robotics for crop monitoring and management. Additionally, the use of data analytics and IoT (Internet of Things) devices can help farmers make more informed decisions about things like crop yields, weather patterns, and pest control. Smart agriculture can help to increase crop yields, reduce water usage, and improve the overall sustainability of farming in India.

Applications of ICTs in Agriculture

i. Providers of information on each crop via online services

ii. Online interaction facility to interact with the nearest Krishi Vigyan Kendra (KVK), State Agricultural University (SAU), Indian Council of Agriculture Research (ICAR) research station, Agriculture and allied departments for advice on current schemes, projects, varieties, etc.

iii. Online interaction facility to interact with the nearest Krishi Vigyan Kendra (KVK), State Agricultural University (SAU), Indian Council of Agriculture Research (ICAR) research station, Agriculture and allied departments for advice on current schemes, projects, varieties, etc.

iv. Details on any possible sources of agricultural credit, as well as information on the terms and conditions of crop insurance.

v. Mechanisms for the exchange of information between farmers, extension workers, and scientists in each district, block, and "village."

vi. There should be a question-and-answer service (similar to the Kisan Call Center) available for each district in the country.

vii. Information available online on the current prices of all commodities sold at the local, state, district, state major, and major national market levels.

viii. Monitoring and information sharing via the Internet for all agricultural development projects taking place across the nation.

ix. Sharing of different Agricultural Technology Management Agency (ATMA) districts' District Agriculture development plans, such as Strategic Research and Tension Plans (SREPs), can be done online.

x. Information available online on the current prices of all commodities sold at the local, state, district, state major, and major national market levels.

xi. Information gathered from the market about the major crops.

xii. Online weather forecasting and its influence on major crops every week, including early warning systems.

xiii. Access to land records through an online service.

xiv. The use of electronic commerce to facilitate direct links between regional manufacturers, merchants, and consumers.

xv. Information on all government and nongovernment organizations working directly or indirectly in the agricultural sector, as well as information on the availability and prices of agricultural inputs such as seeds, fertilizers, pesticides, and machinery, among other things (Chauhan *et al.,* 2016).

ICT for Smart Agriculture

Through the application of a variety of tools, Information and Communication Technology (ICT) plays a significant part in the expansion of agricultural production, which contributes to the achievement of economic sustainability and self-reliance. ICT refers to a wide variety of technological tools and resources that are used to transmit information, store information, create information, share information, or exchange information with other people. The terms ICT and information technology (IT) are frequently confused with one another; however, the meanings of each term change depending on the circumstance. ICT, also known as Information and Communication Technology in agriculture or e-agriculture, is utilized in the agricultural industry in the same way it is used in every other industry. ICT is also an umbrella term that encompasses all forms of digital communication, such as the Internet (including websites, blogs, vlogs, and email), computers, radio, television, webcasting, podcasting, audio and video players, storage devices, mobile phones, telephones, video conferencing, audio conferencing, wireless technologies, Global Positioning System (GPS), Geographic Information System (GIS), intelligent mobile apps, remote sensing, RFID for animal identification, and other similar technologies. It assists farmers, allowing them to improve agricultural practices and contribute to the overall development of rural areas. National e-Governance Plan in Agriculture (NeGP-A), various Kiosks, Krishi Vigyan Kendras, Kisan Call Centres, Agri-Clinics, Common Service Centers, mKisan, Kisan TV, and other similar programs are examples of some of the e-agriculture programs that have been implemented by the Government of India. For agriculture to become more intelligent, the application of ICT is essential. The use of technologies such as the IoT, sensors, GPS, GIS, and AI in agricultural settings is referred to as "smart agriculture." The goal of smart agriculture is to improve both the quality and quantity of crop production while simultaneously reducing the amount of manual labor required. Some of the examples of ICT used in smart agriculture are precision irrigation, precise plant nutrition, climate management using greenhouse, soil, water, moisture, light and temperature sensors, GPS, mobile communications, robots, etc. Through these, farmers can get precise information about their farm, weather, and climate and monitor the crops on their smartphones without even visiting their fields.

Role of ICT in Smart Agriculture

The climatic conditions of different regions have a significant impact on agriculture in those regions. The central idea behind community-supported agriculture (CSA) is to provide a variety of strategies that, when combined, will make food security more robust in the face of climate change by concentrating on potential adaptive and mitigation strategies (Campbell et al., 2014). It is a threat to the farmers' food security and livelihood security that the pattern of rainfall, the frequency of rainfall, the intensity of rainfall, the temperature, the length of the seasons, and the occurrence of drought, flood, hail storms, and snowfall, among other things, are changing. The information and communications technology that enables smart farming will collect data, analyze it, and then deliver the result to the farmers. The farmers can make informed decisions with the help of this information. They can forecast when, where, and how to plant a particular crop, as well as what varieties to use and other relevant information. The primary goal of CSA is to reduce emissions of greenhouse gases while at the same time meeting the demand for food. This must be accomplished while simultaneously increasing the adaptability of small-scale farmers and the efficiency with which agricultural systems use resources. The dissemination of information to farmers about climate-smart agriculture (CSA) technologies, practices, and services can be aided by the use of ICT tools such as mobile phones, radio, televisions, and internet services. There have been a lot of studies done in the past on how ICT enables climate information and agro advisory services, and they have all shown that farmers benefit from these things because they help farmers make better decisions regarding crop management, choose the appropriate technology, and develop more effective marketing strategies (Ospina and Heeks 2012; Kumar and Singh 2012). For example, timely information about monsoons, which is provided by weather forecasting agencies, helps farmers make decisions about when to sow crops to avoid rainfall-related uncertainties, which is especially important in rain-fed areas (Raghuvanshi et al., 2018). The information needed to make an accurate weather forecast is included in all modern smartphones. Some websites provide information on the weather, such as (https://mausam.imd.gov.in/). As a result, farmers can make use of the various ICT tools for CSA practices.

The International Society of Precision Agriculture (https://www.ispag.org/about/definition) adopted the following definition of precision agriculture in 2019: "Precision Agriculture" is a management strategy that "gathers, processes, and analyses temporal, spatial, and individual data and combines it with other information to support management decisions according to estimated variability for improved resource use efficiency, productivity, quality, profitability, and sustainability of agricultural production. "Precision Agriculture" is a management strategy that "gathers, processes, and analyses temporal, spatial, and individual data and combines all. Precision agriculture is practiced with both the crops and the livestock. It uses technologies that automate agricultural operations, practices, and activities and it helps in improving problem diagnosis and making better decisions or performance (FAO, 2022).

Additionally, it helps market the direct products as well as the products with value added. Farmers can get up-to-date information on the prices of various commodities by visiting a website such as https://agmarknet.gov.in/. ICT is useful in both the domestic market and the international market because it also facilitates trade in both directions (export and import). Buying and

Table 18.1: Some of the available e-commerce websites in India

Sl. No	Description	URL
1	Bighaat	https://www.bighaat.com/
2	AgriBegri	https://agribegri.com/
3	Agrowala	https://agrowala.com/
4	Krushikendra	https://krushikendra.com/
5	Farmersstop	https://www.farmersstop.com/
6	Farmagritech	https://farmagritech.com/
7	Farmkart	https://farmkart.com/home
8	Agritell	https://www.agritell.com
9	Agrikendra	https://www.agrikendra.com/
10	Farmkey	https://farmkey.in/
11	Kisanestore	https://www.kisanestore.com/
12	Iffcobazar	https://www.iffcobazar.in/
13	Agribazaar	http://www.agribazaar.co/
14	Agricart	https://www.agricart.com/
15	Farmkey	https://farmkey.in/product-category/pesticides

Source: http://www.agricindia.com/blog/Agriculture-e-commerce-website-in-india/42

selling agricultural products is made easier by the many different kinds of e-commerce platforms that are currently available. Table 18.1 provides some of the e-commerce websites that are available in India.

In addition, there are expert systems that assist farmers in finding solutions to issues that are associated with their crops and livestock. A decision-making support system of this kind also assists farmers in selecting the most advantageous available options, as well as in the management of their crops and livestock. It increases the probability, frequency, and consistency of making good decisions, has an additive effect on the knowledge of many domain experts, facilitates real-time, low-cost expert-level decisions by the nonexpert, enhances the utilization of most of the available data, and frees the mind and time of the human expert to enable him or her to concentrate on more creative activities. All of these benefits come together to make it possible for the human expert to focus on more creative endeavors (Rani et al., 2011). The following is a list of some examples of expert systems:

a. Wheat Pakistani Expert System—Khan et al., (2008) established a web-based expert system for wheat crops in Pakistan.

b. Mango Expert System—Prasad et al., (2006) formulated an expert system viz., AMRAPALIKA for diagnosing 14 different pests including 8 diseases and 6 insects in the Indian mango variety. The expert system is developed for important diseases like Powdery mildew, Black spot, Anthracnose, Red rust, Die back, Bacterial spot, Sooty mold and Malformation, and insects like Shoot-borer, Red ants, White ants, Mealy bug, Mites, and Fruit fly.

c. Rice Expert System—In India, Sarma et al. (2010) developed an expert system to diagnose and manage the diseases occurring in rice crops.

d. Tomato Expert System—A web-based tomato crop expert information system was developed by Babu et al., (2010) in India.

e. Rapeseed-Mustard Expert System—Vinod et al., (2008) developed an image-based rapeseed-mustard disease expert system in India.

There are also many different kinds of agricultural kiosks, each of which can be of assistance to farmers by providing them with the most recent information available. The Information Kiosk is an innovative concept that provides farmers, including those who are illiterate, with access to information through animations, graphics, and other forms of audiovisual and visual communication (Ahmad and Haneef, 2020). The goal of the AGRIKIOSK project in Pakistan is to educate farmers on relevant topics pertaining to agriculture in their native language(s) and to facilitate their interaction with agricultural professionals and one another through the utilization of communication technologies that are up-to-date and appropriate for the digital age. This will be achieved through the incorporation of digitized video, audio, graphics, text, websites, and software applications that were developed to cater to the requirements of the rural population (Afzal et al., 2015). The formulation and evaluation of government policy on farm management, disease and pest management, agro-environmental resource management, disaster management, marketing strategies, and other related topics became more effective as a result of the proliferation of ICT tools in the agricultural sector. This was because the tools were able to collect more data more quickly. Because there are so many different kinds of tools, such as GIS and GPS, the government is able to precisely locate and manage any kind of problem that may arise in the fields of agriculture and the allied sector. The process of transforming conventional agricultural operations into intelligent agricultural operations can also be aided by these tools.

Different ICT Tools Used in Smart Agriculture

In the past, the most common forms of ICT tools utilized by farmers included television, radio, the internet, email, and smartphones. Because of advances in technology, wide varieties of new tools have become available to farmers. These tools enable farmers to better manage and keep track of their crops and livestock. The following is a list of some of the more cutting-edge ICT tools that are used in agriculture:

Sensors

a. **Optical Sensors in Agriculture:** In agriculture, optical sensors are used to estimate and predict the properties of the soil and crop by analyzing the amount of reflected light on the growing parts of the crop in real time. This allows for more accurate data collection. It gives the tools for analysis the ability to regulate the amount of nitrogen that is required. In addition to these applications, it is studied to determine the vigor of the crop and assists farmers in regulating the levels of moisture in the air and soil so that damp conditions are avoided.

b. **Location Sensors in Agriculture:** The name itself suggests that it can ascertain the range, distance, and even height of any position within the specified area. In order to accomplish this goal, they make use of GPS satellites, which assist in enhancing the assessment of particular sections of the plot and provide farmers with a deeper understanding of the land that is suitable for cultivation.

c. **Mechanical Soil Sensors for Agriculture**: It does this by applying resistive forces to the soil in order to get a reading on how resistant the soil is. They do this by utilizing load cells or strain gauges to calculate the amount of force that the roots apply in order to take up water. In the end, using this information, they optimize the tilling methods as well as the water that is introduced.

d. **Electrochemical Sensors**: They analyze the pH level of the soil as well as the quality of the soil. Altering the pH level with it is possible, as is continuing with the same practices in order to keep the pH level where it is currently at throughout the subsequent phases of the lifecycle of any crop. It is utilized in farming that takes place both inside and outside of greenhouses. As a result, they help bring about farming practices that are sustainable and kind to the environment, and they also maximize revenue.

e. **Dielectric Soil Moisture Sensors**: These determine the amount of moisture that is present in the soil. They are also referred to as moisture sensors, and they are installed in conjunction with rain check locations across the farm in order to monitor the conditions of the soil in terms of moisture when the level of vegetation is low.

Drones

In the past decade, drones have found applications in a variety of industries, and now, after undergoing significant development, they are being put to use in the agricultural industry. The human eye is unable to pick up on certain heat signatures, light frequencies, or imagery feedback; however, drones equipped with specialized cameras can do so. These drones can provide necessary field information. They help collect data that is precise and accurate, as well as analyze the growing pattern of crops and the areas that are under stress. The findings of these analyses contribute to the development of treatment plans that are both more environmentally friendly and require fewer resources to carry them out. They are also helpful to crop growers in making decisions that are both effective and efficient. Through the use of multispectral cameras, they can assist growers in better understanding their crops and generate vegetation index maps of fields. The use of drones has many potential benefits, including the early detection of problems, the provision of a comprehensive view of a crop while it is growing in the field, the more effective application of inputs, and the reduction of negative effects on the environment. In addition to these applications, they are utilized in the mapping of terrain, the inspection of solar panels, the monitoring of livestock, the examination of soil and fields, the planting of seeds, the spraying of crops, the mapping and surveying of crops, the monitoring of irrigation, the evaluation of crop health, and other similar applications. As a result, the use of drones in agriculture may have a bright future.

Artificial Intelligence (AI)

AI is the leading technology solution for improving productivity, and agriculture is gradually becoming digitalized with a focus on the following three areas: Agriculture robotics, soil and crop monitoring, and predictive analytics. Farmers who are on the cutting edge of innovation are increasingly turning to robotics, soil sampling, and sensor technology to collect data, which is then stored in their farm management systems. Because of this, they are able to more effectively process and analyze the data, and ultimately, their crops. Depending on the type of industry, the type of technology, and the industry itself, many different kinds of AI technologies can be used in agriculture. The service segment is the type that is growing at the fastest rate, while the technology that is growing at the fastest rate is predictive analytics, and the largest application category in industry is precision farming. Therefore, depending on the requirements of the technology, AI may be applied in agriculture; however, despite their high cost, these systems provide accurate data and assist in making the most efficient use of resources.

Geographic Information System (GIS)

The application of digital tools has ushered in a new era of innovation in the sector of agriculture. The practice of farming has undergone a sea change as a direct result of the much technological advancement, among which geospatial technologies are included. Image data and nonimage data with a spatial context are required for the identification of problem areas in cropland precisely, as well as for monitoring and management in the agricultural industry. GIS, along with its component tools and analytic modules, as well as the data gathered by its supporting technologies such as remote sensing and GPS, all contribute to the provision of information visualization that is both straightforward and easy to understand. This, in turn, helps in making decisions that are driven by data for the purpose of increasing crop productivity. Because of advancements in technology, the application of GIS in agricultural settings is gradually expanding on a day-to-day basis. The evaluation of land suitability and the planning of land use, the management of water resources, the management of soil health and fertility, the evaluation and intervention of biotic and abiotic damage, crop monitoring and yield prediction, precision farming, biomass assessment, supply chain management, and other applications are some of the most common and emerging uses of GIS.

Global Positioning System (GPS)

The GPS is a positioning and navigation system that is based on a satellite network. GPS makes it possible to determine positional information by providing a location's latitude, longitude, and elevation. Farmers and researchers are able to more easily obtain reliable information on the identification of fields and water bodies, the mapping of field boundaries, and areas of the field that are infested or problematic thanks to the location information that is collected by GPS receivers. With the help of these data, one can better comprehend the connection between the location and several other characteristics that exist both inside and outside the confines of the given field. A field that has been meticulously mapped in this way makes it possible to apply fertilizers, nutrients, herbicides, weedicides, pesticides, and water for irrigation in a site-specific manner. As a result, it contributes to the enhancement of crop productivity as well as the reduction of input costs, both of which are the most essential features of precision agriculture.

Conclusion

The agricultural and allied industries can benefit from the use of a variety of tools that are included in ICT. It directs people in making the most accurate decision possible and assists them in selecting appropriate options from among the many that are available. Many different technologies, such as GIS, GPS, remote sensing, use of drones, smartphone apps, kiosks, internet, computers, radio television, and so on, provide farmers with the most up-to-date information and make it easier for them to practice precision agriculture. In addition to these, it assists in the establishment of climate-smart agricultural practices, which is a topic of great importance at present. The implementation of ICT tools in agriculture has made the policy planning process for the government regarding agriculture and allied industries more effective. As a result, it is possible to conclude that tools related to ICT can also be used in agriculture and allied sectors to transform conventional agriculture into smart agriculture.

References

Ahmad, T., and Haneef, R. Information kiosk: Innovations in information and communication technology for strengthening rural community. *Indian Farmers' Digest.* 53, no. 3, (2020): 39-41.

Afzal, M., Zubairim, J.A., Akram, A., and Hina, A. AGRIKIOSK – A Digital Tool to Reach Farmers and Rural Communities. *International Journal of Applied Engineering Research.* 10, no. 15, (2015): 35650-56.

Babu, M.S.P., Murty, N.V.R. and Narayana, S.V.N.L. A web based tomato crop expert information system based on artificial intelligence and machine learning algorithms. *International Journal of Computer Science and Information Technologies.* 1, no. 1, (2010): 6-15.

Campbell, B.M., Thornton, P., Zougmoré, R., Asten, P. and Lipper, L. Sustainable intensification: What is its role in climate smart agriculture. *Current Opinion in Environmental Sustainability.* 8, (2014): 39-43.

Chauhan, N.B., Patel, J.B. and Vinaya Kumar, H.M. 2016. Innovative ICT Models for Sustainable Agricultural Development. National Seminar on Contemporary Innovations for Quantum Extension in Agricultural Development. 75-80. (https://www.researchgate.net/profile/VinayaKumar/publication/324803468_Innovative_ICT_Models_for_Sustainable_Agricultural_Development/links/5ae31e630f7e9b28594a58bf/Innovative-ICT-Models-for-Sustainable-Agricultural-Development.pdf).

FAO (2022) The state of the food and agriculture 2022 – Leveraging automation in agriculture for transforming agri food systems. Rome, FAO. https://doi.org/10.4060/cb9479en

https://www.ispag.org/about/definition Accessed on 28th December, 2022.

http://www.agricindia.com/blog/Agriculture-e-commerce-website-in-india/42Accessed on 29th December, 2022.

https://www.bisinfotech.com/top-5-sensors-used-in-smart-farming/ Accessed on 30th December, 2022.

https://www.folio3.ai/blog/drones-in-agriculture/ Accessed on 30th December, 2022.

Khan, F.S., Razzaq, S., Irfan, K., Maqbool, F., Farid, A., Illahi, I., and Ulamin, T. 2008. Dr. Wheat: A Web-based Expert System for Diagnosis of Diseases and Pests in Pakistani Wheat. Proceedings of the World Congress on Engineering Vol. I.

Kilima, F.T.M., Tarimo, A.J.P., Johnsen, F.H., Nchimbi-Msolla, S., Mbaga, S., Sesabo, J., Abdallah, J.M., and Iranga, G. 2013. The Impact of Agricultural Research on Poverty and Income Distribution: A Case Study of Selected On-farm Research Projects at Sokoine University of Agriculture, Morogoro, Tanzania. *Tanzania Journal of Agricultural Sciences.* 12, no. 1, (2013): 1-9.

Kumar, A. and Singh, K.M. 2012. Role of ICTs in rural development with reference to changing climatic conditions. In: Singh KM, Meena MS (eds.) ICT for agricultural development under changing climate. Narenda, New Delhi.

Ospina, A.V. and Heeks, R. 2012. ICT-enabled response to climate change in rural agricultural communities. Strategy brief 2, Centre for Development Informatics, University of Manchester, Manchester.

Prasad, R., Ranjan, K.R. and Sinha, A.K. 2006. AMRAPALIKA: An expert system for the diagnosis of pests, diseases, disorders in Indian mango. *Knowl.-Based Syst.* 1, no. 1, (2006): 9-21.

Raghuvanshi, R., Raj, S. and Bhattacharjee, S. 2018. Climate smart agriculture and advisory services: Approaches and implications for future. Discussion Paper 1, Center for Agricultural Extension Innovations, Reforms and Agripreneurship (CAEIRA), Rajendranagar, Hyderabad.

Rani, P.M.N., Rajesh, T. and Saravanan, R. Expert systems in agriculture: A review. *Journal of Computer Science and Applications.* 3, no. 1, (2011): 59-71.

Sarma, S.K., Singh, K.R. and Singh, A. An expert system for diagnosis of diseases in Rice plant. *International Journal of Artificial Intelligence.* 1, no. 1, (2010): 1-6.

Vinod, K., Arvind, K., Premi O.P. and Manoj, K. An expert tool for fertilizer management of rapeseed mustard. *Journal of Oilseed Research.* 21, (2004): 130-33.

19

Improving Water Retention and Nutrient Uptake

Vivek Kamble[1], Princy Thakur [2], and Meraj Ahmed[3]

[1]*Department of Soil Science and Agricultural Chemistry, Lovely Professional University, Phagwara, Punjab, India*

[2,3]*Department of Soil Science and Agricultural Chemistry, Lovely Professional University, Phagwara, Punjab, India*

Introduction

Water retention is an important soil hydraulic property that affects how soil operates in an ecosystem and how it is managed. Hydrology, agronomy, meteorology, ecology, environmental protection, and many other soil-related professions employ soil water retention data. Some soil survey programs measure water retention. A soil survey can estimate water retention from other soil properties. However, these measurements are problematic during project design and in large-scale implementations (Rawls et al., 1991; Wo sten et al., 2001) since agricultural land needs rise periodically. Agriculture is using dry land. Food self-sufficiency and becoming a world food barn by 2045 necessitate agricultural land expansion (Mulyani and Agus, 2017). Water concerns afflict farming in this agroecosystem (Syam et al., 1996; Mulyani and Agus, 2017). Because of soil composition, much water stays in the earth. Tang et al. (2022).

For plant growth, soil water retention variables indicate how well the soil holds water. The dirt provides most of the plant's water. Water in root layer dirt pores is absorbed by plants. Thus, soil water retention is the key factor affecting plant development and productivity. Plants can take in the soil pores' water between the field capacity and the point where they always wilt. Field capacity occurs when the matrix potential and gravitational potential match. The most significant component for plant growth is soil water (Tang et al., 2022, Deng et al., 2016, Su et al., 2020). The soil's physical features determine plant water availability. The amount of water available depends on clay minerals, organic matter, absorbed cations, soil texture and structure, and water availability. Thus, to prevent plants from experiencing water or groundwater stress, soil physical attributes must be investigated to determine groundwater availability (Kirkham 2023). Infiltration is crucial to groundwater availability. The amount of water that can penetrate the soil depends on its physical features, such as soil texture, organic matter, bulk density, porosity, aggregate stability/stability, and water content. Thus, soil water availability depends on its physical qualities, particularly its ability to absorb and transfer water. The physical properties of saturated soil affect how water travels through it, which affects runoff and infiltration (Sauer and Logsdon 2002).

Role of Water in Plant Health

Life needs water. It is a solvent, temperature buffer, and metabolite in cells. Plants and microbes get water from their surroundings, unlike animals. Microbes that live on land plants get water from the plant and/or water vapor from the atmosphere, while land plants get water from the soil. Water availability greatly affects plant diversity and the composition of microbial communities (Lau and Lennon, 2012; Blazewicz et al., 2014; Jonas et al., 2015; Taketani et al., 2017). Plants have evolved a reliable system to achieve and sustain water homeostasis in response to a wide range of environmental variables. According to Wheeler and Stroock (2008) research, the water movement that occurs from the roots to the shoots is controlled by a negative pressure (Wheeler and Stroock, 2008). The main force that makes a plant take in water is the soil-root osmotic difference. In normal, nonsalty soils, osmosis has a small effect on water retention, but in saline soils, it may have a much bigger effect, making it harder for glycophytic to get water. Even in salty soil, it seems that halophytes can keep the appropriate gradient. In farming, a soil water potential of 15 bars (1.5 MPa) was thought to be the limit for water plants could get and the constant wilting coefficient. This number is the average of 10–20 bars of data from many tests on permanent wilting (Slatyer, 1967). The plants and evaporation slowly use up the water in the soil at field capacity until the soil's water potential hits about 15 bars (1.5 MPa). The water in the earth between 11.0 and 1.5 MPa is soil water that can evaporate. Later, Tardieu et al. (1992) came up with a new way of thinking about dirt, water, plants, and the atmosphere as all being part of the same continuum (Tardieu et al., 1992). The osmotic equilibrium and the dynamic balance between the tissue water potential and the soil water potential are required for wilting to occur, which is caused by a loss of turgor in the leaves. However, in contrast to non-transpiring plants, transpiring plants' roots can take water from the soil at a lower soil water potential (Tardieu et al., 1992) (Figure 19.1).

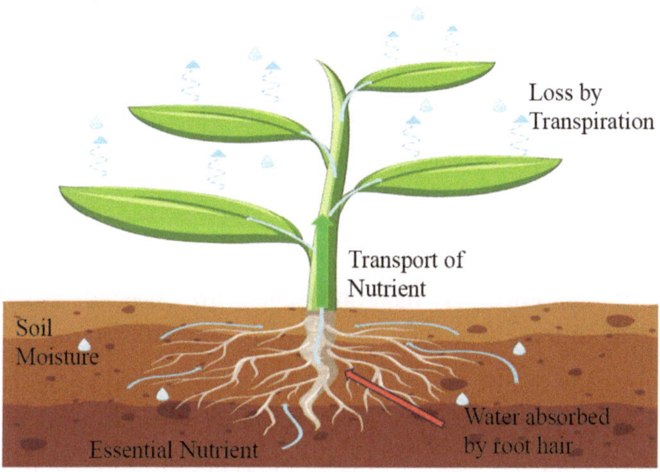

Figure 19.1: Role of water in plant health.
Source: Tardieu et al., 1992

Growth-Required Nutrients

Like all other forms of life, plants require nourishment to thrive. There are 17 nutrients plants cannot grow without (Figure 19.2). The atmosphere and groundwater supply carbon, hydrogen, and oxygen. Soil minerals and soil organic matter (SOM), as well as organic and inorganic fertilizers, provide the remaining 14 important elements: nitrogen, phosphorus, potassium, calcium, magnesium, sulfur, iron, zinc, manganese, copper, boron, molybdenum, chlorine, and nickel (Bhatla et al., 2018).

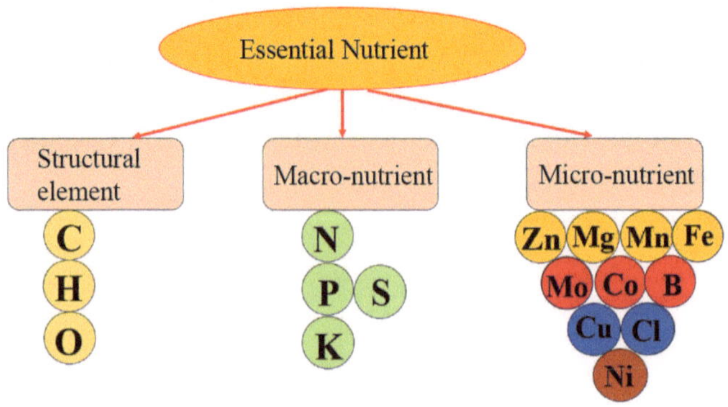

Figure 19.2: Essential Nutrients required for plant growth.
Souce: Pandey, 2018

- Primary nutrients, or macronutrients, are needed most. Carbon, hydrogen, nitrogen, oxygen, phosphorus, and potassium.
- Secondary nutrients: These are needed in moderate amounts. Calcium (Ca), Magnesium (Mg), and Sulphur (S) are secondary nutrients.
- Micronutrients are needed in small amounts. Boron, chlorine, copper, iron, manganese, molybdenum, and zinc are micronutrients (Pandey (2018)).

Water and Nutrient Uptake Issues

The world's desire for water is always going up because of things like a growing population, more irrigated farmland, and economic growth. Even though surface water (lakes, rivers, and ponds) can meet this demand around the world, there are big differences between regions that cause water stress in some places (Wada et al., 2010). The idea of rainwater gathering (RWH) is not new, and it is used in many parts of the world to deal with water shortages and flooding in cities (Mays et al., 2013). It is a major source of groundwater refilling that can be used for both drinking and nondrinking purposes, like irrigation and farming (Texas WDB., 2005; Vieira et al.,2014; Belmeziti et al., 2013; Liaw and Chirang 2014; Campisano et al., 2013; Hajani et al., 2014). History shows that growing interest in collecting rainwater started in the 1980s. This method is now getting more attention and is popular in both the west and east of the world (Hussain et al., 2019). There are two main types of methods used to add water to the ground artificially: direct methods and indirect methods. Direct methods include spreading on the surface and spreading below the surface. Indirect methods include induced methods and methods that change the groundwater (Reddy et al., 2019). Surface spreading techniques, which involve flooding, ditch and furrows, recharge basins, runoff conservation structures (bench terracing, contour bunds and contour trenches, gully plugs, nalah bunds, check dams, percolation ponds), and stream modification/augmentation are aimed at increasing the contact area and residence time of surface water over the soil to enhance the infiltration and to augment the groundwater storage in phreatic aquifers (Reddy et al., 2019). Sub-surface techniques, such as injection wells (also called "recharge wells"), gravity head recharge wells, and recharge pits and shafts, are used to refill deeper aquifers that are covered by layers that don't let water through (Reddy et al., 2019; Ministry of water resources, 2000). Parent material mineral composition and weathering processes affect soil pH. In humid conditions, soil acidification happens for a long period as water moves laterally or downward through the soil, while in dry environments, soil weathering and leaching are less intensive, and soil pH is often neutral or alkaline (Bloom and Skyllberg, 2012). Chemical and physical characteristics of the soil, such as the parent material and minerals that are already in the soil, the amount of organic matter, the depth to bedrock, sand, or gravel, and the soil's permeability, water-holding capacity, and drainage, can affect the availability of nutrients. The abundance of nutrients is also affected by how the environment is and what the crop is like. The amount of water in the soil is important not only to give the crop the water it needs but also to break down nutrients and make them available to the plant. When there is too much water in the soil, however, oxygen (O_2) levels drop, and carbon dioxide (CO2) levels rise. Roots need O_2 to grow and take in nutrients, but too much CO_2 is bad for them. Fernández and Hoeft (2009) say that temperature is important for controlling the speed of chemical processes in soil that make minerals available (Fernandez and Hoeft., 2009).

Assessing Soil Conditions and Water Needs

Soil Testing for Water Retention Capacity

Annual soil testing helps determine fertilizer needs to maximize productivity and prevent arthropod pests and disease. pH and macronutrients including potassium, phosphorus, calcium, and magnesium must be tested. Soil testing can measure sulfur, but forecasting plant demands is tricky. As other crops have shown shortages, boron, iron, manganese, molybdenum, and zinc levels should also be checked. To ensure nutrient absorption, soil pH may need to be changed. Soil pH affects many biological and chemical processes. Extreme pH can cause nutrient shortages and sickness (Baligar et al., 2001).

Assessing Malnutrition and Inadequate Dietary Intake

The bines, leaves, and cones of a hop plant all contribute to the plant's substantial biomass output. Hop plants, like any plant, need a balanced diet to produce a high yield. Hop may not be getting enough of some nutrients or too much of others. When numerous production factors contribute to the same symptom in a plant, it might be challenging to determine which one is at fault. In this section, we go over the common signs of nutrient deficiency and the known interactions nutrients have with illnesses and insect pests. The recommendations for fertilization are out of the scope of this pest management manual. The published literature provides recommendations that vary greatly depending on factors like production region, variety, irrigation method, soil type, and production objectives. Readers who want advice tailored to their specific region and set of circumstances should consult with professionals in that area. (Gent et al., 2015).

Boron: Boron deficiency can delay shoot emergence, stunt, distort, and crinkle young leaves, and yellow and die shoot tips. Due to lobe formation problems, afflicted leaves may be tiny, brittle, and fluffy-tipped. Acid and sandy soils are especially deficient. Boron deficiency may cause red crown rot (Gupta and Solanki 2013).

Calcium: Young tissues and developing sites show calcium-deficient symptoms first. Similar symptoms to boron shortage include yellowing growth tips, limited leaf development, and leaf margin yellowing and death. Excess calcium can impair nutritional intake and deplete positively charged ions including ammonium, magnesium, and potassium (Johnson and Urju, 1989).

Iron. Young leaves show yellowing between veins and green veins due to iron shortage. Iron deficiency is most frequent in alkaline soils, but manganese solubility and uptake in severely acidic soils (pH 5.7 or less) can produce it. (Gent et al., 2015).

Magnesium. Older leaves show magnesium symptoms as yellowing between leaf veins, which can lead to mortality and defoliation. Magnesium deficits are frequent in acid soils or when potassium is overapplied. (Gent et al., 2015).

Manganese: Manganese is scarce in alkaline soils and harmful in acidic ones. Manganese insufficiency causes juvenile leaf yellowing and white speckling. At soil pH below 5.7, manganese accumulates in plant tissues, preventing iron uptake and causing iron shortage.

Nitrogen: Nitrogen deficit causes stunting, poor development, and yellowing, especially on older leaves. Nitrogen-deficient cones are smaller. Excess nitrogen fertilizer can cause powdery mildew, Verticillium wilt, spider mites, and hop aphids. Nitrogen inputs should match crop demands, especially when a pest is present (Gent et al., 2015).

Molybdenum: Older leaves first show yellowing and white speckling from molybdenum deficits. Acidic hop (pH 5.7 or less) has deficiencies. Since affected plants are yellow, molybdenum insufficiency can be mistaken for nitrogen deficit.

Phosphorus: Lower leaves show deficiency symptoms as downcurved, dark-green, dull leaves. Bines is frail. Cones may turn brown. Hop plants low in phosphorus can decrease yield even without symptoms, according to English studies.

Sulfur: Deficient plants feature limited growth, wiry stems, and withering younger leaves. High-leaching soils may lack sulfur.

Zinc: Deficient plants display sluggish growth, short lateral branches, and poor cone development. Small, malformed, yellow, curling, brittle leaves. Damaged plants may perish. Zinc deficits are typical in central Washington soils with a pH above 7.5. Zinc relieves Apple mosaic virus symptoms (Gent et al., 2015).

Enhancing Soil Structure and Composition

Soil Amendments for Improved Water Retention

Both synthetic and natural soil amendments are important because they help create a water storage reservoir for plants in the topsoil layers, where roots typically grow. Soil physical qualities can be enhanced by crop residues, leading to increased water retention and reduced weed growth. Soil humus content, rainfall infiltration, moisture conservation, water storage capacity, and cation exchange capacity were all shown to increase after wheat wastes were introduced to the soil. When wheat waste decomposes, it releases nutrients that can be used by succeeding crops. Improved N recovery from subsoil and reduced soil crusting are just two of the benefits of gypsum's usage as a global soil amendment. Improved water retention in soil and increased water-use efficiency are the results of gypsum's increased water infiltration. Soil microbial activity can be accelerated and water use efficiency (WUE) improved by adding qemisoyl and humic acid, according to a 2011 study. Applications of humic acid at 10 and 15 kg ha1 improve maize production by 11.50% and 19.11% over control, respectively, according to research published by Verlinden et al., 2010. Soil microbial activity and water retention are both improved when irrigation is combined with humic acid, and the use of humic acid in irrigation increases the efficiency with which water is used. Some additives to soil can boost drought tolerance in plants and WUE. Biochar is one type of amendment used to alleviate drought stress in plants. It is considered a drought-resistant material because of the 80% carbon content because of carbon's ability to save water and nutrients for plants. Biochar's various benefits make it a great addition to soil. The water-holding capacity of the soil is also increased, and microbial interaction and activity are enhanced. When applied alongside conventional fertilizers, biochar has been shown to increase yields. Biomass from many kinds of plants can be used to create biochar. Lantana camara is an invasive plant that has a destructive effect on grazing lands and depletes soil nutrients throughout most of the subtropical world. Because of its beneficial effect on soil physicochemical properties, *Lantana camara* can be removed from damaged regions and converted into compost, making it one of the most preferred organic additions. Compost, an external supply of Soil Organic Matter (SOM), can be mixed into the ground to offset the multiple ways in which organic matter is lost from the soil. This, in turn, aids in the retention of soil nutrients (Batool et al., 2015).

Organic Matter and its Impact on Nutrient Availability

SOM stores nutrients and binds the soil together. Overgrazing and converting grasslands to tilled farms have reduced SOM levels in some soils over the last century. This has reduced soil fertility, increased fertilizer demands, and soil erosion in some regions. SOM is also a critical indication of soil quality and a sink for carbon dioxide (CO_2) and other greenhouse gases, nutritional quality, soil microbe availability, and long-term adaptation of plants or microorganisms to persistent nutritional circumstances may affect SOM decomposability and rate. Soil structure and mineralogy stabilize SOM over time. Change capacity in nutrient management and soil pH affects cation and anion exchange capabilities. Exchange capacity is the soil's ability to retain and supply nutrients to crops. Cation Exchange Capacity (CEC) is larger than Anion Exchange Capacity (AEC) because most soils have a net negative charge. High-CEC soil buffer pH changes for the better. Clay-rich and organic-rich soils have higher CEC and buffering capabilities than silty or sandy soils (Rawls et al., 2003). Plant nutrients depend on soil pH. Nutrient Management illustrates numerous nutrients' ideal pH levels. Except for Phpsporous, which is most available in a pH range of 6 to 7, macronutrients (N, K, Ca, Mg, and S) are more available in a pH range of 6.5 to 8, while most micronutrients (B, Cu, Fe, Mn, Ni, and Zn) are more available in a pH range of 5 to 7. Outside these ideal ranges, plants receive fewer nutrients. Soil microbe activity is highest near neutral pH; however, pH ranges vary by species. Very acid soils (less than 5) reduce microbial activity and population (McCauley et al., 2009).

Managing Soil Compaction and Drainage Issues

In the broader topic of soil management, soil compaction is a particularly relevant issue. Loss of organic matter, crusting, slaking, hard-setting, and anaerobes are also parts of the land degradation "syndrome." Recent reports from the United Nations and other international organizations have highlighted the role that soil compaction can play in causing soil erosion, nutrient depletion, and pollution (Hartemink, 2008). Compressive forces applied to the soil by the wheels of tractors, trailers, and harvesters; the passage of tillage implements through the soil (especially powered rotary equipment); and the weight of livestock and other animals on the ground are the primary causes of compaction (Tardieu et al., 1992).

Efficient Irrigation Techniques for Water Retention

Irrigation methods are typically scheduled around other farming tasks when water is abundant. Most farmers, for instance, swap out their furrow irrigation systems every 12 to 24 hours since it's the most time- and labor-efficient way to do it. When water is scarce, you need to reevaluate basic habits to get the most out of what's available. After all, besides the land itself, water is a cultivator's most precious resource. Trading off management and labor for more efficient water use makes sense. There are good arguments for conserving water even in times of plenty. Excessive water use can lead to runoff that washes away valuable topsoil and fertilizer. Deep percolation and leaching of nitrates, nitrites, and other agricultural chemicals are the results of excessive irrigation. The amount of chemicals that aquifers carry every day includes these pollutants. You and other producers, as well as those who care about having access to clean water, stand to profit from self-regulation on the part of the growing community (Levidow et al., 2014).

Approaches to Less Water Utilization

Scheduled irrigation, conservation methods including deficit irrigation and tillage, and the use of more effective irrigation systems can all lead to greater water savings. Furrow irrigation is less effective than sprinkler and drip irrigation systems. While employing some of these techniques may incur additional expenses, the potential for increased crop value may be worth the price (Shock et al., 2013). The following are some of the simple and common approaches to reducing the utilization of water:

- Do not overwater!
- Evapotranspiration (ET)-based irrigation scheduling.

- Water tension or soil moisture content can be used to plan irrigation times.
- Implement deficit irrigation
- Prepare your irrigated land area
- Use conservation tillage techniques.
- Manage surface irrigation with caution.

Drip Irrigation Systems: Pros and Cons

1. Drip irrigation (Figure 19.3) uses plastic tubing near plants' roots to slowly, and evenly apply low-pressure water.
2. Drip irrigation improves chemical application. Only the crop root zone is irrigated, reducing nitrogen leaching losses and optimizing fertilizer use. Less insecticides may be used.

Additional Advantages of Drip Irrigation

- Drip systems can be designed for fields with strange shapes, uneven topography, or soil texture.
- Drip systems function well where other irrigation systems fail due to excessive infiltration, water puddling, or runoff.

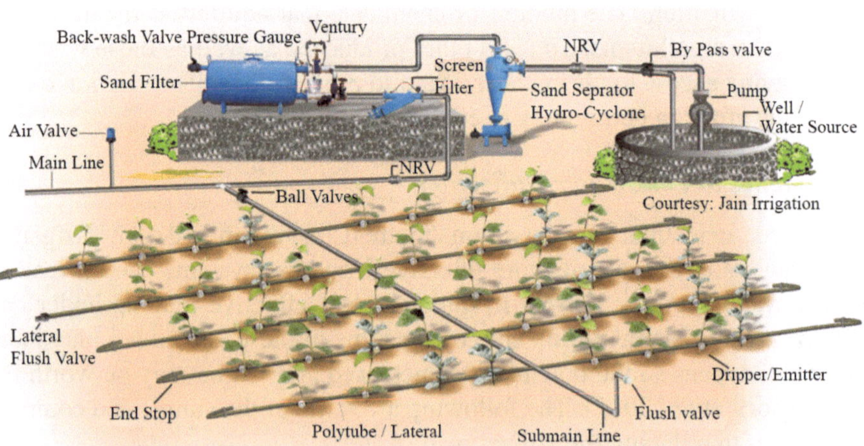

Figure 19.3: Drip irrigation system.
Source: Shock, 2006
Source: Common Wikimedia

- Drip irrigation helps save water. Drip irrigation allows precise nutrition administration and reduces evaporation, runoff, and deep percolation.
- Reduce fertilizer and nitrate expenditures. Plants can benefit from better nutrient timing.
- Drip irrigation systems can keep wheel traffic rows dry enough for tractor use at any time. Apply herbicides, insecticides, and fungicides on time.
- Avoid leaking or clogging drip tape or tubing. Silt in irrigation water can clog drip emitters. Chemical deposits or algae in the tape can potentially clog the emitters (Shock 2006).

Disadvantages of Drip Irrigation System

Initial investment cost is a drawback. Surface drip irrigation (SDI) systems are costly. Field size, shape, water source location, and automation level affect cost per acre.

System capacity: SDI allows frequent, low-volume water and nutrient applications. The irrigation district's water supply schedule may not allow producers to make use of this system attribute. In certain circumstances, a water storage tank or reservoir may be needed, increasing system costs.

Management time: Management of SDI requires more time than conventional irrigation systems, especially in the first 2 years when the learning curve is severe. SDI systems require periodic chlorination and acid injection, which other systems do not.

Short driplines: Limit dripline lengths for homogeneity. Dripline diameter, emitter flow rate, land slope, and emitter spacing determine the maximum dripline length. Large fields may need to be separated into irrigation zones, which may increase mainline and sub-main costs.

Installation: Installing an SDI system needs specialized equipment, is laborious, and accounts for a large percentage of its initial cost. After installation, run water through the driplines to check for leaks and open the soil flow passage to the tape diameter.

Stiff design: SDI systems must be well-designed, implemented, operated, and maintained regardless of cost. Design decisions are irreversible.

Clogged emitter: Emitter clogging plagues SDI and other drip irrigation systems. Before developing the system, a water quality test might reveal issues that may affect its design, performance, and maintenance.

Rodents: SDI systems often struggle to irrigate row crops due to rodents. Gophers and field mice chew on drip lines, which can cause serious issues.

Seed growth: Water stress may impede seed germination and early growth because of the subsurface SDI system's dry soil. In sandy soils, little water rises in the soil profile.

Soil salinity: SDI in arid places may enhance soil salinity above the driplines. This problem may take a long time to develop and is unlikely in places with enough precipitation to transport the salts down to the soil profile.

Dripline alignment: When drip lines are put parallel to crop rows, as is customary, it might be difficult to keep them aligned from season to season (Payero and Yonts, 2005).

Sprinkler Irrigation System: Distributing Water Efficiently

Sprinkler irrigation evenly distributes water by overlapping sprinklers. Water should infiltrate where it falls without runoff or erosion. The water distribution model of the sprinkler, the layout and spacing between sprinklers, and the wind, especially its speed, determine water application in set systems. The sprinkler's water distribution model depends on its type, nozzle number and design, and operational pressure (Keller and Bliesner, 1990). The sprinkler system must cooperate to distribute water evenly (Tarjuelo 2000; Montero 2001a; Ortega 2004b). Thus, solid-set systems Portable systems with plot-dispersed laterals are less uniform and efficient. Solid-set systems require less labor but cost more. Buried pipelines safeguard solid-set sprinkler irrigation systems. Pipes in surface solid-set systems may be damaged by temperature, solar radiation, blows, etc. These systems can be reused in other plots after the crop cycle. Al, PE, and polyvinyl chloride (PVC) pipes on the surface can be employed in solid-set systems. All of them demand more labor than a permanent PVC system with underground pipes. Sprinkler irrigation requires pumping, pipelines, sprinklers, ditches, electricity, labor, maintenance, and water. The sprinkler irrigation system's price relies on its design, materials, and automation. The design, layout, size, distance from the water supply, and pumping requirements also affect this cost. (Esquiroz and Puig, 2001). It is found that sector number and size greatly affect investment costs. The multiplicity of design options necessitates finding the lowest total cost, including investment and operation expenditures (Esquiroz and Puig, 2001).

Automation and Smart Irrigation Systems

Automation is crucial to human life in the 21st century. Automation controls appliances automatically. It saves energy, and time, and provides comfort. Industrial automation and control machines are expensive and unsuitable for farm applications. so, we design a low-cost smart irrigation technology for Indian farmers (Sahu and Behera, 2015). This controlled the water motor automatically and selected the pipe water flow direction using a soil moisture sensor. Finally, communicate the farm field's motor operation and water direction to the user's phone and email. A smart irrigation system automatically controls the entire irrigation system, even when the farmer is not around, and sends messages to the farmer about farm field information and changes in operation (Yuan et al., 2004) and which requires no worker and uses less water than the previous three methods. The moisture sensor Comparator (LM393) transforms analog soil moisture sensor data into discrete. Two 5-cm copper wire soil probes can be immersed in the soil under test. The circuit outputs soil conductivity voltage. Soil moisture determines the resistance between the probes. RASPBERRY-Pi controls irrigation and sends data to the registered mobile number through the Internet. Raspberry-pi Python programming automates message sending. The farmer always updates his field status through text messaging or Gmail if he is far away. Raspberry-pi checks which direction of the agricultural field is dry, opens the electromagnetic valve, and turns on the pump. High-priced devices regulate smart watering systems. However, we employ low-cost hardware that Indian farmers can buy and use. Using fewer sensors makes the smart irrigation system cheaper (Abreu and Pereira 2002).

Implementing Sustainable Water Management Strategies

Rainwater Harvesting and Reuse

Rainwater harvesting is easy, inexpensive, and beneficial. Rainwater can augment brackish groundwater or polluted surface water in the rainy season. In droughts and dry wells, it's a good option. Rainfall is uncontrollable. In dry or semi-arid climates, using rainfall efficiently is vital. Rainwater that might otherwise evaporate or run off is valuable. Many communities worldwide are reaching their water resource limits due to contamination of groundwater and surface waters and population growth. They must use "new" resources like rainwater harvesting (RWH). Rainwater collecting is again a valuable water resource. Rainwater is currently used alongside "conventional" water supply technology in rural and urban areas. Arid and semi-arid areas, small coral and volcanic islands, and distant human settlements benefit from Rain Water Harvesting (RWH) (Nandi and Gonela 2002). There are many and varied benefits to storing and utilizing rainwater for household use:

1. **Rising water demand** Water shortages affect groundwater tables and reservoirs. Many piped water systems fail. Rainwater is another option.
2. **Water availability** Lakes, rivers, and shallow groundwater availability fluctuate. Rainwater can be collected and stored for residential usage during droughts. Rainwater can also help improve river and other surface water quality during the rainy season.
3. **Collection and storage close to use** Traditional sources are distant. Collecting and storing water near homes increases water accessibility and health. It also boosts ownership.
4. **Water quality** Industrial, human, and mineral pollutants can damage water supplies. Rainwater is usually good (Worm 2006).

Mulching and Its Effect on Soil Moisture Conservation

The gardener's use of straw and leaves as mulch may have come from the German word "molsch," which signifies soft decay. Mulching prevents runoff, soil loss, weed infestation, and water evaporation. Thus, it retains soil moisture, controls temperature variations, improves soil physical, chemical, and biological qualities, adds nutrients, and boosts crop development and production. Under rainfed conditions, mulching increases production by 50–60% (Telkar et al., 2017). The following are the benefits of organic mulching:

- Mulch deflects sunlight that heats soil. This cools the soil and prevents evaporation. Hot, dry climates require this.
- Weeds cannot grow under mulch because it blocks light. Mulches minimize soil erosion by shielding it from wind and water.
- Mulches reduce runoff and promote soil absorption. Crops benefit from soil water.
- Organic mulches also benefit the soil. Organic debris from these mulches loosens the soil.
- Organic matter feeds beneficial earthworms and other soil microorganisms, making the soil permeable.
- Root growth, water infiltration, and soil water-holding capacity improve.
- Decaying organic debris provides plant nourishment.
- Even soil temperature. Cleans feet, allowing field access even when damp (Patil et al., 2013).

Conclusion

Ecosystem functioning and agricultural management depend on soil water retention. It affects plant growth, nutrient availability, soil structure, and water management. Hydrologists, agronomists, ecologists, and environmentalists must understand soil water retention. While measuring water retention is difficult, research and technology are improving our understanding of this important soil property. For plant health, soil water retention affects plant water availability, development, and productivity. The delicate balance between soil water potential and plant osmotic processes controls water uptake, affecting plant growth and yield. Soil management practices are crucial to plant growth because water retention and nutrient availability are interconnected. Drip and sprinkler systems conserve water and maximise crop yields. Rainwater harvesting, mulching, and soil amendments improve soil moisture conservation and ecosystem resilience.

References

Abreu, V. M. and Pereira, L. S. 2002. Sprinkler irrigation systems design using ISAMim. p. 022254.

Baligar, V. C., Fageria, N. K., and He, Z. L. Nutrient use efficiency in plants. *Communications in soil science and plant analysis*, 32, no. 7-8, (2001): 921-50.

Batool, A., Taj, S., Rashid, A., Khalid, A., Qadeer, S., Saleem, A. R., and Ghufran, M. A. Potential of soil amendments (Biochar and Gypsum) in increasing water use efficiency of Abelmoschus esculentus L. Moench. *Frontiers in plant science,* 6, (2015): 733.

Belmeziti, A., Olivier, C., and Bernard, D.G. A new methodology for evaluating potential for potable water savings (PPWS) by using rainwater harvesting at the urban level: The case of the municipality of Colombes (Paris region). *Water,* 5, (2013): 312–326.

Bhatla, S. C., A. Lal, M., Kathpalia, R., and Bhatla, S. C. Plant mineral nutrition. *Plant physiology, development and metabolism,* (2018): 37-81.

Blazewicz, S.J., Schwartz, E., and Firestone, M.K. Growth and death of bacteria and fungi underlie rainfall-induced carbon dioxide pulses from seasonally dried soil. *Ecology,* 95, (2014): 1162-72.

Bloom, P. R. andSkyllberg, U. 2012. Soil pH and pH buffering. In: Huang PM, Li Y, Sumner ME, editors. Handbook of Soil Sciences: Properties and Processes. (2nd ed). Boca Raton, FL: CRC Press; 19-14. ISBN: 9781439803059.

Campisano, A., Gnecco, I., Modica, C. and Palla, A. Designing domestic rainwater harvesting systems under different climatic regimes in Italy. *Water Sci. Technol.*, 67, (2013): 2511-18.

Deng, L., Wang, K.B., Li, J.P., Zhao, G.W. and Shangguan, Z. Effect of soil moisture and atmospheric humidity on both plant productivity and diversity of native grasslands across the Loess Plateau, China. *Ecol. Eng.,* 94, 2016: 525-31. DOI: https://doi.org/10.1016/j.ecoleng.2016. 06.048.

Esquiroz, O., and Puig, J. 2. "Optimización económica de instalaciones de riego en parcela." Proc., XIX Congreso Nacional de Riegos, Asociacion Española de Riegos Y Drenajes, Zaragoza, Spain (in Spanish).

Fernández, F. G., and Hoeft, R. G. Managing soil pH and crop nutrients. *Illinois agronomy handbook,* 24, (2009): 91-112.

Gent, D. H., Sirrine, J. R., Darby, H. M., O'Neal, S. D., and Walsh, D. B. 2015. Nutrient management and imbalances. Field Guide for Integrated Pest Management in Hops. pp 98-100. Moxee, WA, USA: Washington Hop Commission.

Gupta, U., and Solanki, H. Impact of boron deficiency on plant growth. *International Journal of Bioassays,* 2, no. 7, (2013): 1048-50.

Hajani, E. and Rahman, A. Reliability and cost analysis of a rainwater harvesting system in peri-urban regions of greater Sydney, Australia. *Water,* 6, (2014): 945-60.

Hartemink, A.E. Soils are back on the global agenda. *Soil Use and Management*, 24, (2008): 327-30.

Hussain, F., Hussain, R., Wu, R.S. and Abbas, T. Rainwater Harvesting Potential and Utilization for Artificial Recharge of Groundwater Using Recharge Wells. *Processes,* 7, (2019): 623.

Johnson, R. S., and Uriu, K. Mineral nutrition. Peach, Plum and Nectarine: Growing and Handling for Fresh Market. Oakland, University of California, Division of Agriculture Resource, (1989): 68-81.

Jonas, J.L., Buhl, D.A., and Symstad, A.J. Impacts of weather on long-term patterns of plant richness and diversity vary with location and management. *Ecology,* 96, (2015): 2417-32.

Kirkham, M. B. 2023. Principles of soil and plant water relations. Elsevier. DOI: https://doi.org/10.1016/B978-0-12-409751-3.X5000-2

Lau, J.A. and Lennon, J.T. Rapid responses of soil microorganisms improve plant fitness in novel environments. *Proc. Natl Acad. Sci. USA*, 109, (2012): 14058-62.

Levidow, L., Zaccaria, D., Maia, R., Vivas, E., Todorovic, M., and Scardigno, A. Improving water-efficient irrigation: Prospects and difficulties of innovative practices. *Agricultural Water Management*, 146, (2014): 84-94.

Liaw, C.H. and Chiang, Y.C. Dimensionless analysis for designing domestic rainwater harvesting systems at the regional level in northern Taiwan. *Water,* 6, (2014): 3913–33.

Mays, L., Antoniou, G.P. and Angelakis, A.N. History of water cisterns: Legacies and lessons. *Water*, 5, (2013): 1916-40.

McCauley, A., Jones, C., and Jacobsen, J. Soil pH and organic matter. *Nutrient Management Module,* 8, no. 2, (2009): 1-12.

Ministry of Water Resources 2000. Guide on Artificial Recharge to Groundwater; Delhi, India: Ministry of Water Resources.

Montero, J., Tarjuelo, J., and Carrión, P. "SIRIAS: A simulation model for sprinkler irrigation. Part II: Calibration and validation of the model." *Irrig. Sci.*, 202, (2001a): 85-98.

Mulyani, A., dan Agus, F., Kebutuhan dan ketersediaanlahancadanganuntukmewujudkancita-citaindonesiasebagai lumbung pangan dunia tahun 2045, *Analisis Kebijakan Pertanian*, 15 no. 1, (2017): 1-17. DOI: https://doi.org/10.21082/akp.v15n1.2017

Nandi, S., and Gonela, V. Rainwater harvesting for domestic use: A systematic review and outlook from the utility policy and management perspectives. *Utilities Policy*, 77, (2022): 101383.

Pandey, N. Role of plant nutrients in plant growth and physiology. *Plant nutrients and abiotic stress tolerance,* (2018): 51-93.

Patil, S., Tushar, K.S. and Bhalerao S.A. Mulching: A soil and water conservation practice. *Research Journal of Agriculture and Forestry Sciences,* 1, no. 3, (2013): 26-29.

Rawls, W. J., Pachepsky, Y. A., Ritchie, J. C., Sobecki, T. M., and Bloodworth, H. Effect of soil organic carbon on soil water retention. *Geoderma,* 116, no. (1-2), (2003): 61-76.

Rawls, W.J., Gish, T.J., and Brakensiek, D.L. Estimating soil water retention from soil physical properties and characteristics. *Adv. Soil Sci.,* 16, (1991): 213-34.

Reddy, G.S., Sandhya, S., and Hanumanthu, R. The need of artificial recharge techniques for sustainable development of ground water resources. *Int. J. Geol. Earth Sci. 5*, (2019): 131-45.

Sahu, C. and Behera, P. 2015. DzA Low-Cost Smart Irrigation Control Systemdz, IEEE Sponsored ʹnd International Conference on Electronics and Communication System (ICECS 2015).

Sauer, T. J., and Logsdon, S. D. Hydraulic and physical properties of stony soils in a small watershed. *Soil Science Society of America Journal*, 66, no. 6, (2002): 1947-56.

Shock, C. C. 2006. Drip irrigation: An introduction.

Shock, C. C., Shock, B. M., and Welch, T. 2013. Strategies for efficient irrigation water use.

Slatyer, R.O. 1967. Plant Water Relationships. London: Academic Press.

Su, F. L., Wang, F. W., Li, Z., et al. J. Predominant role of soil moisture in regulating the response of ecosystem carbon fluxes to global change factors in a semi-arid grassland on the Loess Plateau. *Sci. Total Environ.,* 738 (2020), Article 139746. DOI: https://doi.org/10.1016/j.scitotenv.2020.139746

Syam, A., K. Kariyasa, E. SUjitno dan Z. Zaini. 1996. ProsidingLokakaryaEvaluasi Hasil Penelitian Usahatani Laban Kering, 1997. Pusat Penelitian Tanah dan Agroklimat. Badan Penelitian dan PengembanganPertanian, DepartemenPertanian, Bogor.

Taketani, R.G., Lanconi, M.D., Kavamura, V.N., Durrer, A., Andreote, F.D., and Melo, I.S. Dry season constrains bacterial phylogenetic diversity in a semi-arid rhizosphere system. *Microb. Ecol.,* 73, (2017): 153-61.

Tang, K., Li, F. Y., Jaesong, S., et al. 2022. Soil water retention capacity surpasses climate humidity in determining soil organic carbon content but not plant production in the steppe zone of Northern China, Ecological Indicators, Volume 141, ISSN 1470–160X. DOI: https://doi.org/10.1016/j.ecolind.2022.109129.

Tardieu, E., Zhang, J., and Davies, W.J. What information is conveyed by an ABA signal from maize roots in drying field soil? *Plant Cell Environ.,* 15, (1992): 185-91.

Tarjuelo, J. M., Ortega, J. F., Montero, J., and de Juan, J. A. "Modeling evaporation and drift losses in irrigation with medium size impact sprinklers under semi-arid conditions." Agric. *Water Manage.,* 433, 2000: 263-84.

Telkar, S. G., Kant, K., Pratap, S., and Solanki, S. Effect of mulching on soil moisture conservation. *Biomolecule reports*, 9, (2017): 1-4.

Texas Water Development Board (2005). The Texas Manual on Rainwater Harvesting, 3rd ed.; Texas Water Development Board: Austin, TX, USA.

Verlinder, G., Coussens, T., Vliegher, A. De., Baert, G. and Haesaert, G. Effect of humic substances on nutrient uptake by herbage and on production and nutritive value of herbage from sown grass pastures. *Grass and Forage Science.* 65, no. 1, (2010): 133-44.

Vieira, A.S., Beal, C.D.; Ghisi, E., and Stewart, R.A. Energy intensity of rainwater harvesting systems: A review. *Renew. Sustain. Energy Rev.*, 34, (2014): 225-42.

Wada, Y., van Beek, L.P., Van Kempen, C.M., Reckman, J.W.T.M., Vasak, S., and Bierkens, M.F. Global depletion of groundwater resources. *Geophys. Res. Lett.*, 37, (2010): L20402.

Wheeler, T.D. and Stroock, A.D. The transpiration of water at negative pressures in a synthetic tree. Nature, 455, (2008): 208-12

Wo sten., J.H.M., Pachepsky, Ya. A., and Rawls, W.J. Pedotransfer functions: Bridging the gap between available basic soil data and missing soil hydraulic characteristics. *J. Hydrol.*, 251, 2001: 123-50.

Worm, J. 2006. AD43E Rainwater harvesting for domestic use (No. 43). Agromisa Foundation.

Yuan, G., Luo, Y., Sun, X, and Tang, D. DzEvaluation of a crop water stress index for detecting water stress in winter wheat in the North China Plain, dzAgricult. *Water Manag.,* 64, no. 1, (2004): 29-40.

20

Mitigating Climate Change (CC) by Arresting Soil Organic Carbon

Princy Thakur[1], Meraj Ahmed[2], Rodrick Lepcha[3], and Akshita Barthwal[4]

[1–3]Department of Soil Science and Agricultural Chemistry, Lovely Professional University, Phagwara, Punjab, India

[4]R&D Bayer Crop Science Ltd, Thane, Maharashtra, India

Introduction

The changing climate and its variability are a big concern for the human being. The recurrent droughts and floods, which are the results of this changing climate seriously hamper the livelihood of billions of people who depend on land for most of their needs. According to the United Nations Framework Convention on Climate Change (UNFCCC) (UNFCC; 1992), "climate change is the change that can be attributed directly or indirectly to human activity that alters the composition of the global atmosphere and which is in addition to natural climate variability observed over comparable time periods" The Intergovernmental Panel on Climate Change (IPCC report, 2022) issued a cautionary statement regarding the imminent attainment of the 1.5°C threshold by the global climate system within the forthcoming 20-year period. It further emphasized that only implementing exceedingly stringent reductions in carbon emissions, commencing at present, would serve as a viable measure to avert an impending ecological catastrophe.

Droughts: A time in which circumstances are significantly drier than average is referred to as a drought.

Floods: A flood is an abnormally large outpouring of water (or, on very rare occasions, other fluids) that covers normally dry ground.

Livelihoods: The term "livelihood" refers to a collection of activities that are carried out during one's lifetime and are considered necessary to one's day-to-day existence. Obtaining necessities such as water, food, fodder, medication, shelter, and clothing are examples of such activities.

Ecological catastrophe: A catastrophic occurrence that occurs in the natural environment as a direct result of activities carried out by humans is referred to as an environmental disaster or ecological disaster.

Global Scenario of Climate Change

It was reported that the Greenland Ice Sheet holding enough ice to eventually raise global sea levels by 7 m may likewise be established below 2°C while suffering of the pan-tropical mass bleaching of the coral reefs (Robinson et al., 2012; Frieler, 2013) could be minimized if the global temperature stays well below 2°C and some remnants of the world's coral reefs can be saved. In another study, the prices for main staples may increase between 30% (for rice in an "optimistic" scenario) and 100% (for maize in a baseline scenario) by 2050 for the declining crop yields due to climate change. Hence, the main challenges for agriculture in this century will be to increase the supply of food, improve its distribution and access, and enhance the resilience of the whole food system while reducing greenhouse gas (GHG) emissions as well as getting rid of air and water pollution from agriculture and land use, avoiding biodiversity and habitat losses, and phasing out unsustainable water withdrawals. Besides the high seasonal temperatures, (Battisti and Naylor 2009) beyond what has been already noted in the last 50 years, may become further widespread in several Mesoamerican and South American locations in the remaining period of the century. Furthermore, the changes in forest cover in the Amazon basin affect the flux of moisture to the atmosphere, regional convection, and hence regional rainfall (Roy and Avisaar, 2002).

As per the United Nations Report of FAO published in 2016; Climate change affects nutrition status and dietary choices through its impacts on food security, diseases, water safety, sanitation, livelihoods, and caregiving. It amplifies the impact of droughts, floods, and storms and exposes large numbers of people especially the poor and most vulnerable to the risk of undernutrition following extreme climate events. The seasonal patterns of inadequate food availability and access, a major cause of undernutrition among poor rural communities, are accentuated by climate change, which has impacts also on livelihood security and on intra-family food distribution, affecting in turn the nutrition status of children and women in particular (Wijesinha et al., 2013). It was estimated that wheat grown under the high levels of CO_2 expected by 2050, wheat grain had 9 % less zinc, 5 % iron, and 6 % less protein, while losses in rice were 3 %, 5 %, and 8 %, respectively, compared to expected yields without climate change. Maize would suffer similar losses of nutrients; soybeans would not lose protein but would contain less zinc and iron. Food safety may be compromised by an increase in food-borne pathogens, as well as contamination or chemical changes that increase the prevalence of toxic compounds in food. For example, the presence of algal surface blooms will contaminate drinking water and shellfish with cyanotoxins (Paerl and Huisman, 2009), while higher temperatures and humidity increase the risk of mycotoxin contamination of stored cereals and pulses (Paterson and Lim, 2010). In addition, the changes in patterns of plant and animal diseases may lead to increased use of potentially harmful agricultural chemicals.

Soil organic carbon (SOC) serves as a significant indicator for assessing the health of soil and land due to its ability to include several intrinsic soil characteristics and exhibit a strong response to aboveground landscape dynamics. This includes factors such as land-use change and processes associated with land degradation. The preservation or addition of SOC is of utmost importance in both prospective climate change mitigation and adaptation, owing to its significant contribution to agricultural and land production. Soil has the potential to serve as a significant carbon (C) sink owing to its capacity for storing C, which is often higher than that of plants. Additionally, soil has a rather stable collection of different organic and inorganic C components. The potential impacts of climate change on agricultural productivity, in conjunction with the recent expansion of agricultural areas due to population growth and rising food demand, have generated heightened attention toward soil carbon sequestration and global assessments of SOC stocks. The Kyoto Protocol also recognizes the need to take agricultural soils into account when addressing environmental concerns.

Soil Carbon Sequestration as a Measure Against Climate Change

What Causes Climate Change?

The intervention of humans (anthropogenic climate change) and nature are the main and important drivers of climate change. The concentration of GHGs and aerosols, changes in solar radiation and land surface properties disturb the energy balance of the climate system, i.e., in altering the balance of incoming and outgoing energy in the earth-atmosphere system which is an important factor of the potential climate change mechanism (IPCC, 2007).

Some of the prominent and utmost important natural drivers of climate change are continental drift, volcanoes, ocean currents, the earth's tilt, comets, and meteorites. Industrialization, deforestation, and pollution have greatly increased concentrations of water vapor, carbon dioxide, methane, and nitrous oxide, all GHGs that trap heat near the earth's surface. The persistence of these gases in the atmosphere is very high, which means that even if such emissions were eliminated today, it would not immediately stop global warming. There are many uncertainties in deducing the consequential climatic effects. Agriculture, forestry, and land-use change account for about one-fifth of global GHG emissions (FAO, 2016) which is becoming a major factor in climate change.

Impact of Climate Change on Agriculture

The effect of climate change on agriculture is a wide area of concern. It includes the impact of climate change on soil, pest and disease incidence, crop productivity, etc. Soils are important to food security (Pimentel, 2006), and climate change has the potential to threaten food security through its effects on soil properties and processes (Brevik, et al., 2013).In past decades some expectations that increased atmospheric CO_2 would lead to increased plant productivity (Coughenour and Cheng, 1997) coupled with increased C sequestration by soil, which means there is increased plant growth. Organic matter is important for many soil properties, including structure formation and maintenance, water-holding capacity, cation exchange capacity, and for the supply of nutrients to the soil ecosystem (Brevik et al., 2013). Increased temperature is likely to hurt C allocation to the soil, leading to reductions in SOC and creating a positive-feedback in the global C cycle (increased temperatures lead to increased CO_2 release from soils to the atmosphere, which leads to more increase in temperature) as global temperatures rise (Gorissen et al., 2004 and Wan et al., 2011). In a study of soils in a semi-arid steppe, it was observed (Link et al. 2003) that soil warming and drying led to a 32% reduction in soil cover over 5 years, a much more rapid reduction in soil C than reductions that have been observed due to increased tillage. Besides, the modeling of C responses to climate change in Canada predicted small increases in aboveground biomass in forest and tundra ecosystems but larger decreases in soil and litter pools, for an overall increase in atmospheric carbon (Price, et al., 1999). When CO_2 enrichment increases the soil C:N ratio, decomposing organisms in the soil need more N, which can reduce N mineralization because of this climate change and anthropogenic activities, many of our world's soils have become or are expected to become more susceptible to erosion by wind and/or water (Gill et al., 2002). In a study in Australia, the increased rainfall due to climate change could lead to significant increases in runoff, with amplification greater in arid areas (up to five times more runoff than the percentage increase in rainfall) than in wet and temperate areas (twice as much runoff as the percentage change in rainfall) (Chiew et al., 1995). Greater runoff would be expected to cause increased erosion. Water erosion models in the United Kingdom predicted that a 10% increase in winter rainfall could increase annual soil erosion by as much as 150% during wet years, but that long-term averages of soil erosion would show a modest increase over current conditions (Favis Mortlock and Boardman, 1995). Li et al., (2011) predicted changes in water erosion of 5% to 195% for conventional tillage and 26% to 77% for conservation tillage in China's Loess Plateau region, while Zhang et al., (2004) predicted increased erosion in Oklahoma, USA of 19% and 40% under conservation and conventional tillage, respectively by 2056–2085 as compared to average annual erosion in the period from 1950 to 1999. The linkage between crop yield and climate variability has been confirmed multiple times. Increases in atmospheric CO_2, heat waves, droughts, and flooding will likely hamper the yield potential of the crops. The changes in crop production related to climatic variables will possibly have major influences on regional as well as global food production (Abraha and Savage, 2006).

Indirect climate impacts include increased competition from weeds, expansion of pathogens and insect pest ranges, and other alterations in crop agroecosystems. With the changes in temperature and humidity levels, the populations of the insects may expand their geographic range, and expose animals and humans to diseases to which they have no natural immunity (Abraha and Savage, 2006). Insects are cold-blooded organisms; their body temperature is approximately the same as that of the environment. Therefore, temperature is probably the single most important environmental factor influencing insect behavior, distribution, development, survival, and reproduction. Insect life stage predictions are most often calculated using accumulated degree days from a base temperature and biofix point. Some researchers believe that the effect of temperature on insects largely overwhelms the effects of other environmental factors (Bale et al., 2002). The fossil records some researchers (Bale et al., 2002) conclude that the diversity of insect species and the intensity of their feeding have increased historically with increasing temperature. A growing number of pests and diseases could lead to higher and even unsafe levels of pesticide residue and veterinary drugs in local food supplies. The effect of climate change on agriculture or more precisely on insect pests and diseases of agricultural crops is multidimensional. The magnitude of this impact could vary with the type of species and their growth patterns. (Chattopadhyay et al., 2011). Some classes of pesticides (pyrethroids and spinosad) are less effective in controlling insects at higher temperatures (Musser and Shelton, 2005). Research has shown that host plants such as wheat and oats become more susceptible to rust diseases with increased temperature; but some forage species become more resistant to fungi with increased temperature (Coakley et al., 1999). Physiological changes in host plants may result in higher disease resistance under climate change scenarios, and host resistance to disease may be overcome more quickly by more rapid disease cycles, resulting in a greater chance of pathogens evolving to overcome host plant resistance. Fungicide and bactericide efficacy may change with increased CO_2, moisture, and temperature. Changes in temperature and precipitation regimes due to climate variation may alter the growth stage, development rate, and pathogenicity of infectious agents, and the physiology and resistance of the host plant (Mboup et. al., 2012). Harvell et al., (2002) considered the consequences of warmer temperatures on host-pathogen interactions and concluded that there will be three main effects: (1) increases in pathogen development rate, transmission, and generations per year, (2) increases in overwintering of pathogens, and (3) changes in host susceptibility to infection. Climate change is a threat to livestock production because of the impact on the quality of feed crops and forage, water availability, animal and milk production, livestock diseases, animal reproduction, and biodiversity. It has been reported that global demand for livestock products is expected to double by 2050, mainly due to improvements in the worldwide standard of living.

How Is Agriculture Becoming the Contributor to Climate Change?

Agriculture nowadays, has a significant share of the GHG emissions that are causing climate change; 17% directly through agricultural activities, and an additional 7% to 14% through land use changes (Xu et al., 2021). A study in Finland on the effect of crops on GHG fluxes from soils showed that the organic soils in their study were a net source of CO_2 for all cropping systems studied, with average annual emissions ranging from 1500 to 7500 kg CO_2 ha^{-1} (Martikainen, et al., 2002). Carbon is also naturally emitted from soils as CO_2 and CH_4 gases due to microbial respiration, with the form of the Carbon gas depending on the oxygen status of the soil system. Well-aerated soils are dominated by CO_2 emissions, while anaerobic conditions are associated with CH_4 generation. Agriculture accounts for about 47% of annual global anthropogenic emissions of CH_4, the main anthropogenic source of soil-derived methane is rice (*Oryza sativa*) production (Heilig et al., 1994). Different vegetation growing on the same soil will cause differences in CH_4 emission or consumption. In a study soil under forest vegetation might act as a net sink of CH_4 while the same soil in a nearby field planted with maize (*Zea mays*) was essentially CH_4 neutral and a third field of the same soil planted with grass cover was a net source of CH_4 to the atmosphere. Various methods of irrigation, tillage, and soil management lead to the production of N_2O, and the use of manure contributes to both CH_4 and N_2O emissions. Clearing space for agriculture (e.g., deforestation) is also a contributor to carbon emissions and land degradation. In a study, it was found that livestock is also becoming a major contributor to climate change. IPCC reported that agricultural land use contributes 12% of the global GHG emissions with deforestation at about 18%, and animal agriculture at 14%.

Mitigation Strategies for Climate Change

Within the vast expanse of scholarly discourse, one encounters a trinity of paramount methodologies when it comes to addressing the pressing issue of climate change mitigation. Initially, one must acknowledge the utilization of conventional mitigation endeavors, which encompass an array of decarbonization technologies and techniques. These endeavors are specifically designed to curtail the emission of CO_2, thereby mitigating its detrimental impact on the environment. Notable examples of such techniques include the adoption of renewable energy sources, the practice of fuel switching, the pursuit of efficiency gains, the integration of nuclear power, and the implementation of carbon capture, storage, and utilization methodologies. The vast majority of these technological solutions have been well-tested and have an acceptable degree of risk that has been handled (Bustreo, 2019). The following includes ways in order to mitigate climate change effects:

Carbon Sinks

A carbon sink refers to any entity that exhibits a net uptake of carbon from the atmosphere, surpassing its carbon emissions. Examples of carbon sinks include vegetation, ocean, and soil.

Agriculture: Agroforestry is a sustainable agricultural practice that entails the deliberate integration of trees or woody plants within agricultural systems, alongside crops or livestock. Trees function as carbon sinks by actively sequestering CO_2 from the atmosphere and subsequently storing carbon within their biomass and soil. Agroforestry systems have the potential to enhance carbon sequestration, alongside offering additional advantages such as enhanced soil fertility, biodiversity, and microclimate regulation.

Cover crops: Cover crops, also known as noncash crops, are primarily cultivated to provide soil coverage during periods when main crops are not actively growing. These ecological functions encompass the mitigation of soil erosion, enhancement of soil quality, and carbon sequestration facilitated by the proliferation of plant biomass. The carbon contained within cover crops is incorporated into the soil organic matter upon decomposition.

Traditional Tillage Practices

The potential to liberate carbon, which is stored in the soil, into the atmosphere in the form of carbon dioxide (CO_2). No-till and reduced tillage systems aim to minimize soil disturbance by retaining crop residues on the field. This practice aids in the preservation of soil structure and organic matter, thereby mitigating carbon loss and facilitating carbon sequestration.

a. **Residue management:** The practice of crop residue management involves the retention of crop residues on agricultural fields, as opposed to their burning or removal. This approach has been found to have a positive impact on the soil's organic carbon content. Over a period of time, the residues undergo decomposition, resulting in the addition of carbon to the soil and thereby enhancing its overall quality.

The Implementation of Rotational Grazing Systems

The strategic rotation of livestock across various grazing areas has been shown to enhance the overall well-being of grasslands and promote the sequestration of carbon. Effective management practices are essential in mitigating the negative impacts of overgrazing, such as soil degradation and carbon loss.

a. **Implementation of composting techniques**: The adoption of organic farming practices has the potential to enhance the levels of SOC content. Organic farming is characterized by the deliberate avoidance of synthetic fertilizers, instead emphasizing the adoption of practices that aim to improve soil fertility and facilitate carbon sequestration.

b. **Wetland Restoration**: Within certain agricultural landscapes, the implementation of wetland restoration initiatives can be undertaken to facilitate the capture of carbon. Wetlands serve as carbon sinks through the accumulation of organic matter and the facilitation of anaerobic conditions, which impede decomposition processes and foster carbon storage.

c. **Biochar:** Biochar is a certain type of charcoal that is made by heating organic materials while there isn't much oxygen around. The incorporation of biochar into soil has the potential to augment carbon sequestration by its durable carbon composition and capacity to enhance soil characteristics.

d. **Perennial crops:** It exemplified by specific varieties of grains, fruits, and vegetables, exhibit extended periods of growth, and possess more extensive root systems. These attributes enable them to effectively sequester a greater amount of carbon in the soil when compared to annual crops.

e. **Enhanced Nutrient Management**: The implementation of effective fertilizer utilization and nutrient management strategies has the potential to optimize plant growth while simultaneously mitigating the release of nitrous oxide, a highly potent GHG. The successful implementation of these practices necessitates a synergistic approach involving educational initiatives, policy frameworks, and financial incentives. The incorporation of carbon sinks within agricultural practices not only serves as a means of mitigating climate change but also holds promise for enhancing soil health, water retention, biodiversity, and overall sustainability within agricultural systems.

f. **Ocean:** The oceans play a critical role in regulating the global carbon cycle by absorbing and storing large amounts of carbon dioxide. However, it's important to note that human activities, such as the burning of fossil fuels, have led to an increase in atmospheric CO_2 levels, which has consequences for the oceans as well. Rapid carbon absorption by the oceans can lead to ocean acidification, which can harm marine ecosystems and disrupt the balance of ocean carbon sinks.

Carbon Sequestration Management

Crop rotation is a sustainable agricultural practice that involves growing different crops in the same field over a sequence of planting seasons. This practice offers several benefits for carbon sequestration and overall soil health, making it an important component of carbon sequestration management. The following points indicate how crop rotation contributes to carbon sequestration:

a. **Organic Matter Input:** Different crops have varying root structures, growth rates, and residues. By incorporating diverse crops into a rotation, a wider range of organic materials is added to the soil in the form of crop residues, root exudates, and other organic matter. This organic matter serves as a food source for soil microorganisms, promoting their activity and leading to the accumulation of organic carbon in the soil.

b. **Increased Root Biomass**: Different crops have different root systems that vary in depth, thickness, and chemical composition. Deep-rooted crops can penetrate deeper into the soil, bringing organic matter with them and increasing the potential for carbon storage at greater depths.

c. **Reduced Soil Erosion**: Crop rotation can include cover crops and plants with strong root systems that help prevent soil erosion. Reduced erosion means that the topsoil, which contains a significant amount of organic carbon, remains intact and is not lost to wind or water erosion.

d. **Enhanced Microbial Diversity**: Crop rotation can lead to a more diverse microbial community in the soil. Different crops release various types of root exudates and residues, which provide a range of nutrients for different microbes. This diversity enhances soil health and carbon cycling processes.

e. **Nitrogen Fixation**: Some crop rotation systems include leguminous plants that have the ability to fix atmospheric nitrogen into the soil. This nitrogen fixation not only improves soil fertility but also contributes to the buildup of organic matter in the form of nitrogen-rich residues.

f. **Reduced Soil Disturbance**: Crop rotation can reduce the need for intensive tillage, especially if cover crops are included in the rotation. Reduced tillage minimizes the disruption of soil structure and the exposure of organic matter to the atmosphere, which can lead to better carbon retention in the soil.

g. **Weed and Pest Management**: By rotating crops, farmers can disrupt the life cycles of pests and pathogens that may target specific crops. This can reduce the need for chemical pesticides, which can have environmental impacts. A healthier, biodiverse soil ecosystem resulting from crop rotation can also help naturally suppress some pests.

h. **Improved Water Management**: Some crops are more efficient at utilizing water than others. By rotating crops with different water requirements, farmers can better manage water resources and reduce the risk of soil degradation due to water stress.

Overall, crop rotation contributes to carbon sequestration by promoting the accumulation of organic matter in the soil, improving soil structure, enhancing microbial activity, and reducing soil erosion. These benefits not only mitigate climate change by sequestering carbon but also support sustainable agricultural practices by maintaining soil health and fertility.

Other Strategies

a. **Carbon Trading**: Carbon trading, also known as emissions trading or cap-and-trade is a market-based approach to mitigating climate change by reducing GHG emissions. It is a tool used to incentivize companies and nations to reduce their CO_2 and other GHG emissions cost-effectively. Here's how carbon trading works in the context of addressing climate change. Carbon trading is one of the market-based mechanisms that can contribute to achieving emissions reduction targets and addressing climate change. It has been implemented in various forms at regional, national, and international levels, and its effectiveness depends on factors such as the design of the trading system, the stringency of emission caps, and the level of transparency and oversight.

b. **Emissions Cap and Allowances**: A regulatory authority, often a government or an international body, sets an overall limit (or cap) on the total amount of GHG emissions that a particular sector, industry, or country can release into the atmosphere. This cap is usually based on established emission reduction targets.

c. **Allowance Allocation**: Under a carbon trading system, emissions allowances are distributed among participants. Each allowance typically represents the right to emit a specific amount of CO_2 or its equivalent. These allowances can be freely allocated to participants or auctioned off, generating revenue for the regulatory authority.

d. **Trading and Marketplaces**: Participants who emit more GHGs than their allocated allowances can purchase additional allowances from participants who have surplus allowances. This creates a marketplace for trading emissions allowances. The price of allowances is determined by supply and demand dynamics, and it can fluctuate over time.

e. **Incentive for Emission Reduction**: Companies that can reduce their emissions below their allocated allowances can sell their surplus allowances to others. This creates a financial incentive for companies to invest in emission reduction technologies and practices, as they can generate revenue by selling unused allowances.

f. **Flexible Compliance**: Carbon trading offers flexibility for companies to choose how they achieve emissions reductions. Some companies may find it more cost-effective to invest in emission reduction projects, while others might choose to purchase allowances to comply with the emission cap.

g. **Innovation and Technology Adoption**: Carbon trading encourages the development and adoption of cleaner technologies and practices. Companies seek innovative solutions to reduce emissions in order to lower their compliance costs and potentially profit from selling surplus allowances.

h. **Emission Reduction Transparency**: Companies participating in carbon trading must accurately measure, report, and verify their emissions. This transparency helps ensure that actual emissions reductions are achieved and that the cap is adhered to.

i. **International Cooperation**: Carbon trading can facilitate international cooperation in addressing climate change. Countries with binding emission reduction commitments can trade allowances with each other, allowing for more efficient and cost-effective emission reductions on a global scale.

j. **Carbon Offsets**: Some carbon trading systems allow for the use of carbon offsets, which are emissions reduction projects that occur outside the capped sector. These projects generate carbon credits, which can be used to offset emissions within the capped sector. Carbon offset projects can include reforestation, renewable energy projects, and methane capture from landfills.

k. **Complexities and Criticisms**: While carbon trading has the potential to drive emissions reductions, it also faces criticisms and challenges. These include concerns about the potential for market manipulation, the accuracy of emissions reporting, and the risk of allowing emissions to simply be "traded" without significant reductions occurring.

Conclusion

The increasing interest in the establishment and preservation of SOC contents and stocks in relation to soil health and carbon dioxide (CO_2) mitigation has captured the attention of a diverse range of stakeholders. This includes policymakers, nongovernmental organizations (NGOs), and land managers. In order to effectively promote the implementation of SOC sequestration practices in managed soils, it is imperative to incorporate methodologies that are both dependable and precise while also being economically viable. These methodologies should enable the quantification of alterations in SOC stocks and the anticipation of SOC reactions under varying management strategies, climatic patterns, and soil conditions. The sink potential of managed habitats, such as forests, farmland, and wetlands, can be enhanced through the adoption of prudent land use strategies and the implementation of suitable forestry, agricultural, and pasture management practices. The intentional manipulation of biological processes has the potential to expedite the process of CO_2 sequestration, enhance soil health, augment food production for the population, and promote the sustainability of land as a valuable resource. However, it is essential to adopt an integrated structural approach to mitigating climate change for management systems to achieve success.

References

Abraha, M.G. and Savage, M.J. Potential impacts of climate change on the grain yield of maize for the midlands of KwaZulu-Natal, South Africa. *Agricultural Ecosystem Environment.* 115, (2006): 150-60.

Bale, J.S., Masters, G.J., Hodkinson, I.D., Awmack, T.M., et al. Herbivory in global climate change research: Direct effects of rising temperatures on insect herbivores. *Global Change Biology.* 8, (2002): 1-16

Battisti, D. S., and R. L. Naylor. Historical warnings of future food insecurity with unprecedented seasonal heat. *Science.* 323, (2009): 240-44.

Brevik, E.C. Soils and Human Health—An Overview. In *Soils and Human Health*, pp. 29-56.

Bustreo, C. How fusion power can contribute to a fully decarbonized European power mix after 2050. *Fusion Eng. Des.* 146, (2019): 2189-193. DOI: https://doi.org/10.1016/j.fusengdes.2019.03.150

Coakley, S.M., Scherm, H. and Chakraborty, S. Climate Change and Disease Management. *Ann. Rev. Phyto.* 37, (1999): 399-426.

Chattopadhyay, C., Bhattacharya, B.K., Kumar, V., Kumar, V. and Meena, P. D. Impact of Climate Change on Pests and Diseases of Oilseeds Brassica - the Scenario Unfolding in India. *Journal of Oilseed Brassica.* 2, no. 2, (2011): 48-55.

Chiew, F.H.S., Whetton, P.H., McMahon, T.A. and Pittock, A.B. Simulation of the impacts of climate change on runoff and soil moisture in Australian catchments. *J. Hydrol.* 1, no. 167, (1995): 121-47.

Coughenour, M.B. and Chen, D.-X. Assessment of grassland ecosystem responses to atmospheric change using linked plant-soil process models. *Ecol. Appl.* 7, (1997): 802-27.

FAO. 2016. The agriculture sectors in the Intended Nationally Determined Contributions: Summary. Rome.

Favis-Mortlock, D. and Boardman, J. Nonlinear responses of soil erosion to climate change: A modeling study on the UK South Downs. *Catena.* 25, (1995): 365-87.

Frieler, K. Limiting global warming to 2 degrees Celsius is unlikely to save most coral reefs. *Nature Climate Change.* 3, (2013): 165-70. DOI: http://doi10.1038/Nclimate1674

IPCC (Intergovernmental Panel on Climatic Change) 2007. Climate Change: The Physical Science Basis. Extracts from the IV Assessment Report. Survey of the Environment 2007, The Hindu, pp147-55.

Gill, R.A., Polley, H.W., Johnson, H.B., Anderson, L.J., Maherali, H. and Jackson, R.B. Nonlinear grassland responses to past and future atmospheric CO2. *Nature*, 417, (2002): 279-82.

Gorissen, A., Tietema, A., Joosten, N.N., Estiarte, M., Peñuelas, J., Sowerby, A., Emmett, B.A., Beier, C. Climate change affects carbon allocation to the soil in shrublands. *Ecosystems.* 7, (2004): 650-61.

Harvell, C.D., Mitchell, C.E., Ward, J.R., Altizer, S., Dobson, A.P., Ostfeld, R.S. and Samuel, M.D. Climate warming and disease risks for terrestrial and marine biota. *Science.* 296, (2002): 2158-62.

Heilig, G.K. The greenhouse gas methane (CH4): Sources and sinks, the impact of population growth, possible interventions. *Popul. Environ.* 16, (1994): 109-37.

Li, Z., Lui, W.-Z., Zhang, X.-C. and Zheng, F.-L. Assessing the site-specific impacts of climate change on hydrology, soil erosion, and crop yields in the Loess Plateau of China. *Climate. Change.* 105, (2011): 223-42.

Link, S.O., Smith, J.L., Halverson, J.J., and Bolton, H., Jr. A reciprocal transplant experiment within a climatic gradient in a semiarid shrub-steppe ecosystem: Effects on bunchgrass growth and reproduction, soil carbon, and soil nitrogen. *Global. Change Biology.* 9, (2003): 1097-105.

Martikainen, P.J., Regina, K., Syväsalo, E., et al. 2002. Agricultural Soils as a Sink and Source of Greenhouse Gases: A Research Consortium (AGROGAS). In *Understanding the Global System, the Finnish Perspective*; Käyhkö, J., Talve, L., (eds.). Finnish Global Change Research Programme FIGARE: Turku, Finland, pp. 55-68.

Mboup, M., Bahri, B., Leconte, M., De Vallavieille-Pope, C., Kaltz, O. and Enjalbert, J. Genetic structure and local adaptation of European wheat yellow rust populations: The role of temperature-specific adaptation. *Evol. Appl.* 5, (2012): 341-52.

Musser, F. P. and Shelton, A. M. The influence of post-exposure temperature on thetoxicity of insecticides to *Ostrinianubilalis* (Lepidoptera: Crambidae). *Pest Management Science*. 61, (2005): 508-10.

Paerl, H. and Huisman, J. Climate change: A catalyst for global expansion of harmful cyanobacterial blooms. *Environmental Microbiology Reports*. 1, no. 1 (2009): 1-95.

Paterson, R. & Lim, N. How will climate change affect mycotoxins in food? *Food Research International*, 43, (2010): 1902-14.

Pimentel, D. Soil erosion: A food and environmental threat. *Environ. Dev. Sustain. 8*, (2006): 119-37.

Price, D.T., Peng, C.H., Apps, M.J., Halliwell, D.H. Simulating effects of climate change on boreal ecosystem carbon pools in central Canada. *Journal of Biogeography.* 26, (1999): 1237-48.

Robinson, A., Calov, R. and Ganopolski, A. Multistability and critical thresholds of the Greenland ice sheet. *Nature Climate Change,* 2, (2012): 429-32. DOI: http://doi:10.1038/nclimate1449

Wijesinha, B. R., Kennedy, G., Dirorimwe, C. and Muehlhoff, E. Considering seasonal variations in food availability and caring capacity when planning complementary feeding interventions in developing countries. *International Journal of Child Health and Nutrition,* 2, no. 4, (2013): 335-52.

Wan, Y., Lin, E., Xiong, W., Li, Y., and Guo, L. Modeling the impact of climate change on soil organic carbon stock in upland soils in the 21st century in China. *Agricultural. Ecosystem. Environment.* 141, (2011): 23-31.

Xu, X., Sharma, P., Shu, S., Lin, et al. Global greenhouse gas emissions from animal-based foods are twice those of plant-based foods. *Nature Food.* 2, no. 9, (2021): 724-32.

Zhang, X.C., Nearing, M.A., Garbrecht, J.D. and Steiner, J.L. Downscaling monthly forecasts to simulate impacts of climate change on soil erosion and wheat production. *Soil Science. Society. American. Journal.* 68, no. 9, (2004): 1376-85.

21

Harmonizing Nature's Symphony Unleashing Agricultural Potential through Livestock-Crop Integration

Abhijit Das[1], Sahin Aktar Munshi[2], and Bikram Barman[3]

[1]*Department of Agricultural Economics and Extension, School of Agriculture, Lovely Professional University, Phagwara, Punjab, India*

[2]*ICAR-Indian Agricultural Research Institute, New Delhi, Delhi, India*

[3]*ICAR-Indian Agricultural Research Institute, New Delhi, Delhi, India*

Introduction

In the face of a growing global population and increasing pressures on natural resources, the quest for sustainable and efficient agricultural practices has become highly significant. The integration of livestock with cropping systems and grasslands stands out as a holistic approach that not only addresses these pressing challenges but also unlocks a myriad of economic benefits. This chapter delves into the profound significance of integrating livestock with cropping systems and grasslands in modern agriculture, exploring how this synergistic combination enhances resource utilization, promotes resilience, and fosters a thriving agricultural sector.

The roots of integrated farming practices can be traced back to ancient agricultural traditions, where farmers instinctively recognized the advantages of combining crop cultivation and animal husbandry (Devendra and Thomas, 2002). Over time, this integrated approach evolved and adapted to different cultural and geographical contexts. In traditional mixed farming systems, livestock provided essential services such as draft power (Gomiero et al.,

2011), manure for fertilization, and protein-rich food. As agriculture progressed with technological advancements, there was a shift toward specialized monoculture and intensive livestock farming (Herrero et al., 2013). However, the unintended consequences of these systems, including soil degradation, environmental pollution, and declining rural livelihoods, paved the way for a resurgence of interest in integrated farming methods (FAO, 2020).

This chapter aims to shed light on the economic significance of integrating livestock with cropping systems and grasslands in the context of modern agriculture. By delving into the various aspects of this approach, the chapter aims to:

1. Illuminate the advantages of adopting integrated farming practices in optimizing resource utilization and improving overall farm productivity.
2. Examine the historical context and evolutionary trajectory of integrated farming, highlighting lessons learned from past practices and how they inform contemporary agricultural models.
3. Explore the environmental advantages of such systems, showcasing their potential to reduce the carbon footprint, conserve biodiversity, and promote sustainable land management.

By interpreting the significance of this integrated approach, the chapter aims to inform policymakers, agricultural practitioners, researchers, and consumers about the potential of livestock-crop integration to foster sustainable agriculture, ensuring food security and environmental stewardship for generations to come.

Livestock-Crop Integration: Enhancing Agriculture through Synergy

Concept Of Integrating Livestock with Cropping Systems and Grasslands

Livestock-crop integration, often referred to as mixed farming or integrated farming, is a holistic agricultural approach that combines crop production with livestock rearing (Devendra and Thomas, 2002), creating a synergistic and sustainable system. This concept recognizes the inherent interdependence between plants and animals, where each component contributes to the overall productivity and resilience of the farm. In integrated systems (Herrero et al., 2013), the outputs and by-products of one enterprise become valuable inputs for the other, leading to improved resource efficiency and waste reduction (Gomiero et al., 2011).

Various Forms of Integration

Mixed Farming

Mixed farming represents the most traditional and straightforward form of livestock-crop integration. In this system, farmers grow crops and raise livestock on the same piece of land. The crop residues and by-products, such as straw and husks, serve as feed for animals, while the animal manure enriches the soil, benefiting subsequent crop growth (Kremen and Miles, 2012). The integration of crops and livestock closely optimizes resource utilization and minimizes waste (Herrero et al., 2009).

Agro-Pastoral Systems

Agro-pastoral systems involve the combination of crop cultivation and grazing livestock. While the crops are grown in separate fields, the livestock are allowed to graze on the stubble after harvest or during fallow periods (Lin et al., 2015). This integration ensures efficient utilization of grasslands and crop residues while facilitating natural weed and pest control by grazing animals.

Silvopastoral Systems

Silvopastoral systems take integration to another level by incorporating trees, either fruit-bearing or timber, into the mix. Livestock graze under the shade of trees, benefiting from the forage provided by grasses and tree leaves. In return, the animals help fertilize the soil and control undergrowth, promoting healthy tree growth. The presence of trees also mitigates soil erosion and contributes to biodiversity conservation.

Benefits of Integration

Enhanced Resource Utilization

Integrating livestock with cropping systems and grasslands optimizes the utilization of resources on the farm. Crop residues, which were once considered waste, become valuable feed for livestock (Kremen and Miles, 2012), reducing the need for additional animal feed imports. Livestock plays a vital role in nutrient recycling by converting organic waste into fertilizer, enriching the soil, and benefiting future crops (Lin et al., 2015). This efficient use of resources leads to cost savings for farmers.

Waste Recycling

Livestock-crop integration allows for effective waste recycling on the farm. The manure produced by livestock is utilized as organic fertilizer, enriching the soil with essential nutrients (Herrero et al., 2013). This reduces the need for synthetic fertilizers, which can be costly and environmentally harmful. The integration of grazing animals in agro-pastoral and silvopastoral systems helps manage grasslands and crop residues, turning potential waste into valuable forage (Herrero et al., 2013).

Risk Mitigation

Integrated farming systems offer a degree of risk mitigation by diversifying income streams for farmers. The combination of crop and livestock enterprises provides a buffer against market fluctuations (Lin et al., 2015). If one enterprise faces challenges due to adverse weather or market conditions, the other may continue to generate income, ensuring greater financial stability for the farm.

Resource Utilization in Integrating Crop-Livestock Farming

Optimizing the Use of Land, Water, and Nutrients

The integration of crop and livestock farming offers a strategic approach to optimize the use of crucial resources such as land, water, and nutrients (Valbuena et al., 2012). By combining these two components (Rufino et al., 2007), farmers can achieve greater efficiency and sustainability in agricultural practices.

Land utilization: Integrated farming allows for complementary land use, where crop fields can be rotated with grazing pastures (Chivenge et al., 2015). Crop residues and by-products can be fed to livestock, efficiently utilizing land that would otherwise lie fallow. The integration of pasture-based grazing systems also enhances the utilization of grasslands, promoting sustainable forage production.

Water utilization: Livestock-crop integration reduces water wastage by ensuring that water is used efficiently for both purposes (Chivenge et al., 2015). For instance, water used for cleaning livestock areas can be repurposed for irrigation in crop fields, minimizing water losses and improving overall water use efficiency on the farm.

Nutrient utilization: Livestock play a crucial role in the nutrient cycle within integrated farming systems. Animal manure, rich in nitrogen, phosphorus, and other essential elements, can be utilized as organic fertilizer for crops

(Valbuena et al., 2012). This reduces the dependence on synthetic fertilizers and prevents nutrient imbalances in the soil, ultimately leading to improved crop growth and yields (Rufino et al., 2007).

Nutrient Cycling and Its Impact on Soil Fertility

Nutrient cycling is a fundamental ecological process in integrated crop-livestock systems that influences soil fertility positively. This process involves the continuous flow of nutrients through different components of the farm ecosystem (Valbuena et al., 2012), ensuring their efficient use and recycling.

Organic matter decomposition: Livestock contribute to nutrient cycling by converting plant residues (Chivenge et al., 2015), crop by-products, and animal waste into organic matter. As these organic materials decompose, they release nutrients back into the soil, enriching it with essential elements required for healthy plant growth.

Manure as fertilizer: Animal manure is a valuable source of organic nutrients, including nitrogen, phosphorus, and potassium. When applied to croplands, it replenishes nutrient levels in the soil, promoting nutrient availability to plants. Additionally, organic matter in manure improves soil structure and water retention, further enhancing soil fertility (Rufino et al., 2007).

Reduced nutrient leaching: Nutrient cycling in integrated systems reduces the risk of nutrient leaching into water bodies. As nutrients are efficiently utilized by crops and pastures, there is less excess runoff (Rufino et al., 2007), which can otherwise lead to water pollution and eutrophication in lakes and rivers.

The Role of Livestock in Converting Crop Residues and By-Products

Livestock plays a crucial role in transforming crop residues and by-products into valuable resources through various processes:

Crop residue consumption: Livestock, such as cattle and sheep, can graze on crop residues after harvest, utilizing plant materials that would otherwise go to waste (Valbuena et al., 2012). This not only provides a cost-effective feed source but also helps manage residue disposal and pest control.

By-product utilization: Crop by-products, such as straw and husks, can be fed to livestock as supplementary feed, converting them into high-quality animal products like meat and milk. This adds value to the by-products and reduces the need for additional feed purchases (Rufino et al., 2007).

Manure production: Livestock consume crops and forage and produce nutrient-rich manure (Chivenge et al., 2015). This organic waste serves as an essential resource for fertilizing crops and enriching the soil, thus completing the nutrient cycle within the integrated farming system.

Environmental Benefits of Integrating Crop-Livestock Farming

Environmental Advantages of Integrated Systems

Integrating crop and livestock farming offers significant environmental advantages that contribute to sustainable land management and reduced ecological impacts. By harmonizing these two components (Gomiero et al., 2011), farmers can mitigate greenhouse gas emissions and soil erosion, leading to a more environmentally friendly agricultural system.

Reduced greenhouse gas emissions: Integrated farming systems are associated with lower greenhouse gas emissions compared to conventional monoculture or intensive livestock farming. The adoption of practices such as rotational grazing and manure management in integrated systems reduces methane and nitrous oxide emissions, both potent greenhouse gases.

Data from a meta-analysis published in the journal *Global Change Biology* indicates that integrated crop-livestock systems can reduce greenhouse gas emissions by up to 30% when compared to conventional intensive livestock systems (Herrero et al., 2013).

Soil erosion control: The combination of crop fields and pastures in integrated systems helps mitigate soil erosion. The presence of perennial vegetation, such as grasses in pastures and cover crops in rotation with cash crops, stabilizes the soil, reducing the risk of erosion caused by wind and water. This leads to better soil health and fertility over the long term (Gomiero et al., 2011).

Potential for Biodiversity Conservation and Habitat Creation

Biodiversity conservation: Integrated crop-livestock systems promote biodiversity by creating diverse habitats for various plant and animal species. The integration of pastures, trees, and crops provides a variety of niches that support a wide range of flora and fauna. Integrated crop-livestock systems supported higher biodiversity, including beneficial insects, birds, and soil organisms, compared to monoculture systems (Herrero et al., 2013).

Habitat creation: The presence of diverse vegetation in integrated systems provides habitats for wildlife and beneficial insects. Trees, hedgerows, and grasslands in agroforestry and silvopastoral systems offer nesting sites and food sources for birds and small mammals. This habitat creation contributes to the conservation of native species and helps restore ecological balance in agricultural landscapes. Research conducted by researchers at the University of California, Davis, demonstrated that the inclusion of trees in integrated systems increased bird diversity (Lin et al., 2015) and abundance, positively impacting pest control and ecosystem services.

Challenges and Limitations in Adopting Integrated Farming Practices

Potential Obstacles and Constraints

Knowledge and education: One of the primary challenges in adopting integrated farming practices is the lack of knowledge and awareness among farmers. Traditional farming methods may have been ingrained in their practices for generations, making it difficult to transition to a new and integrated approach. Proper education, training, and extension services are crucial to familiarize farmers with the benefits and techniques of integrated farming.

Investment costs: Integrating livestock with cropping systems may require initial investment costs, such as setting up infrastructure for livestock management, implementing rotational grazing, and adopting agroforestry practices. Small-scale farmers, in particular, may find these upfront costs challenging to bear.

Market access and demand: Farmers adopting integrated farming practices may face challenges in accessing markets for their specialized products. The demand for organic produce and grass-fed meat may be limited in certain regions, leading to uncertainties in marketing and pricing.

Land availability and size: The availability and size of land can be a constraint for farmers interested in integrating crop and livestock enterprises. Small landholdings might limit the scope for rotational grazing or setting up dedicated pastures for livestock.

Ways to Overcome These Challenges and Promote Wider Adoption

Capacity building and training: Government agencies, NGOs, and agricultural institutions can play a vital role in providing training and capacity-building programs for farmers interested in adopting integrated farming practices. These programs should focus on the benefits, technical know-how, and best practices of integrated systems.

Financial support and incentives: Governments can offer financial support and incentives, such as subsidies for initial infrastructure setup or grants for adopting sustainable practices. These incentives can alleviate the financial burden and encourage farmers to adopt integrated farming practices.

Market development and linkages: Creating strong market linkages for products from integrated farming systems is essential. Governments and industry associations can facilitate market access, connect farmers with buyers, and promote niche marketing to capitalize on the demand for sustainable and ethically produced products.

Knowledge sharing and networking: Establishing farmer networks and knowledge-sharing platforms can enable the exchange of experiences and best practices. Farmers can learn from successful adopters of integrated farming and gain insights into overcoming challenges.

Policy support: Governments should formulate supportive policies that promote integrated farming practices. This may include regulations on organic certification, support for sustainable agriculture, and the inclusion of integrated systems in agricultural extension programs.

Research and development: Investment in research and development can lead to the improvement of integrated farming practices, including the development of integrated crop-livestock models suited to specific agroecological regions. Research can also provide evidence-based data on the economic and environmental benefits of integrated farming, further encouraging wider adoption.

Conclusion

In the pursuit of feeding a growing global population sustainably, integrated farming stands as a beacon of hope, offering a pathway to balance overall prosperity with environmental stewardship. Governments, agricultural institutions, and farmers must collaborate to provide education, training, and financial support to encourage wider adoption of integrated farming practices. By doing so, we can forge a future where agriculture thrives in harmony with nature, ensuring food security, environmental resilience, and the well-being of generations to come. Embracing the importance of integrating livestock with cropping systems and grasslands is a collective responsibility, and its realization will pave the way for a more sustainable and prosperous agricultural landscape.

References

Chivenge, P., Mabhaudhi, T., Modi, A. T., and Mafongoya, P. The potential role of neglected and underutilised crop species as future crops under water scarce conditions in sub-Saharan Africa. *International Journal of Environmental Research and Public Health*, 12, no. 6, (2015): 5685-711. DOI: http://dx.doi.org/10.3390/ijerph120605685

Devendra, C., and Thomas, D. Crop–animal systems in Asia: Importance of livestock and characterisation of agro-ecological zones. *Agricultural Systems*, 71, no. (1-2), (2002): 5-15. DOI: https://doi.org/10.1016/S0308-521X(01)00032-4

Food and Agriculture Organization of the United Nations. 2020. World Livestock: Transforming the livestock sector through the Sustainable Development Goals. FAO.

Gomiero, T., Pimentel, D., and Paoletti, M. G. Environmental impact of different agricultural management practices: Conventional vs. organic agriculture. *Critical Reviews in Plant Sciences*, 30, no. (1-2), (2011): 95-124. DOI: https://doi.org/10.1080/07352689.2011.554355.

Herrero, M., Havlík, P., Valin, H., et al. Biomass use, production, feed efficiencies, and greenhouse gas emissions from global livestock systems. *Proceedings of the National Academy of Sciences*, 110, no. 52, (2013): 20888-93.

Herrero, M., Thornton, P. K., and Gerber, P. Livestock, livelihoods and the environment: Understanding the trade-offs. *Current Opinion in Environmental Sustainability*, 1, no. 2, (2009): 111-20. DOI: https://doi.org/10.1016/j.cosust.2009.10.003.

Kremen, C., and Miles, A. Ecosystem services in biologically diversified versus conventional farming systems: Benefits, externalities, and trade-offs. *Ecology and Society*, 17, no. 4, (2012): 40. DOI: https://doi.org/10.5751/es-05035-170440.

Lin, B. B., Philpott, S. M., and Jha, S. The Future of Urban Agriculture and Biodiversity-Ecosystem Services: Challenges and Next Steps. *Basic and Applied Ecology*, 16, (2015): 189-201. DOI: http://dx.doi.org/10.1016/j.baae.2015.01.005.

Rufino, M. C., Tittonell, P., van Wijk, M. T., Castellanos-Navarrete, A., Delve, R. J., de Ridder, N., and Giller, K. E. Manure as a key resource within smallholder farming systems: Analysing farm-scale nutrient cycling efficiencies with the NUANCES framework. *Livestock Science*, 112, no. 3 2007: 273-87. DOI: https://doi.org/10.1016/j.livsci.2007.09.011

Valbuena, D., Erenstein, O., Homann-Kee Tui, S., Abdoulaye, T., Claessens, L., Duncan, A. J., and van Rooyen, A. F. Conservation agriculture in mixed crop–livestock systems: Scoping crop residue trade-offs in Sub-Saharan Africa and South Asia. *Field Crops Research*, 132, (2012): 175-84. DOI: http://dx.doi.org/10.1016/j.fcr.2012.02.022.

22

Recycling Wastes to Ameliorate Soil Biology

Anushka Mishra[1], Princy Thakur[2], Meraj Ahmed[3], Neeraj Singh Parihar[4], and Raghu Nandan Singh Khatana[5]

[1,2]Department of Soil Science and Agricultural Chemistry, Lovely Professional University, Phagwara, Punjab, India

[2–4]Department of Soil Science and Agricultural Chemistry, Lovely Professional University, Phagwara, Punjab, India

[5]Department of Soil Science, School of Agriculture, Suresh Gyan Vihar University, Jaipur, Rajasthan, India

Introduction

The process of recycling involves the conversion of waste materials into reusable products or materials and aims to reduce the consumption of fresh raw materials, minimize energy use, and mitigate trash creation. The primary objective of recycling is to effectively preserve finite resources, safeguard the natural environment, and mitigate the volume of garbage deposited in landfills or subjected to incineration (Lienig et al., 2017). Recycling refers to the systematic transformation of discarded materials into fresh resources and tangible products. The aforementioned term often encompasses the process of extracting energy from waste materials. The recyclability of a material is contingent upon its capacity to regain the characteristics it has in its initial form. There exists an alternative approach to garbage disposal that has the potential to save resources and mitigate the release of greenhouse gases. Additionally, it can mitigate the waste of potentially valuable resources and diminish the use of pristine raw materials, therefore reducing energy consumption, air pollution resulting from incineration, and water pollution stemming from landfill practices. The promotion of environmental sustainability is achieved by the removal of raw material input and the redirection of waste output within the economic system. The increasing volumes of trash have given rise to several environmental challenges, including air pollution, water pollution, soil contamination, and biodiversity loss. The emission of diverse greenhouse gases from solid waste landfills also contributes to the phenomenon of global warming. At now, the predominant techniques for solid waste disposal are open dumping and landfilling. Various alternative techniques, such as pyrolysis, gasification, anaerobic digestion, composting, and bio-methylation, are used to extract valuable resources from solid waste materials (Griffin et al., 2018).

Prior to exploring the recycling of garbage, it is essential to have an understanding of the complex soil ecology that resides under the Earth's surface. The soil is a multifaceted composition of many species such as bacteria, fungi, protozoa, nematodes, and earthworms. Each of these organisms fulfills essential functions in processes such as nutrient cycling, decomposition, and enhancement of soil structure. The combined efforts of those involved contribute to the creation of a conducive ecosystem that facilitates the development of plants and the maintenance of life. The flora and fauna present on the planet Earth have played a crucial role in the provision of sustenance, the enhancement of well-being, and the maintenance of ecological balance. The soil plays a crucial role in facilitating plant development by providing a habitat for essential organisms. Soil microbes have a significant role in ecosystem dynamics through their contributions to plant nutrition, plant health, soil structure, and soil fertility. Soil microorganisms play a crucial role in the biogeochemical cycles. Microorganisms stimulate plant development via the enhancement of nutrient accessibility (Meena and Meena 2017).

Agricultural waste refers to the byproducts created as a result of various agricultural activities. The sources of organic waste in agricultural systems include a variety of components, including manure and other byproducts generated by farms, poultry houses, and slaughterhouses. Additionally, harvest waste, fertilizer run-off from fields, pesticides that infiltrate water, air, or soils, and the drainage of salt and silt from fields contribute to the waste pool (Figure 22.1 and Table 22.1 (De Corato et al., 2018; Esparza et al., 2020). The composting process of recycling biomass waste can result in either the rationalization and reduction of residual biomass or a source of humified soil organic matter (SOM), minerals, and beneficial microbial communities that are associated with soil and plants. These resources are especially valuable for crops that require improved nutrition, vegetative vigor, health, and productivity (Zaccardelli et al., 2010).

The production of organic waste has emerged as a significant environmental issue, prompting widespread worldwide concern. These issues may be effectively mitigated by the implementation of sustainable solutions that include the recycling and composting of organic waste materials, such as vermicompost. Earthworms are capable of producing vermicast through the decomposition of biodegradable materials. The term "vermicomposting" was introduced to encompass the entire process of vermicompost production. This process enhances the bioavailability of essential nutrients by facilitating the degradation and digestion of raw materials. Vermicast also contains substances that promote plant growth, including ethylene, auxin, and gibberellin, as well as enzymes such as cellulose, nitrogenase, and phosphatase (Aslam and Ahmad, 2020; Aslam et al., 2020). The agricultural industry is responsible for the production of around 23.7 million metric tons of food daily worldwide. This sector plays a significant role in contributing to greenhouse gas emissions, accounting for almost 21% of the total emissions (Gerber et al., 2013). The rise in agricultural output is exerting an influence on the environment, resulting in adverse effects on soil, air, and water resources. In contemporary times, a pressing worldwide concern is the need to mitigate environmental damage. Nowadays, a pressing worldwide concern revolves around the imperative to mitigate environmental degradation. This necessitates the adoption of more ambitious and expeditious methods to effectively accomplish the specified targets during the upcoming 10 years (Global Sustainable Development Report, 2019) published by the United Nations.

Emerging technologies use formulations consisting of microbial consortia, minerals, and organic compost-based chemicals. Among the several options available, it is recommended to use the humic fractions derived from compost for two purposes: Either as unrefined compost teas (CTs) or as customized combinations of humic acids, fulvic acids, and the humus-like fraction known as humin. In recent years, there has been a growing interest in the use of soluble organic molecules and mineral nutrients in the phytosanitary market. These substances have been found to have significant impacts on various metabolic processes of plants, owing to their unique molecular structures (Khaled and Fawy, 2011; Morard et al., 2011). Significant progress has been made in the management of agricultural biomass waste on-site by using a differentiated sorting approach for food waste and municipal garbage. This approach aims to separate the various organic components, such as green/brown, wet/dry, solid/liquid, and so on. The use of biomass wastes in local production cycles via landfilling, on-farm composting for compost production and liquid derivatives (such as CTs and mixes of humic acids), and anaerobic digestion for the generation of biogas, biomethane, and digestate, poses a novel challenge. According to De Corato et al. (2018), on-farm composting offers a range of environmental advantages in comparison to landfilling and anaerobic digestion. These benefits include reduced greenhouse gas emissions and decreased leachate formation.

Table 22.1: Classification of wastes

Type of wastes	Source of wastes	Treatment procedure	Final product (applied to soil)	References
Organic agro-waste	Agriculture and agro-produce-based industries	Composting, vermicomposting anaerobic digestion, thermal treatment	Compost, vermicompost, biogas plant slurry, biochar	Hossain et al., (2018), Baiyeri*et al.* (2019)
Inorganic industrial wastes	Thermal power plants and steel industries, construction site waste	Direct application	Fly ash, bottom ash, steel slag	Antonkiewicz et al. (2020)
Mineral inorganic waste	Coal mining waste, metallurgical waste, slaughter-house waste	Chemical treatment, Pulverization	Soil additives, bone meal	Sisay (2019)
Nonhazardous inorganic waste	Glass, lime, fertilizer, glass, ceramic, and brickkiln waste	Pulverization	Waste mica, gypsum	Biswas et al. (2018), Basak (2019)
Hazardous waste	Metallurgical, tannery, and galvanizing industrial waste	Chelation to remove heavy metal	Inorganic complex	Ayilara et al. (2020)

Source: Meraj

Agricultural Wastes

Agricultural wastes include many forms of organic matter generated from agricultural activities, such as crop leftovers, weeds, leaf litter, sawdust,

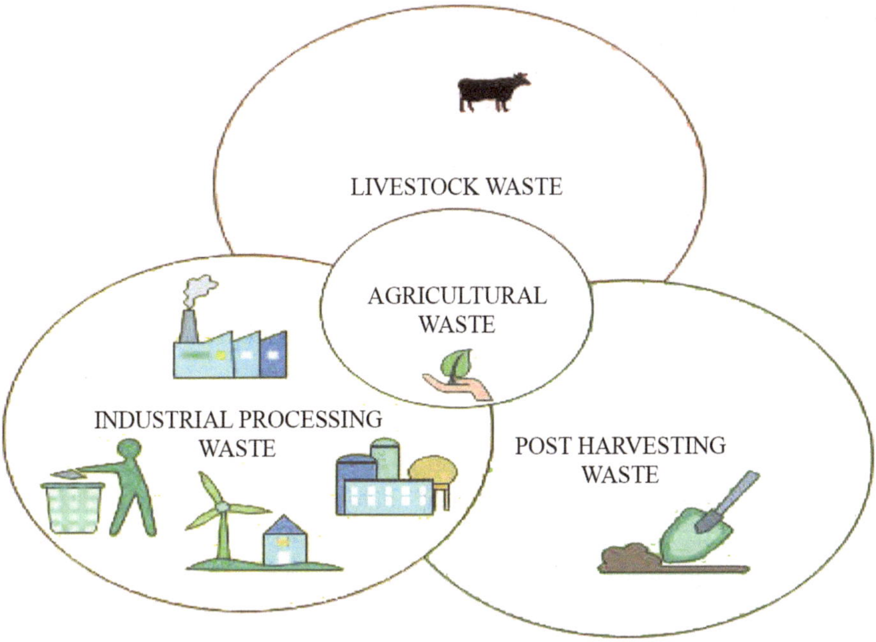

Figure 22.1: Types of wastes.
Source: Meraj

forest trash, and cattle manure. Livestock manure is consistently favored by researchers as a feedstock for earthworms and as a bulking substrate for vermicomposting, among the diverse range of agricultural wastes available. Livestock manure is widely regarded as a viable organic supplement for improving the vermicomposting process due to its affordability, ready accessibility, adequate nutritional composition, and optimal carbon-to-nitrogen ratio. Crop leftovers, leaf litter, and cattle manure are also used in composting and mulching processes.

Municipal Solid Wastes

Waste materials originating from residential households mostly consist of kitchen trash, paper waste, and food waste, ultimately culminating in the formation of municipal solid waste (MSW). Vermicomposting may be a very successful method for managing household trash due to its ability to efficiently decompose easily biodegradable materials that are often non-toxic and contain a substantial amount of nutrients.

Industrial Wastes

Industrial wastes are generated by several sources such as sugar mills, textile industries, food processing facilities, pulp and paper mills, tanneries, palm oil mills, and milk processing units. The results of many experiments consistently demonstrate that vermicompost obtained from various industrial waste sources displays pathogen-free characteristics, contains substantial amounts of plant nutrients, and may be efficiently used as an organic fertilizer. Vermicomposting has the potential to efficiently eliminate heavy metals from industrial waste.

Mechanisms of Recycling of Wastes

Ex-situ *recycling*

Composting: The composting process has two distinct stages, namely the breakdown phase and the building-up phase. Breakdown refers to the process of decomposing biodegradable compounds, such as organic waste, into degraded components that exist in simpler forms, such as amino acids. These deteriorated substances further transform into nitrates and nitrites. The building-up phase involves the development of intricate humic acids, which ultimately transform into a refined humus substance inside the composting material during the final phases. The aforementioned activities are executed by aerobic microflora, such as actinomycetes, Azotobacter, and Nitrosomonas (Meena et al., 2018). Composting, also known as bio-composting, is a biologically driven process that is both natural and environmentally sustainable. Through the activity of indigenous microorganisms that inhabit organic feedstocks, it facilitates the conversion of highly biodegradable organic compounds into stable organic matter. The bio-oxidative transformation occurring in the solid phase is contingent upon adequate oxygenation of the biomass, achieved by organizing it into piles or windrows and subjecting it to forced air. This process is facilitated by the intricate interactions between the substrate and bacteria. The generation of thermal energy inside the central mass is a crucial aspect of the composting process, especially in relation to the sanitization achieved by pasteurization. This thermal energy serves as the driving force for subsequent stages of the composting process. The first stage, referred to as the "thermophilic" phase, is marked by a rise in temperature above 60°C and reaching levels as high as 75–78°C for a minimum duration of 5–6 consecutive days. This process enables the comprehensive sterilization of biomass by effectively removing or deactivating seeds and pathogenic microorganisms that pose risks to human, animal, and plant health. The second phase, known as "mesophilic," is characterized by a gradual decrease in temperature, during which a majority of the carbon molecules become more susceptible to microbial degradation

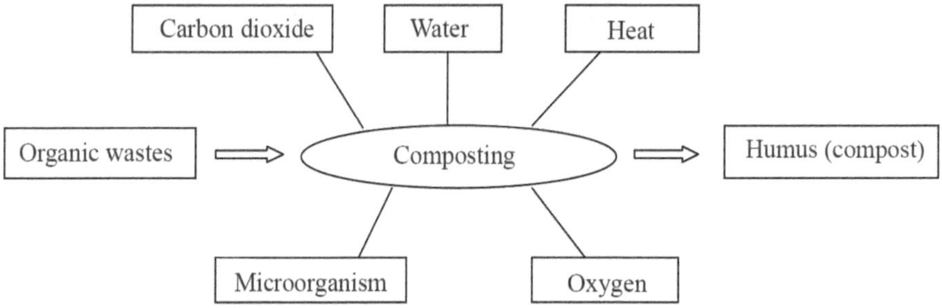

Figure 22.2: Schematic of composting.
Source: Sartore et al., 2018

inside the windrow (Niladri et al., 2019). The perceived challenge at hand may be effectively overcome by using on-farm composting techniques, which are increasingly being used in organic farming systems globally. In addition to industrial composting systems, there are on-farm technologies that provide safer and more sustainable methods for producing high-quality compost. These technologies are particularly successful in utilizing agricultural wastes and associated co/by-products within their original location (Scotti et al., 2016). The utilization of on-farm composting technology allows for the recycling of raw agricultural wastes (Zaccardelli et al., 2010) and refined co/by-products obtained from biofuel production chains (De Corato et al., 2015). This practice addresses the challenge of biomass disposal while simultaneously providing organic amendments that enhance soil quality and promote plant health. Compost may significantly contribute to the bioeconomy, promoting optimal environmental sustainability in organic farming systems via the conversion of agricultural waste into valuable resources. A notable illustration (Figure 22.2) is provided by the global dissemination of biodegradable plastics composed of starch-based polymeric composite material (Sartore et al., 2018).

Vermicompost: Vermicomposting is a very efficient process, surpassing traditional composting techniques in terms of speed, while the latter often needs a lengthier period of around 3 months. The process of converting organic waste yields a high-quality final product known as vermicompost. This vermicompost is composed of earthworm castings and nutrients that are enclosed by mucus, which is created during the composting process when the organic material is combined with fine soil particles within the digestive tracts of earthworms. Plants possess the capability to efficiently and gradually uptake macro and micronutrients from the soil, leading to the attainment of an optimal amount (Meena et al., 2015). Vermicomposting is a straightforward technique that has exceptional qualities, ensuring that there is no harm to plant life. The technique involves the use of earthworms to facilitate the decomposition of organic waste by vermicomposting, resulting in the stabilization of the waste in conjunction with microbes. Aerobic bacteria are responsible for the decomposition of organic waste, which then undergoes digestion by earthworms. Within the earthworm's digestive system, the organic waste is thoroughly mixed with coelomic fluid and mucus. Furthermore, the aforementioned substance undergoes processing by the microorganisms present in the gastrointestinal tract, hence further augmenting its antibacterial characteristics. The vermicast is combined with decomposed material that is devoid of any pathogens. The earliest step of composting involves the use of thermophilic bacteria in a process known as thermophilic digestion. The vermicompost demonstrates an augmented abundance of beneficial microorganisms, as shown by studies conducted by Ansari (2012); Meena et al. (2017), and Buragohain et al. (2017).

This product serves as a viable substitute for chemical fertilizers in agriculture, effectively enhancing soil health through the promotion of a robust soil microbial community. Additionally, vermicomposting facilitates the efficient recycling of minerals, thereby contributing to a safe environment and improved human health (Asma et al., 2016). At a significant magnitude, there is limited use of the aforementioned earthworms in the practice of vermicomposting (Datta et al., 2017), mostly due to variations in their capacity to decompose organic material. *Epigeic species*, such as *Eisenia foetida* and *Eudrilus eugeniae*, are often used in the vermicomposting process to facilitate the conversion of food waste, agricultural wastes, and manures into a very beneficial soil amendment known as vermicompost. The (C:N) ratio in vermicompost reaches a range of 14 to 20 at the stage of maturity, as reported by (Meena and Yadav, 2014). The physical, chemical, and biological qualities of soil are enhanced and restored when vermicompost is used as an organic amendment (Hussain and Abbasi 2018; Yadav et al. 2017). Vermicomposting may be done in several ways. The vermicomposting devices may be set up in tanks, boxes, containers, baskets, drums, and pits in the soil. The depth of the units should be less than 1 m or 1 f. These should be supplied with proper drainage at the bottom (with suitable holes). The other dimensions can vary depending on the appropriateness and amount of vermicompost output necessary.

In situ ***recycling:*** Green manures refer to crops, particularly forage or leguminous plants, cultivated primarily to obtain leafy components that contribute to soil conservation efforts. The practice of using green leaves and twigs obtained from trees, bushes, and herbs gathered from external sources is also referred to as green leaf manuring. The practice of cultivating green manure crops or using green leaves as a kind of organic fertilizer has been shown to have many beneficial effects on soil quality. These include improvements in soil structure, increased water retention capacity, and a reduction in soil erosion. It effectively mitigates the multiplication and development of unwanted vegetation. In addition, it also contributes to the reclamation of alkaline soils and serves to mitigate the impact of root-knot nematode infestations. Green manures provide significant promise as a beneficial supply of nitrogen and organic matter. In rice-based cropping systems, dhaincha (*Sesbania* sp.) is considered a particularly ideal green manure crop for regions with water logging and excessive rainfall. In contrast, sunhemp (*Crotalaria juncea*) exhibits suitability for rainfed regions. Green gram, black gram, fodder cowpea, and horse gram are often used as green manure crops in tropical climates. Niger and cockscomb have been chosen as green manure crops because of their high potash content. Typically, a green manure crop 45 to 60 days old can collect around 100 kg of nitrogen per hectare. This nitrogen accumulation is equivalent to the quantity of nitrogen that is typically provided to crops via the use of mineral fertilizers (Chatterjee et. al., 2017).

The application of mulch materials to the root rhizosphere has many benefits, including the suppression of weed development, enhancement of water infiltration, improvement of soil water retention, preservation of surface soil structure, promotion of drought tolerance, and protection against erosion and nutrient leaching. The use of biomulch expedited the process of agricultural waste decomposition and facilitated the cycling of nutrients. The mechanism of action involves the promotion of the inherent process of biological deterioration. The use of bio-mulches has the potential to enhance the levels of SOM, increase the capacity of soils to retain water and nutrients, particularly in soils that are prone to leaching, and stabilize the pH of the soil. It has the potential to serve as a source of both macro and micronutrients (Chatterjee et. al., 2017).

Effects of Recycled Organic Wastes on Crop Growth

Improved Soil Structure and Texture

The enhancement of soil structure via the use of recycled organic waste is achieved by enhancing the process of aggregation and increasing the ability to retain water. This, in turn, results in improved aeration and facilitates the penetration of roots. This facilitates the availability of nutrients and contributes to the general health of plants.

Enhanced Nutrient Availability

Organic waste has a significant abundance of essential elements such as nitrogen, phosphorus, and potassium, which are gradually liberated throughout the process of decomposition. This phenomenon facilitates a consistent supply of vital nutrients to crops, hence fostering enhanced and healthier development.

Increased Microbial Activity

The presence of organic waste in soil fosters the development of a wide range of microbial communities. These communities play a crucial role in several ecological processes, including nutrient cycling, disease control, and decomposition of organic materials. The microbial activity seen in this context is known to have a significant role in enhancing soil fertility.

Carbon Sequestration

The process of converting organic waste into soil serves as a means of carbon sequestration, so aiding in the mitigation of climate change via the retention of atmospheric carbon dioxide. Additionally, it contributes to the augmentation of soil organic carbon levels, hence positively impacting soil health and agricultural output.

Reduced Soil Erosion

The incorporation of organic waste into soil will have a positive impact on soil structure, leading to a reduction in erosion via the enhancement of soil particle binding. This practice serves to safeguard the uppermost layer of soil, so mitigating the potential loss of nutrients by surface water runoff.

Water Retention and Drought Resistance

Soils treated with organic waste have enhanced water-holding capacity, resulting in reduced irrigation requirements and improved resilience to drought in crops. This method is particularly advantageous in areas characterized by limited access to water resources.

Suppressed Plant Diseases

Several organic wastes include advantageous bacteria that may aid in the suppression of soil-borne pathogens, resulting in improved crop health and decreased occurrence of diseases.

Enhanced Biodiversity

The promotion of biodiversity in the soil ecosystem is facilitated by the use of recycled organic waste, hence fostering a harmonious and robust agricultural milieu. Enhanced biodiversity has the potential to positively impact pest control practices and enhance agricultural productivity.

Lower Input Costs

The integration of recycled organic wastes into agricultural systems results in the reduction of synthetic fertilizer usage and chemical inputs. Hence, it generates economic efficiencies for farmers and fosters the adoption of sustainable farming methodologies.

Improved Food Security

Recycled organic waste plays a significant role in bolstering food security via its potential to augment soil fertility, enhance nutrient availability, and elevate crop production. This practice effectively enhances agricultural productivity and fortifies resistance against fluctuating environmental circumstances (Eden, et al., 2017).

Impact of Recycling of Waste on the Soil Microbial Population

The soil microflora, which includes bacteria, fungi, mycorrhiza, and actinomycetes, as well as the soil fauna, may be categorized into three groups: Microfauna (such as Nematodes and protozoa), mesofauna (for example, ascarids and enchytraeus), and macrofauna (including earthworms, termites, and big arthropods). The classification of these organisms may also be characterized based on soil food webs, including microflora and microfauna responsible for the decomposition of organic waste, mesofauna that prey upon them, and macrofauna that feed on mesofauna (Roger et al., 2010). The combination of green manuring and agricultural residue has been shown to have a significant and favorable impact on the biological properties, namely in terms of dehydrogenase and alkaline phosphatase activities (Pathak et al., 2015). The dehydrogenase activity was enhanced by rice straw compost and the required dosage of inorganic fertilizer, as compared to the recommended quantity of fertilizer (Goyal, et al., 2009). The absorption of crop residues led to a significant increase in dehydrogenase activity in silty clay loam soil (Ramchandra, 2011) and the existence of greater functional diversity within the bacterial communities in the soil. The study's findings indicate that vermicomposting was successful in significantly improving the quality of distilled grape marc. As a result, the manufacturing process yielded a final product exhibiting enhanced bacterial diversity and functional attributes and being abundant in essential nutrients. Therefore, this final product may serve as an effective stimulant for plant growth and/or as a natural additive for soil enhancement (Gómez-Brandon et al., 2020).

Metabarcoding techniques to investigate the presence of bacteria from the phylum Bacteroidetes, as well as the classes Flavobacteria and Spingobacteriales, in soils supplemented with different types of compost. The results indicated that these bacterial groups exhibited significantly higher levels of abundance in soils supplemented with poultry litter compost (PLC) compared to those supplemented with dairy manure compost (DMC) or left unamended (Neher et al., 2020). Several bacterial genera, including *Cellulomonas*, *Solirubacter*, Alerer*y*throbacter, and *Massilia*, as possible bio-indicators indicating heightened aerobic decomposition. The family *Nitrosomonadaceae* was seen to be linked with the process of ammonia oxidation, whereas bacterial genera *Nitrospira* and *Reyranella* are connected with nitrite oxidation and nitrate reduction, respectively (Kim et al., 2020).

Future Aspects of Recycling Waste to Enhance Soil Biology (Table 22.2)

a. The potential future developments in soil recycling may prioritize the precise control of soil microorganisms. Gaining insight into the distinct functions performed by diverse microbial populations and their interrelationships has the potential to provide deliberate interventions aimed at achieving particular agricultural goal.

b. Synthetic biology techniques might potentially be utilized to manipulate microorganisms in a manner that promotes the improvement of nutrient cycling, disease resistance, and organic matter breakdown within soil systems, thereby leading to an overall enhancement of soil health.

c. The use of recycling has the potential to make a substantial contribution to bioremediation endeavors, whereby the utilization of specifically engineered flora or microorganisms aids in the remediation of contaminated soils via the degradation or immobilization of hazardous compounds. Phytomining and phytoremediation are two distinct processes that include the use of plants for various purposes. The use of plants for the purpose of extracting precious metals (known as phytomining) or eliminating pollutants (known as phytoremediation) from soil has the potential to advance via the identification and manipulation of hyper-accumulator plant species and their associated microorganisms.

d. The use of nanotechnology in the field of soil management. The use of nanoparticles has the potential to facilitate the delivery of nutrients, regulate water availability, and provide precise therapies to soil microorganisms, so fundamentally transforming our interactions with the soil ecosystem.

Table 22.2: Different types of compost feedstock used in recycling waste products and their effect on soil health

Compost feedstock	**Experimental conditions**	**Effect on soil health**	**Reference**
Green wastes	Compost was applied at three different rates: 5 kg/each tree, 10kg/each tree and 15 kg/each tree	The high level of compost application amount (15 kg/tree) had the greatest effect on soil improvement. Soil pH decreased to 7.28–7.45, soil organic matter content reached more than 35 g/kg. Soil total nitrogen, soil available phosphorus, and soil available potassium increased by 25–28%, 200–400%, and 80–177%, respectively. Additionally, the soil microbial structure was changed such that bacterial abundance increased by 12–13%.	Tong, et al., (2018)
Mix of food wastes, animal bedding and manure	Compost was incorporated in soil for 12 years (33% by volume) with the use of a backhoe, and annual top-dressing with mulch	Compared to unamended soils, improvements were found in aggregate stability (72.41%), available water-holding capacity (0.22%), total organic matter (8.43%), potentially mineralizable nitrogen (27.53 mg/kg), active carbon (1022.47 mg/kg), and reduction in bulk density (0.89 g/cm3).	Sax, et al., (2017)
Sheep manure and wheat straw	Continuous application of compost for 5 years in the proportion of 60:40 (volume basis)	The amended soil by compost increased the SOM and water-soluble C fraction in parallel with an increase in microbial parameters (microbial biomass C, adenosine triphosphate (ATP), basal respiration and dehydrogenase).	Jindo, et al., (2016)
Dairy waste	Dairy waste compost was applied at a rate of 100 T/ha for 5 years	The dairy waste compost increased organic carbon by 143 and 54% compared to ammonium sulfate and liquid dairy waste treatments, respectively, applied at the same available N level (200 kg N/ha), whereas the C pool was enhanced by 115%.	Habteselassie et al., (2006)
Digestates and compost	Digestate and compost were applied at a rate of 100 m^3/ha for four years	Increased pH of the soil and improved biological soil activity (e.g., enzymatic activities).	Fuchs, *et al.*, (2014)

Source: Meraj

e. Recent developments in sensor technology and data analytics have the potential to provide accurate and targeted application of recycled organic materials. This approach aims to align the specific requirements of the soil with suitable organic inputs, therefore, reducing unnecessary waste.

f. The phenomenon of soil-plant signaling refers to the communication and exchange of information between soil and plants. Enhanced comprehension of the complex communication mechanisms between plants and soil microorganisms has the potential to facilitate the development of methods aimed at improving nutrient absorption, stress resilience, and overall plant productivity.

g. The advancement of soil health sensors has the potential to enable farmers to make well-informed choices on recycling techniques and the optimization of organic matter inputs. These sensors would provide real-time monitoring of soil conditions, allowing for adjustments depending on the dynamic nature of the soil.

h. Circular agriculture models refer to a concept in which agricultural practices are designed to mimic natural ecosystems, creating a closed-loop system that minimizes waste and maximizes resource efficiency. This approach aims to move away from the linear. In the forthcoming recycling strategies, there is a potential to combine agricultural, animal, and food waste systems into circular frameworks, to enhance nutrient cycle efficiency and reduce adverse environmental consequences.

i. Climate-resilient soil practices are a set of techniques and strategies aimed at enhancing the ability of soil to withstand and adapt to the impacts of climate change. These practices are designed to improve soil health.

j. Recycling systems have the potential to respond to climate change by integrating heat- and drought-resistant microorganisms, plants, and organic substances to sustain soil quality amidst evolving environmental circumstances.

k. The potential future implications of incorporating recycling practices into soil biology provide promising prospects for enhancing agricultural sustainability, bolstering soil vitality, and safeguarding food security amidst the dynamic shifts occurring in our global landscape.

Conclusion

Recycling waste to ameliorate soil biology presents a win-win solution for both the environment and agriculture. By embracing composting, mulching, and vermicomposting, and exploring innovative techniques for nonorganic waste recycling, our soil ecosystems can be rejuvenated. Promoting these sustainable practices not only reduces waste pollution but also fosters healthier soil higher crop yields, and enhanced biodiversity. As stewards of the land, it is our responsibility to embrace recycling and embrace a harmonious relationship with the soil beneath our feet. Through such actions, we can pave the way for a greener and more sustainable future.

References

Ansari, A.A. 2012. Permutations and combinations of organic waste-vermitechnology. p. 76. Germany: Lambert Academic Publishing.

Antonkiewicz, J., Popławska, A., Kołodziej, B., Ciarkowska, K., Gambus, F., Bryk, M. and Babula, J. Application of ash and municipal sewage sludge as macronutrient sources in sustainable plant biomass production. *J Environ Manage* 264, (2020): 110450. DOI: https://doi.org/10.1016/j.jenvman.2020.11045

Aslam, Z. and Ahmad, A. Effects of vermicompost, vermi-tea and chemical fertilizer on morpho-physiological characteristics of maize (*Zea mays* L.) in Suleymanpasa District, Tekirdag of Turkey. *Journal of Innovative Sciences,* 6, no. 1, (2020): 41-46. DOI: https://doi.org/10.17582/journal.jis/2020/6.1.41.46

Aslam, Z., Ahmad, A., Idrees, M., et al. Comparative analysis of nutritional sources on the morphophysiological characteristics of mung bean(*Vigna radiata*). *International Journal of Food Science and Agriculture,* 4, no. 3, (2020): 314-22. DOI: https://doi.org/10.26855/ijfsa.2020.09.011

Asma, F., Ahmad, L. and Singh, P. Vermicomposting: An effective technique to recycle waste into valuable organic fertilizer: a review. *J Pure Appl Microbiol.* 10, no. 2, (2016): 1-7.

Ayilara, M.S., Olanrewaju, O.S., Babalola, O.O. and Odeyemi, O. Waste management through composting: Challenges and potentials. *Sustainability,* 12, no. 11, (2020): 4456.

Baiyeri, K.P., Chukwudi, U.P., Chizaram, C.A. and Aneke, N. Maximizing rice husk waste for Daucus carota production. *Int J Recycl Org Waste Agric.* 8, no. 1, (2019):399-406.

Basak, B.B. Waste mica as alternative source of plant-available potassium: Evaluation of agronomic potential through chemical and biological methods. *Nat Resour Res.* 28, (2019): 953-65. DOI: https://doi.org/10.1007/s11053-018-9430-3.

Biswas, D.R., Ghosh, A., Ramachandran, S., Basak, B.B. and Moharana, P.C. Dependence of thermaland moisture sensitivity of soil organic carbon decomposition on manure composition in an inceptisol under a 5-year-old maize-wheat cropping system. *J Geophys Res-Biogeo.* 123, (2018): 1637-50. DOI: https://doi.org/10.1029/2017JG004329.

Buragohain, S., Sharma, B., Nath, J.D., Gogaoi, N., Meena, R.S. and Lal, R. Impact of ten years of bio-fertilizer use on soil quality and rice yield on an inceptisol in Assam, India. *Soil Res.*, 2017.

Chatterjee, R., Gajjela, S. and Thirumdasu, K.R. Recycling of Organic Waste for Sustainable Soil Health and Crop Growth. *International Journal of Waste Resources.* 7, (2017): 3.

Datta, R., Baraniya, D., Wang, Y.F., Kelkar, A., Moulick, A., Meena, R.S., Yadav, G.S., Ceccherini, M.T. and Formanek, P. Multi-function role as nutrient and scavenger off reeradical in soil. *Sustain MDPI.* 9, (2017): 402. DOI: https://doi.org/10.3390/su9081402.

De Corato, U., De Bari, I., Viola, E. and Pugliese, M. Assessing the main opportunities of integrated biorefining from agro-bioenergy co/by-products and agro-industrial residues into high value-added products associated to some emerging markets: A review. *Renew Sustain Energy Rev*. 88, (2018): 326-46. DOI: http://doi.10.1016/j.rser.2018.02.041.

De Corato, U., Pane, C., Bruno, G.L., Cancellara, F.A. and Zaccardelli, M. Co-products from a biofuel production chain in crop disease management. *A review. Crop Prot*. 68, (2015): 12-26. DOI: http://doi.10.1016/j.cropro.2014.10.025.

Eden, M., Gerke, H. H. and Houot, S. Organic waste recycling in agriculture and related effects on soil water retention and plant available water: A review. *Agron. Sustain. Dev*. 37, (2017): 11. DOI: https://doi.org/10.1007/s13593-017-0419-9.

Esparza, I., Jiménez-Moreno, N., Bimbela, F., Ancín-Azpilicueta, C. and Gandía, L.M. Fruit and vegetable waste management: Conventional and emerging approaches. *J Environ Manag*. 265, (2020): 110510. DOI: http://doi.10.1016/j.jenvman.2020.110510.

Fuchs, J., Fliessbach, A., Mader, P., Weibel, F., Tamm, L., Mayer, J. and Schleiss, K. Effects of Compost on Soil Fertility Parameters in short-, mid-and long-term field experiments. *Acta Horticultiure.*1018 (2014): 39-46.

Gerber, P.J., Steinfeld, H., Henderson, B., Mottet, A., Opio, C., Dijkman, J., Falcucci, A. and Tempio, G. 2013. Tackling climate change through livestock – A global assessment of emissions and mitigation opportunities. Food and Agriculture Organization of the United Nations (FAO), Rome.

Global Sustainable Development Report. 2019. The Future is Now – Science for Achieving Sustainable Development. New York: United Nations.

Gómez-Brandon, M., Aira, M., Santana, N., Perez-Losada, M. and Dominguez. J. Temporal Dynamics of Bacterial Communities in a Pilot-Scale Vermireactor Fed with Distilled Grape Marc. *Microorganisms* 8, (2020): 642.

Goyal, S., Singh, D., Suneja, S. and Kapoor, K.K. Effect of rice straw compost on soil microbiological properties and yield of rice. *Indian Journal of Agricultural Research*. 43, no. 4, (2009): 263-68.

Griffin, S., Sarfraz, M., Farida, V. and Jawad, M. No time to waste organic waste: Nanosizing converts remains of food processing into refined materials. *Journal of Environmental Management*. 210, (2018): 114e21.

Habteselassie, M.Y., Miller, B.E., Thacker, S.G., Stark, J.M. and Norton, J.M. Soil Nitrogen and Nutrient Dynamics after Repeated Application of Treated Dairy-Waste. *Soil Sci. Soc. Am. J.* 70, (2006): 1328-37.

Hossain, S.S., Mathur, L. and Roy, P.K. Rice husk/rice husk ash as an alternative source of silica in ceramics: A review. *J Asian Ceramic Soc*. 6, no. 4, (2018): 299-13.

Hussain, N. and Abbasi, S.A. Efficacy of the vermicomposts of different organic wastes as "clean" fertilizers: State of the art. *Sustainability* 10, (2018): 1205.

Jindo, K., Chocano, C., De-Aguilar, J.M., González, D., Hernandez, T. and García, C. Impact of Compost Application during 5 Years on Crop Production, Soil Microbial Activity. Carbon Fraction, and Humification Process. *Commun. Soil Sci. Plant Anal*. 47, (2016): 1907-19.

Kim, N., Zabaloy, M.C., Riggins, C.W., Rodríguez-Zas, S. and Villamil, M.B. Microbial Shifts Following Five Years of Cover Cropping and Tillage Practices in Fertile Agroecosystems. *Microorganisms* 8, (2020): 1773.

Khaled, H. and Fawy, H.A. Effect of different levels of humic acids on the nutrient content, plant growth, and soil properties under conditions of salinity. *Soil Water Res*. 6, no. 1, (2011): 21–29. DOI: http://doi.10.17221/4/2010-SWR.

Lienig, J. and Bruemmer, H. Recycling Requirements and Design for Environmental Compliance. *Fundamentals of Electronic Systems Design,* (2017). Pp. 193-218. DOI: https://doi.org/10.1007/978-3-319-55840-0_7.

Meena, H. and Meena, R.S. Assessment of sowing environments and bio-regulators as adaptation choice for cluster bean productivity in response to current climatic scenario. *Bangladesh J Bot.* 46, no. 1, (2017): 241-44.

Meena, R.S., Kumar, V., Yadav, G.S. and Mitran, T. Response and interaction of *Bradyrhizobium japonicum* and *Arbuscular mycorrhizal* fungi in the soybean rhizosphere: A review. *Plant Growth Regul.* 84, (2018): 207-23.

Meena, R.S., Meena, P.D., Yadav, G.S. and Yadav, S.S. Phosphate solubilizing microorganisms, principles and application of microphos technology. *J Clean Prod* 145, (2017): 157-8.

Meena, R.S., Yadav, R.S., Meena, H., Kumar, S., Meena, Y.K. and Singh, A. Towards the current need to enhance legume productivity and soil sustainability worldwide: A book review. *J Clean Prod.* 104, (2015): 513-5.

Meena, R.S. and Yadav, R.S. Phonological performance of groundnut varieties under sowing environments in hyper arid zone of Rajasthan, *India. J Appl Nat Sci.* 6, no. 2 (2014): 344-8.

Morard, P., Eyheraguibel, B., Morard, M. and Silvestre, J. Direct effects of humic-like substance on growth, water, and mineral nutrition of various species. *J Plant Nutr.* 34, no. 1, (2011): 46-59. DOI: http://doi.10.1080/01904167.2011.531358.

Neher, D.A., Limoges, M.A., Weicht, T.R., Sharma, M., Millner, P.D. and Donnelly, C. Bacterial Community Dynamics Distinguish Poultry Compost from Dairy Compost and Non-Amended Soils Planted with Spinach. *Microorganisms* 8, (2020): 1601.

Niladri, P., Utpal, G. and Gourab, R. 2019. Composting. In: Organic Fertilizers–History, Production, and Applications. P 19. Edited by *IntechOpen*. DOI: http://doi.10.5772/intechopen.88753.

Pathak, D.V., Gaur, R.K., Yadav, P.K. and Yadav, S.S. Effect of green manuring and residue incorporation on soil properties and seed yield of rapeseed mustard. *International Journal of Farm Sciences*. 5, no. 4 (2015): 70-7.

Ramchandra Effect of summer crops and their residue management on yield of succeeding wheat and soil properties. *Journal of the Indian Society of Soil Science*. 59, no. 1, (2011): 37-42.

Roger-Estrade, J., Anger, C., Bertrand, M. and Richard G. Tillage and soil ecology: Partners for sustainable agriculture. *Soil and Tillage Research*. 111, no. 1, (2010): 33-40.

Sartore, L., Schettini, E., de Palma, L., Brunetti, G., Cocozza, C. and Vox, G. Effect of hydrolyzed protein-based mulching coatings on the soil properties and productivity in a tunnel greenhouse crop system. *Sci Tot Environ*. 645, (2018): 1221-29. DOI: http://doi.10.1016/j.scitotenv.2018.07.259.

Sax, M.S., Bassuk, N., Van-Es, H. and Rakow, D. Long-term remediation of compacted urban soils by physical fracturing and incorporation of compost. *Urban For Urban Green*. 24, (2017): 149-56.

Scotti, R., Pane, C., Spaccini, R., Palese, A.M., Piccolo, A., Celano, G. and Zaccardelli, M. On-farm compost: A use full tool to improve soil quality under intensive farming systems. *Appl Soil Ecol.* 107, (2016): 13-23. DOI: http://doi.10.1016/j.apsoil.2016.05.004.

Sisay, A. The principal role of organic fertilizer on soil properties and agricultural productivity-a review. *Agri Res Tech.* 22, no. 2 (2019): 556192. DOI: https://doi.org/10.19080/ARTOAJ.2019.22.556192.

Tong, J., Sun, X., Li, S., Qu, B. and Wan, L. Reutilization of Green Waste as Compost for Soil Improvement in the Afforested Land of the Beijing Plain. *Sustainability* 10, (2018): 2376.

Yadav, G. S., Lal, R., Meena, R. S., Babu, S., Das, A., Bhomik, S. N., Datta, M., Layak, J. and Saha, P. Conservation tillage and nutrient management effects on productivity and soil carbon sequestration under double cropping of rice in north eastern region of India. *Ecological Indicators,* (2017). 105: 303-15. Available from: https://www.sciencedirect.com/science/article/abs/pii/S1470160X17305617.

Zaccardelli, M., Villecco, D., Pane, C., Ragosta, G., Palese, A.M. and Celano, G. Realizzazione di un Sistema di compostaggio on farm deiresidui di pomodoro. *Biol. It.* 8, (2010): 63-7 (In Italian).

23

Revival of the Agricultural Economy Future Directions in Agricultural Research and Innovation

***Nidhi Dubey*[1], *Sneha Choudhary*[2], *Sharad Sachan*[3], and *Harshal Avinashe*[4]**

[1,4]*Department of Genetics and Plant Breeding, Lovely Professional University, Phagwara, Punjab, India*

[2]*Department of Agricultural Economics and Extension, Lovely Professional University, Phagwara, Punjab, India*

[3]*Department of Plant Pathology, Lovely Professional University, Phagwara, Punjab, India*

Introduction

Reviving the agricultural economy is a crucial endeavor with far-reaching implications for food security, economic prosperity, and sustainable development. Several key strategies can be adopted to achieve this goal. First, policymakers must prioritize agriculture by providing necessary financial support, infrastructure development, and research and development investments. Adequate credit facilities and subsidies can help farmers access essential resources and modern technology, boosting productivity and efficiency (Asian Development Bank, 2018). Promoting sustainable agricultural practices is vital to ensure long-term viability. Encouraging the adoption of eco-friendly methods such as organic farming, crop rotation, and water-efficient irrigation can enhance environmental resilience while safeguarding natural resources for future generations. Empowering farmers through training and education is essential. Equipping them with the latest knowledge and techniques enhances their capacity to adapt to changing conditions and make informed decisions about crop selection, pest control, and market opportunities. Furthermore, enhancing market linkages is critical to connect farmers with consumers and ensure fair prices for agricultural produce (African Development Bank, 2022). Developing efficient supply chains and reducing post-harvest losses can improve income prospects for farmers and attract more youth to the sector. Lastly, encouraging private sector involvement and investment in agriculture can accelerate growth and innovation. Public-private partnerships can create opportunities for technological advancements, value addition, and market expansion (Liu & Lee, 2018). Reviving the agricultural economy requires a collaborative effort involving governments, farmers, the private sector, and civil society. By addressing challenges, investing in modernization, and promoting sustainable practices, nations can create a thriving agricultural sector that contributes to food security, economic development, and environmental sustainability (Brown, 2021).

The challenge and necessity of reviving the agricultural economy are of paramount importance in today's world, where food security, economic growth, and environmental sustainability are closely interconnected (Carter and Evans, 2020). The agricultural sector has been facing a decline due to various factors, such as rapid urbanization, changing demographics, inadequate investment, and the allure of alternative career opportunities for rural youth. One of the primary reasons for reviving the agricultural economy is the urgent need to ensure food security for a growing global population. As the world's population increases, so does the demand for food. Reviving and strengthening agriculture is crucial to meet this escalating demand and avoid potential food crises. A robust agricultural sector can enhance a nation's self-sufficiency in food production and reduce its reliance on costly imports.

Beyond food security, agriculture plays a vital role in poverty alleviation and rural development. In many developing countries, a significant portion of the population depends on agriculture for their livelihoods (Agriculture and Agri-Food Canada, 2022). Reviving the agricultural sector can create employment opportunities, especially in rural areas, which can help reduce poverty and uplift communities. It also contributes to balanced regional development by promoting economic activities outside urban centers. A revived agricultural economy also has far-reaching economic benefits (Smith, 2022). A prosperous agricultural sector can boost a country's exports, leading to increased foreign exchange earnings. Additionally, a vibrant agricultural economy stimulates growth in related industries, such as food processing, agribusiness, and logistics. This, in turn, generates more jobs and contributes to overall economic development. Furthermore, a focus on revitalizing agriculture aligns with sustainable development goals. Modern agricultural practices, coupled with eco-friendly techniques, can help conserve natural resources and protect the environment (Anderson and White, 2019). Sustainable agriculture can reduce soil degradation, conserve water, and mitigate greenhouse gas emissions, promoting environmental resilience in the face of climate change.

Reviving the agricultural economy poses several significant challenges that need to be addressed to ensure its successful recovery and growth. These challenges include:

i. **Declining Agricultural Productivity:** Over the years, the agricultural sector has witnessed a slowdown in productivity due to factors such as outdated farming practices, limited access to modern technology, and inadequate training and education for farmers.

ii. **Land Degradation and Soil Health:** Land degradation, soil erosion, and nutrient depletion have adversely affected agricultural productivity. Restoring soil health and implementing sustainable land management practices are essential for reviving the sector.

iii. **Water Scarcity and Irrigation:** Water scarcity, inefficient irrigation systems, and uneven distribution of water resources pose significant challenges to agriculture. Implementing efficient water management techniques and promoting water-saving practices are crucial for agricultural revival.

iv. **Climate Change Impacts:** Increasingly unpredictable weather patterns, extreme weather events, and rising temperatures due to climate change pose risks to agricultural production. Adopting climate-resilient farming practices and technologies is necessary to cope with these challenges.

v. **Market Access and Infrastructure:** Inadequate infrastructure, poor transportation, and limited market access hinder farmers' ability to reach consumers and obtain fair prices for their produce. Developing robust market linkages and investing in rural infrastructure are essential for agricultural revitalization.

vi. **Financial Constraints:** Limited access to credit and financial services can impede farmers' ability to invest in modern inputs and technologies, hindering agricultural growth. Ensuring easy and affordable access to credit is crucial for the sector's revival.

vii. **Policy and Governance:** Inconsistent and poorly designed agricultural policies can create uncertainty and discourage investments in the sector. Implementing supportive and coherent policies that promote sustainable practices and protect farmers' interests is vital for agricultural revival.

viii. **Changing Demographics:** The aging farming population and the lack of interest among youth in pursuing careers in agriculture pose challenges in ensuring a skilled and vibrant farming workforce for the future.

ix. **Global Trade Dynamics:** Agriculture is subject to international trade dynamics, and changes in global trade policies can affect domestic agricultural markets. Navigating global trade agreements while safeguarding domestic interests is a complex challenge for agricultural economies.

x. **Socio-Economic Disparities:** Socio-economic disparities between rural and urban areas can hinder agricultural development. Addressing these disparities and ensuring inclusive growth is essential for the revival of the agricultural economy.

Addressing these challenges requires a comprehensive approach that involves sustainable practices, technology adoption, supportive policies, investments in research and development, and empowerment of farmers and rural communities. Collaboration between governments, private sectors, and civil society is crucial in tackling these challenges and achieving a thriving and resilient agricultural economy.

Decline and Factors Affecting the Agricultural Economy

The decline of the agricultural economy has been a significant concern in many regions around the world. Several factors have contributed to this decline, affecting the sector's productivity, profitability, and overall sustainability (Johnson, M., & Smith, A., 2020). Some of the key factors include:

i. **Urbanization:** Rapid urbanization has led to the conversion of agricultural land into urban areas, reducing the overall available land for farming. As cities expand, agricultural activities are pushed to less fertile or remote regions, impacting productivity and increasing transportation costs.

ii. **Changing Demographics:** The aging farming population and the lack of interest among younger generations in pursuing careers in agriculture have resulted in a shortage of skilled and motivated farmers.

This demographic shift hampers the sector's ability to adapt to modern practices and technologies.

iii. **Climate Change:** Increasingly unpredictable weather patterns, extreme events, and shifts in precipitation patterns due to climate change have disrupted traditional farming practices. Farmers are facing challenges related to water availability, crop yields, and pest and disease outbreaks.

iv. **Land Degradation and Soil Erosion:** Unsustainable agricultural practices, such as overuse of agrochemicals and improper land management, have led to soil erosion and degradation. Depleted soils affect crop productivity and long-term sustainability.

v. **Water Scarcity:** Growing water scarcity, driven by increasing demand and changing climate patterns, has become a major challenge for agricultural production. Insufficient water resources can lead to reduced crop yields and even crop failure.

vi. **Market Instability:** Fluctuations in commodity prices and market demands can result in income uncertainties for farmers. Lack of market access and integration can hinder farmers' ability to reach consumers and obtain fair prices for their produce.

vii. **Lack of Access to Finance:** Limited access to credit and financial services prevents farmers from investing in modern inputs, technology, and infrastructure, hindering their ability to improve productivity and competitiveness.

viii. **Policy and Institutional Constraints:** Inconsistent or inadequate agricultural policies, poorly designed subsidies, and cumbersome administrative processes can create uncertainties and deter investments in the agricultural sector.

ix. **Global Trade Dynamics:** Agricultural economies are influenced by international trade agreements and tariff policies, which can impact domestic agricultural markets and create challenges for local farmers to compete on a global scale.

x. **Pest and Disease Outbreaks:** Outbreaks of pests and diseases can devastate crops, leading to significant losses for farmers and disruptions in food supply chains.

Policy Framework for Agricultural Revival

Reviving the agricultural sector is a critical priority for ensuring food security, rural development, and economic growth. A well-crafted and comprehensive policy framework is essential to address the challenges faced by the agricultural economy and create an environment conducive to its revitalization (Harris and Jackson, 2020). This policy framework should encompass various aspects, including sustainable practices, technology adoption, market access, financial support, and institutional reforms.

i. **Sustainable Agricultural Practices:** Promoting sustainable agricultural practices is fundamental for long-term agricultural revival. This includes encouraging the adoption of practices such as organic farming, crop rotation, agroforestry, and integrated pest management (IPM). Governments should provide technical assistance, training, and incentives to farmers who adopt sustainable practices. Financial support, research and development, and extension services should be directed toward the development and dissemination of sustainable agricultural technologies.

ii. **Climate Change Adaptation and Mitigation:** Climate change poses significant challenges to agriculture, affecting crop yields, water availability, and overall productivity. The policy framework should prioritize climate-smart agriculture, which involves the use of climate-resilient crop varieties, water-efficient irrigation techniques, and sustainable land management practices. Governments should support research on climate change impacts on agriculture and develop strategies to mitigate its effects.

iii. **Investment in Agricultural Research and Development:** Investing in agricultural research and development is crucial for fostering innovation and increasing productivity. Governments should allocate adequate funding for agricultural research institutions and collaborate with the private sector and academia to develop and disseminate cutting-edge technologies and best practices. Encouraging public-private partnerships can accelerate technological advancements and enhance the sector's competitiveness.

iv. **Access to Finance and Credit:** Ensuring easy and affordable access to finance and credit is essential for smallholder farmers to invest in modern inputs, equipment, and infrastructure. Governments should establish agricultural credit institutions, microfinance schemes, and farmer cooperatives to facilitate access to credit. Streamlining loan procedures and reducing interest rates for agriculture-related loans can incentivize farmers to invest in their operations.

v. **Market Access and Value Chains:** Enhancing market linkages is crucial for agricultural revival. Governments should invest in rural infrastructure, such as roads and storage facilities, to facilitate the efficient movement of agricultural produce from farms to markets. Encouraging the establishment of farmer-producer organizations and cooperatives can help farmers collectively negotiate better prices and access larger markets. Supporting value-addition initiatives and agribusinesses can create new income streams for farmers and increase the overall value of agricultural products.

vi. **Capacity Building and Extension Services:** Providing farmers with training and extension services is vital for improving their skills, knowledge, and decision-making abilities. Governments should establish robust agricultural extension systems that deliver practical and timely information on best practices, weather forecasts, and market trends. Training programs should focus on modern farming techniques, post-harvest management, and financial literacy to empower farmers with the tools they need to succeed.

vii. **Land Reforms and Tenure Security:** Efficient and equitable land reforms are essential for agricultural revival. Clear and secure land tenure systems can incentivize investment in land improvement and modernization. Governments should streamline land registration processes, protect farmers' land rights, and implement land redistribution programs where necessary to ensure equitable access to land.

viii. **Trade and Export Promotion:** Promoting agricultural exports can provide additional revenue streams for farmers and stimulate economic growth. Governments should facilitate trade agreements, reduce export barriers, and provide market information to farmers and exporters. Supporting the development of quality standards and certification can enhance the competitiveness of agricultural products in the global market.

ix. **Women and Youth Empowerment:** Empowering women and youth in agriculture is crucial for sustainable agricultural development. Governments should implement policies that promote gender equality and create opportunities for women and youth in the agricultural sector. This can include targeted training programs, access to credit, and support for women's cooperatives and youth-led agricultural initiatives.

x. **Institutional Reforms:** Strengthening agricultural institutions is essential for effective policy implementation and coordination. Governments should streamline bureaucratic processes, enhance interagency cooperation, and establish clear roles and responsibilities for various stakeholders. Creating a conducive policy environment and ensuring political commitment to agricultural revival are critical for the success of the policy framework.

Case Studies of Successful Agricultural Revival Initiatives

Case Study 1: The Green Revolution in India

One of the most well-known successful agricultural revival initiatives is the Green Revolution in India, which took place in the mid-1960s and continued into the 1970s. India faced severe food shortages and struggled to feed its growing population. In response, the government and agricultural scientists implemented a series of agricultural reforms aimed at increasing food production. The key components of the Green Revolution included the introduction of high-yielding varieties of rice and wheat, the use of chemical fertilizers, and the expansion of irrigation facilities. These new technologies and practices significantly increased crop yields and transformed India from a food-deficit nation to a food-surplus one. The Green Revolution not only improved food security but also had a profound impact on rural development and poverty reduction. Increased agricultural productivity led to higher incomes for farmers, while surplus food production allowed for more significant agricultural exports, boosting the country's economy (Yogesh, 2023).

Case Study 2: Agricultural Transformation in Brazil

Brazil's agricultural transformation is another notable example of a successful agricultural revival initiative. Starting in the 1970s, Brazil made significant investments in agricultural research and development, modernized its agricultural practices, and expanded its agricultural lands. The development of the Cerrado region in Brazil, a vast savannah with initially poor soil quality, showcased the country's ability to transform marginal lands into highly productive agricultural areas. The adoption of advanced farming techniques, such as no-till farming and precision agriculture, helped conserve soil and water resources while boosting crop yields. Furthermore, Brazil's success in developing a robust agribusiness sector played a pivotal role in its agricultural revival. Investments in infrastructure, transportation, and storage facilities facilitated the efficient movement of agricultural produce from rural areas to urban centers and export markets. As a result of these efforts, Brazil emerged as one of the world's leading agricultural producers and exporters, particularly in soybeans, sugarcane, and beef. The agricultural sector's growth contributed significantly to Brazil's overall economic development and enhanced its global standing in the agricultural trade market (Valdes, 2022).

Case Study 3: Coffee Revival in Rwanda

Rwanda's coffee revival serves as an inspiring case study of how targeted interventions and cooperative efforts can lead to the revival of a key agricultural commodity. In the early 2000s, Rwanda's coffee industry was facing significant challenges, including declining production, poor quality, and low prices. The Rwandan government, in collaboration with international development agencies and coffee cooperatives, initiated a coffee sector revitalization program. The program focused on improving coffee quality through better processing techniques, providing training to farmers, and investing in post-harvest infrastructure. Additionally, efforts were made to promote sustainability and traceability in the coffee value chain, which attracted premium prices in international markets. Specialty coffee certifications and direct trade arrangements with roasters further enhanced the sector's economic viability. Through these initiatives, Rwanda's coffee industry experienced a remarkable revival. Coffee production and exports increased, and coffee farmers' incomes improved significantly. The success of the coffee revival also contributed to poverty reduction and enhanced the overall economic resilience of rural communities in Rwanda (Bhatt, 2015).

Case Study 4: The Ethiopian Agricultural Transformation

In the 2000s, Ethiopia faced severe food insecurity and poverty, with many rural communities relying on subsistence farming. To address these challenges, the Ethiopian government launched the Agricultural Development Led Industrialization (ADLI) strategy in 1994. The ADLI focused on improving smallholder agriculture, enhancing rural infrastructure, and promoting market-oriented farming. Under the ADLI strategy, significant investments were made in irrigation, improved seeds, and extension services. The government also provided subsidies for fertilizers and modern agricultural inputs. These interventions led to increased crop yields and higher incomes for smallholder farmers. Moreover, the expansion of rural infrastructure, such as roads and markets, improved market access and enabled farmers to sell their produce at better prices. As a result, Ethiopia's agricultural sector experienced significant growth, contributing to poverty reduction and economic development (Gates, 2012).

Case Study 5: The Vietnamese Rice Revolution

In the 1980s, Vietnam faced food shortages and a struggling agricultural sector. To boost rice production and achieve food self-sufficiency, the government initiated the Doi Moi policy, which involved agricultural reforms and liberalization measures. The Doi Moi policy encouraged the adoption of modern agricultural technologies, including high-yielding rice varieties and improved irrigation systems. The government also provided credit and technical assistance to farmers, facilitating the adoption of these technologies. As a result of these efforts, Vietnam witnessed a remarkable increase in rice production. It transformed from a rice-importing country to one of the world's leading rice exporters. The success of the Vietnamese rice revolution has played a pivotal role in the country's economic growth and poverty reduction (VIEGO, 2021).

Case Study 6: The Malawi Agricultural Subsidy Program

Malawi faced chronic food insecurity and low agricultural productivity. In response, the government launched the Agricultural Input Subsidy Program (AISP) in 2005. The program aimed to provide smallholder farmers with subsidized fertilizer and improved seeds to increase crop yields. The AISP significantly boosted maize production, the country's staple crop, leading to food surplus and reduced food prices. This had a positive impact on food security, household incomes, and rural development. However, the program also faced criticism for its sustainability and potential negative environmental impacts due to increased fertilizer use. Despite these challenges, the AISP has been instrumental in addressing food insecurity in the short term (Dorward et al., N.A).

Case Study 7: The Israeli Agricultural Innovation

Israel's agricultural sector is a prime example of successful agricultural revival in a challenging environment. With limited water resources and arid conditions, Israel has developed innovative agricultural practices to ensure food security. Through investments in research and development, Israel has pioneered technologies such as drip irrigation, precision agriculture, and greenhouse cultivation. These practices have maximized water efficiency and crop productivity, allowing the country to produce a diverse range of high-value crops. Israel's success in agricultural innovation has not only ensured food security for its population but also positioned it as a global leader in agricultural technology. The export of agricultural know-how and technologies has become a significant source of revenue for the country (Leichman, 2022).

Case Study 8: Bangladesh Experiences Fish Farming Revolution

Bangladesh faced challenges in meeting the protein needs of its growing population. In the 1980s, the government promoted fish farming as a means to address food security and poverty. Bangladesh's fish farming revolution involved training and providing technical support to small-scale fish farmers. Innovative practices, such as integrating fish farming with agriculture and using homestead ponds, were adopted to maximize fish production. The expansion of fish farming led to increased fish consumption and higher incomes for farmers. Fish became an affordable source of protein for millions of people, contributing to improved nutrition and poverty reduction (IFPRI, 2017).

Environmental Impact and Conservation Efforts

The revival of the agricultural economy can have significant environmental impacts, both positive and negative. As agriculture is closely tied to the natural environment, it is essential to consider conservation efforts and sustainable practices to minimize adverse effects and promote ecological resilience. Here are some key environmental impacts and conservation efforts associated with agricultural revival:

Land Degradation and Soil Health

1. **Environmental Impact:** Unsustainable agricultural practices, such as excessive tilling and overuse of chemical inputs, can lead to land degradation, soil erosion, and nutrient depletion. These practices can reduce the land's long-term productivity and negatively impact the environment.
2. **Conservation Efforts:** Implementing sustainable land management practices is crucial for preserving soil health. Techniques like no-till farming, crop rotation, and cover cropping help prevent soil erosion, improve soil structure, and enhance nutrient retention. The use of organic matter, compost, and green manure can further promote soil fertility and biodiversity.

Water Usage and Quality

1. **Environmental Impact:** Agriculture is a major consumer of water resources, and unsustainable irrigation practices can lead to groundwater depletion and water scarcity. Moreover, agricultural runoff containing fertilizers and pesticides can contaminate water bodies, causing pollution and harming aquatic ecosystems.

2. **Conservation Efforts:** Promoting water-efficient irrigation systems, such as drip irrigation and sprinklers, helps conserve water and reduces wastage. Implementing water management techniques like rainwater harvesting and using drought-tolerant crop varieties can improve water use efficiency. Furthermore, adopting IPM practices minimizes the use of chemical pesticides, reducing their impact on water quality.

Biodiversity Loss and Habitat Destruction

1. **Environmental Impact:** The expansion of agricultural lands often results in deforestation and habitat destruction, leading to a loss of biodiversity and threatening wildlife populations.
2. **Conservation Efforts:** Protecting natural ecosystems, establishing wildlife corridors, and promoting agroforestry can help mitigate habitat loss and preserve biodiversity. By integrating trees and crops on farms, agroforestry supports diverse habitats for wildlife while providing economic benefits for farmers.

Chemical Pesticides and Fertilizers

1. **Environmental Impact:** The excessive use of chemical pesticides and fertilizers in agriculture can harm beneficial insects, pollinators, and soil microorganisms, leading to ecological imbalances.
2. **Conservation Efforts:** Adopting IPM) practices is an effective way to reduce reliance on chemical pesticides. IPM involves monitoring pests, using natural predators, and employing cultural practices to manage pest populations. Additionally, organic farming methods minimize the use of chemical inputs, contributing to healthier ecosystems and reduced environmental impact.

Climate Change

1. **Environmental Impact:** Agriculture is a significant contributor to greenhouse gas emissions, primarily through methane from livestock and nitrous oxide from fertilizers. Furthermore, deforestation for agricultural expansion contributes to carbon dioxide emissions.
2. **Conservation Efforts:** Climate-smart agriculture involves practices that reduce emissions and enhance carbon sequestration. This includes using climate-resilient crop varieties, implementing agroforestry systems, and optimizing livestock management to reduce methane emissions.

Conservation efforts that prevent deforestation and promote reforestation also contribute to mitigating climate change.

Genetic Diversity

1. **Environmental Impact:** The focus on high-yielding crop varieties can lead to a reduction in genetic diversity, making agriculture more susceptible to pests, diseases, and climate change.
2. **Conservation Efforts:** Conserving genetic diversity in crops is crucial for ensuring agricultural resilience. Initiatives that promote the cultivation of traditional and locally adapted crop varieties help preserve genetic diversity, making agriculture more adaptable to changing conditions.

Wildlife Conservation

1. **Environmental Impact:** Agriculture expansion can lead to habitat fragmentation and encroachment on natural areas, threatening wildlife populations and contributing to the loss of biodiversity.
2. **Conservation Efforts:** Wildlife-friendly agricultural practices, such as maintaining buffer zones and wildlife corridors, help protect natural habitats and support wildlife conservation. By creating wildlife-friendly landscapes, agricultural revival can coexist with biodiversity conservation.

The revival of the agricultural economy presents both challenges and opportunities for the environment. Environmental impacts such as land degradation, water pollution, and biodiversity loss can be addressed through conservation efforts and the adoption of sustainable agricultural practices. By promoting eco-friendly farming techniques, preserving natural habitats, and enhancing biodiversity, agricultural revival can contribute to a more sustainable and resilient food system. Collaboration between governments, farmers, researchers, and environmental organizations is essential to ensure that agricultural development aligns with environmental conservation, fostering a healthier and more balanced relationship between agriculture and the natural world.

Future Prospects for Reviving Agricultural Economy

The prospects for the reviving agricultural economy are promising yet pose several challenges and opportunities. As nations work toward revitalizing the agricultural sector, several key trends and factors will shape its trajectory:

i. **Technology and Innovation:** Advancements in agricultural technology, precision farming, biotechnology, and data analytics will play a pivotal role in transforming agricultural practices. Smart farming techniques, including the Internet of Things (IoT) and artificial intelligence, will lead to more efficient resource utilization and higher crop yields.

ii. **Sustainable Practices:** The emphasis on sustainable agriculture will continue to grow. Climate-smart practices, organic farming, and regenerative agriculture will gain momentum, as consumers and policymakers prioritize environmentally friendly and socially responsible food production.

iii. **Digitalization of Agriculture:** The integration of digital technologies into agriculture will streamline operations, enhance supply chain transparency, and facilitate access to markets. E-commerce platforms and digital marketplaces will connect farmers directly with consumers and buyers, reducing intermediaries and ensuring fair prices.

iv. **Market Diversification:** Agricultural economies will explore new markets and value chains. The demand for specialty crops, organic produce, and high-value products will create opportunities for farmers to diversify their income sources.

v. **Circular Economy in Agriculture:** The adoption of circular economy principles, such as waste reduction, recycling, and nutrient cycling, will gain prominence in agricultural practices. These efforts will minimize waste and promote resource efficiency.

vi. **Climate Resilience:** With climate change impacting agricultural productivity, investments in climate-resilient crops and practices will be essential. Research on drought-tolerant, heat-resistant, and flood-resistant varieties will help farmers adapt to changing weather patterns.

vii. **Inclusive Growth:** Governments and stakeholders will work toward ensuring inclusive growth in agriculture, empowering smallholder farmers, women, and youth. Training, access to finance, and support for marginalized farming communities will be prioritized.

viii. **Public-Private Partnerships:** Collaborative efforts between governments, private sector entities, research institutions, and civil society will drive agricultural development. Public-private partnerships will be instrumental in technology dissemination, infrastructure development, and market linkages.

ix. **Policy Reforms:** Effective agricultural policies will be critical for fostering growth and sustainability. Governments will need to design supportive policies that incentivize investment in agriculture, encourage research and development, and address social and environmental concerns.

x. **Climate Finance and Green Investments:** International funding and private investments in climate-smart and sustainable agriculture will increase. Climate finance will support agricultural projects that mitigate climate change impacts and promote environmental conservation.

The prospects for the reviving agricultural economy are dynamic and hold significant potential for growth and transformation. Leveraging technology, embracing sustainability, fostering innovation, and ensuring inclusivity will be essential for driving agricultural development. By addressing challenges and capitalizing on opportunities, nations can build a resilient and sustainable agricultural sector that contributes to food security, economic prosperity, and environmental stewardship in the years to come.

Conclusion

The revival of the agricultural economy has been a critical endeavor in recent years, fueled by the recognition of agriculture's pivotal role in ensuring food security, poverty alleviation, and sustainable development. Through comprehensive reforms, technological advancements, and increased investment, countries have made significant strides toward rejuvenating their agricultural sectors. The results have been promising, as evidenced by improved productivity, enhanced income for farmers, and overall economic growth. One of the primary driving forces behind this revival has been the implementation of modern agricultural practices and technologies. The integration of precision agriculture, biotechnology, and smart farming techniques has substantially increased crop yields and resource efficiency. Additionally, the adoption of climate-resilient seeds and farming methods has helped mitigate the impact of changing weather patterns and extreme events, bolstering agricultural resilience. Furthermore, governments and international organizations have recognized the need for supportive policies and financial incentives to invigorate the agricultural sector. Investment in rural infrastructure, such as irrigation systems, roads, and storage facilities, has reduced post-harvest losses and improved market access for farmers. Moreover, the availability of credit and insurance facilities has facilitated access to capital and risk management tools, empowering smallholder farmers to invest in their farms and adopt modern technologies.

The revival of the agricultural economy has not only benefited farmers but also contributed to broader economic development. As the agricultural sector thrives, it generates employment opportunities, particularly in rural areas where job opportunities were previously limited. This, in turn, boosts consumer spending and fosters economic growth in related industries. Environmental sustainability has also been a focal point in the revival process. Farmers are increasingly adopting sustainable agricultural practices to preserve natural resources and reduce environmental impacts. This shift toward eco-friendly practices has positive ramifications on biodiversity, soil health, and water quality. Nevertheless, challenges persist, and the journey toward a fully revitalized agricultural economy remains ongoing. Climate change, water scarcity, and fluctuating commodity prices continue to pose threats. Thus, continuous investment in research and development, coupled with policy support, is crucial to address these challenges.

References

African Development Bank (AfDB). 2022. Unlocking Agricultural Potential for Economic Revival in Africa. Abidjan, Côte d'Ivoire: AfDB.

Agriculture and Agri-Food Canada. 2022. Investing in the Future of Agriculture: A Roadmap for Economic Revival. Ottawa, Canada: Agriculture and Agri-Food Canada.

Anderson, J., & White, E. (2019). Agricultural Investment and the Revival of Rural Economies. *Journal of Rural Development,* 28, no. 1, (2019): 65-80.

Asian Development Bank (ADB). 2018. Agriculture-Led Growth Strategies for Economic Revival in Asia. Manila, Philippines: ADB.

Bhatt, N. 2015. The History of Rwandan Coffee & The Story Behind Epiphanie Muhirwa. Retrieved 26 July, 2023 from https://perfectdailygrind.com/2015/04/the-history-of-rwandan-coffee-the-story-behind-epiphanie-muhirwa/#:~:text=The%20Story%20Behind%20Epiphanie%20Muhirwa,-As%20the%20coffee&text=The%20coffee%20cherries%20are%20hand,Muhirwa%20and%20her%20son%20Samuel.

Brown, L. Agriculture and Economic Recovery: A Comparative Study of Developed and Developing Countries. *Journal of Development Studies,* 35, no. 4, (2021): 589-05.

Carter, S., & Evans, D. Enhancing Agricultural Productivity for Economic Growth and Revival. *Economic Development Quarterly*, 30, no. 2, (2020): 256-70.

Dorward, A., Chirwa, E. & Jayne, T.S. (N.A.). Malawi's Agricultural Input Subsidy Program Experience over 2005-09. https://documents1.worldbank.org/curated/en/304221468001788072/930107812_201408252024422/additional/634310PUB0Yes0061512B09780821387450.pdf

Gates, B. 2012. Ethiopia's agricultural revolution. Retrieved 26 July, 2023 from https://www.gatesnotes.com/Ethiopias-Revolution-in-Agriculture

Harris, P., & Jackson, K. Agricultural Modernization and Its Impact on Economic Development. *Agricultural Economics Review*, 4, no. 3 (2020): 315-29.

IFPRI. 2017. Bangladesh Experiences Fish Farming Revolution. Retrieved 27 July, 2023 from https://bangladesh.ifpri.info/2017/08/bangladesh-experiences-fish-farming-revolution/

Johnson, M., & Smith, A. Agricultural Transformation for Economic Revival: A Case Study of Developing Nations. *Journal of Agricultural Economics*, 25, no. 3, (2020): 102-17.

Leichman, A.K. 2022. Israeli agricultural innovations will keep the world fed. Retrieved 26 July, 2023 from https://www.israel21c.org/israeli-agricultural-innovations-will-keep-the-world-fed/#:~:text=Many%20Israeli%20companies%20enable%20regenerative,crops%20and%20not%20the%20weeds.

Liu, W., & Lee, H. Sustainable Agriculture and Economic Revival: Lessons from East Asia. *Journal of Development Economics*, 42, no. 5, (2018): 791-808.

Smith, R. Agricultural Innovations and Their Impact on Economic Development. *Economic Journal*, 40, no. 2: (2022): 215-31.

Valdes, C. 2022. Brazil's Momentum as a Global Agricultural Supplier Faces Headwinds. Retrieved 25 July, 2023 from https://www.ers.usda.gov/amber-waves/2022/september/brazil-s-momentum-as-a-global-agricultural-supplier-faces-headwinds/

VIEGO. 2021. The In-depth Guide To Vietnam Rice Market. Retrieved 26 July, 2023 from https://viegoglobal.com/vietnamese-rice-the-in-depth-guide-to-vietnam-rice-market/

Yogesh. 2023. Green Revolution in India Advantages, Impacts & Achievements. Retrieved 25 July, 2023 from https://www.studyiq.com/articles/green-revolution-in-india/

Index

About the Contributors

Dr. Haobijam James Watt is currently working as an Assistant Professor at the Department of Agricultural Economics and Extension (AEE), School of Agriculture, Lovely Professional University. He completed his Ph.D. (Agricultural Extension) from Visva-Bharati, Santiniketan, West Bengal, post-graduate (M.Sc. in Agricultural Extension) from Bidhan Chandra Krishi Viswavidyalaya, West Bengal, and graduated from Tamil Nadu Agricultural University, Coimbatore. He has qualified ICAR-AIEEA-UG, ICAR-AIEEA-PG, ICAR-SRF-(PGS) exams, ICAR-ASRB-NET, and UGC-NET (Adult. Edu. /Con.Edu/Andragogy/N.F. Edu). He has been involved in the teaching of U.G. and P.G. courses in Agricultural Extension. He has published several research articles including one book and several book chapters in reputed national as well as international journals. His area of research lies in the field of livelihood, sustainability, entrepreneurship, and other areas.

Dr. Devina Seram is currently working as an Assistant Professor at the Department of Entomology, School of Agriculture, LPU. She completed her Ph.D. (Agricultural Entomology) from Tamil Nadu Agricultural University, Coimbatore with a DST-INSPIRE fellowship, post-graduation (M.Sc. in Agricultural Entomology) from College of Post Graduate Studies (Umiam), CAU, Imphal, Manipur and graduated from Assam Agricultural University with distinction. She received the CAU Merit Scholarship and is a Gold Medallist in her master's degree. She has qualified for ICAR-AIEEA-PG, ICAR-SRF-(PGS), and ICAR-ASRB-NET and also received several awards in her field of expertise including the National Dedication Day Award, Excellence in Research Award, and a few others. She has several research and review papers including book chapters to her publication credit, has guided four master's students, and is currently guiding two students including one Ph.D. scholar. Her research specialization and interests include host plant resistance, insect resistance bioassay, storage entomology, molecular biology, etc.

Dr. Yudhishther Singh Bagal is currently working as an Assistant Professor in the School of Agriculture, Lovely Professional University, Phagwara, Punjab, India. He did his Ph.D. in Agricultural Extension and Communication from Sher-e-Kashmir University of Agricultural Sciences and Technology, Jammu,

J&K, India. He has published several research articles including book chapters in reputed national as well as international journals and books respectively.

Dr. Chongtham Tania is currently working as a Scientist (Senior Scale), Horticulture, ICAR-RC, NEHR, Manipur Centre, Lamphelpat, Imphal. She did her M.Sc. and Ph.D. from BCKV, Nadia, West Bengal. She got selected as ARS in 2013 and qualified ICAR-NET. She has organized and attended several training programs, National and International Seminars, Workshops, etc. She is a member for procurement of seeds, quality planting materials, inputs and agricultural items for implementing various schemes for State Horticulture Department. She has acted as resource person in many awareness programmes. She has published 27 research papers in national and international journals.

Mr. Navneet Singh is currently undergoing B.Sc. (Agri. Hons.) degree from the School of Agriculture, LPU.

Mr. Lakshay Kaushal is currently undergoing B.Sc. (Agri. Hons.) degree from the School of Agriculture, LPU.

Dr. I Yimjenjang Longkumer is currently working as an Assistant Professor, Department of Entomology at, Lovely Professional University, Phagwara (Punjab), since 2022, teaching undergraduate, post graduate students. He obtained B.Sc. (Ag.) in 2015 from School of Agricultural Science, Nagaland, M.Sc. (Ag.) in Entomology from CAU, Imphal, Manipur (2017) and Ph.D. (Genetics and Plant Breeding) in 2020 from Dr. Rajendra Prasad Central Agricultural University, Pusa, Bihar. He is a Life Fellow of Entomological Society of India and has published 20 papers, one book and 10 popular articles.

Dr. Rahat Ashraf is currently working as an Assistant Professor in School of Agricultural Sciences and Technology, RIMT University, Mandi-Gobindgarh, Punjab. He did his Ph.D. in Floriculture and Landscape Architecture from SKUAST, Kashmir. He has published several research articles including book chapters in reputed national as well as international journals and books respectively.

Ms. Prabhjot Kaur is currently working as an Assistant Professor in the School of Agricultural Sciences and Technology, RIMT University, Mandi-Gobindgarh, Punjab. She did her Master's in Agronomy at Khalsa College, Amritsar, India. She has published book chapters and research papers in international scientific journals. She has also participated in several national and international seminars, symposiums/conferences.

Dr. Harshal Avinashe is currently working as an Associate Professor, Genetics and Plant Breeding, Lovely Professional University, Phagwara (Punjab) since

2016, teaching undergraduate, post graduate students. Dr Harshal Avinashe obtained B.Sc. (Ag.) in 2009 from Dr. P.D.K.V, Akola, M.Sc. (Ag.) in Genetics and Plant Breeding in 2011 from AAU, Anand, Gujarat and Ph.D. (Genetics and Plant Breeding) in 2016 from JNKVV, Jabalpur, M.P. He has published 37 refereed research papers, five popular articles and achieved two awards.

Dr. Sharad Sachan is currently working as an Associate Professor, Agricultural Economics, Lovely Professional University, Phagwara (Punjab) since 2015, teaching undergraduate students. He has obtained B.Sc. (Ag.) in 2010, M.Sc. (Ag.) Agricultural Economics in 2012 from CSAUA&T, Kanpur, U.P. and Ph.D. (Agricultural Economics) in 2015 from SVPUA&T, Meerut, U.P. He has published 18 refereed research papers, five popular articles and six patents and achieved one award.

Dr. Sneha Choudhary is currently working as an Assistant Professor, Plant Pathology, Lovely Professional University, Phagwara (Punjab) since 2022, teaching undergraduate and post-graduate students. She has obtained B.Sc. (Hons.) Ag. in 2014 M.Sc. (Ag.) in Plant Pathology in 2017 from CSKHPKV, Palampur, H.P. and Ph.D. (Plant Pathology) in 2022 from SKUAST-J, Jammu, J&K. She has published 15 refereed research papers, three popular articles, one patent, one copyright, received one Gold medal and also awarded with ICAR-SRF.

Dr. Nidhi Dubey is currently working as an Associate Professor, Genetics and Plant Breeding, Lovely Professional University, Phagwara (Punjab) since 2016, teaching undergraduate and post graduate students. She has obtained B.Sc. (Ag.) in 2009 M.Sc. (Ag.) in Genetics and Plant Breeding in 2011 and Ph.D. (Genetics and Plant Breeding) in 2016 from JNKVV, Jabalpur, M.P. She has published 32 refereed research papers, five popular articles and achieved two awards.

Mr. Kartik Sawant has completed M.Sc. (Plant Pathology) in 2023 in the division of Plant Pathology from Lovely professional University, Phagwara. He has published 2 research papers and 4 book chapters.

Dr. Meenakshi Rana is currently working as an Assistant Professor at Lovely Professional University, Phagwara, Punjab, India. She got her Ph.D. degree in Plant Pathology discipline from Govind Ballabh Pant University of Agriculture and Technology, Pant Nagar. She has qualified ICAR NET in Plant Pathology and having teaching experience of more than 5 years. She has published around 50 research papers, review articles and book chapters in national and international reputed journals. Her current areas of research expertise are mycology, biocontrol agents and plant disease management.

Dr. Seweta Srivastava is currently working as an Associate Professor in Lovely Professional University, Punjab. She obtained her Ph.D. degree in Plant Pathology from Banaras Hindu University. She is a recipient of Senior Research Fellowship of ICAR, New Delhi, in Plant pathology. She also qualified ICAR-NET in Plant Pathology. She has published 3 books and more than 140 research papers, short communications and reviews in national and international journals of repute. Her current areas of research expertise are seed pathology, mycology, biocontrol agents and biochemical analysis.

Dr. Devendra Kumar is currently working as an Assistant Professor at the Department of Plant Pathology, School of Agriculture, Lovely Professional University, Punjab. He has completed B.Sc. Agriculture and M.Sc. (Plant Protection) from Chaudhary Charan Singh University, Meerut Uttar Pradesh. He received Gold Medal during his master's degree program. He completed his Ph.D. from Shere-e-Kasmir University of Agricultural Sciences and Technology- Jammu. He has three years of work experience as JRF in DBT Project SRI. He has published various research papers, book chapters and popular articles in various national and international journals.

Dr. Dechan Choskit is currently working as a Research Fellow, CSIR-Indian Institute of Integrative Medicine, Jammu, J&K. She did her Ph.D. in Plant Pathology from Sher-e-Kashmir University of Agricultural Sciences and Technology of Jammu (SKUAST-J). She has published book chapters and research papers in international scientific journals. She has also participated in several national and international seminars, symposiums/conferences.

Dr. Rodrick Lepcha is currently working as an Assistant Professor, Department of Soil Sciences and Agricultural Chemistry, Lovely Professional University (LPU), Punjab. He did his M.Sc. and Ph.D. from BCKV, Mohanpur, West Bengal. He has qualified for the ICAR-NET. He has attended many national seminars, training programmes, and workshops. He published 7 research papers, including book chapters and articles in national and international journals.

Miss Sona Jaison is currently undergoing B.Sc. (Agri. Hons.) degree from the School of Agriculture, LPU.

Dr. Princy Thakur is currently working as an Assistant Professor, Department of Soil Science and Agricultural Chemistry, Lovely Professional University (LPU), Punjab. She did her graduation from CSKHPKV, Palampur, M.Sc. Ph.D. from UBKV, West Bengal. She has been awarded best Ph.D. thesis award. She had qualified for the ICAR-NET. She has attended many national and international seminars, training programmes, and workshops. She has published 10 research papers in national and international journals, including 3 book chapters and several popular articles.

Dr. Boris Huirem is currently working as an Assistant Professor, Department of Horticulture and Food Science, School of Agriculture and Allied Sciences, The Neotia University, WB. He did his Ph.D. in Food Engineering from BCKV, West Bengal. He has been awarded "National Fellowship for OBC-2018" by UGC. He has attended several National and International seminars, training programmes and workshops. He has published 8 publications including research articles, proceedings and book chapters in National and International Journals.

Dr. Pradeepta Kumar Sahoo is currently serving as a Professor and Dean, Department of Food Engineering, Faculty of Agricultural Engineering, BCKV, WB. He did his Ph.D. in Diary and Food Engineering from IIT, Kharagpur. He has been awarded "2006 Seligman Fellowship" by Society of Chemicals industry, University of Cambridge, UK for carrying out his Post-Doctoral research, which has been completed at the Department of Chemical Engineering, Indian Institute of Science, Bangalore. He has attended many National and International seminars, workshops, training programmes. He has published more than 50 publications including research articles, proceedings, book chapters in National and International Journals.

Dr. Richa Bharti is currently working as an Assistant Professor (contractual basis) at Government Degree College, Udhampur. She did her Ph.D. in Plant Breeding and Genetics from Sher-e-Kashmir University of Agricultural Sciences and Technology of Jammu (SKUAST-J). She has more than 5 years of teaching experience and is an author of more than 10 research papers and chapters in national and international journals.

Ms. C. Kanimozhi is currently working as an Assistant Professor with expertise in Horticulture at SRM College of Agricultural Sciences, Kattankulathur campus, SRM Institute of Science and Technology. With an expertise in Floriculture and Landscaping, she has created the Botanical Garden at the SRM College of Agricultural Sciences. Currently, she is involved in three projects as a Principal Investigator (PI) and Co-Principal Investigator (Co-PI). She was awarded the Meritorious Tamil Nadu Agricultural University Merit Scholarship and the SRM Green Ambassador Award for her efforts in promoting greenery in the campus. She has organized numerous national seminars, training programs, and workshops, and has given multiple guest lectures at various schools and institutions. She has authored 8 research papers and articles in both national and international journals. She completed her UG at Horticulture College and Research Institute, Periyakulam, and her PG at Horticulture College and Research Institute, Coimbatore and is also currently pursuing her Ph.D. with specialization in "Therapeutic Horticulture".

Dr. R. Angelin Silviya is currently working as an Assistant Professor (SS&AC) in Department of Soil Science and Agricultural Chemistry (SS&AC) at SRM College of Agricultural Sciences, SRM Institute of Science and Technology. She obtained her M.Sc. and Ph.D. degree from Tamil Nadu Agricultural University, Coimbatore. She received a private scholarship from Kothari Sugars Pvt. Ltd. to pursue M.Sc. and MANF scholarship from UGC to pursue Ph.D. She has published 11 research papers and a book chapter in different national and international journals. At present, she holds two projects as a Co-PI. She has received the Prof. T. Balakrishnan prize for securing the highest cumulative credit points in PG from TNAU, the Dr. P. Murugesa Boopathi Endowment Award for the best Ph.D. thesis from TNAU and the Best Women Researcher award from PEARL foundation.

Dr. S. Anandhi is an Assistant Professor of Horticulture specializing in Floriculture and Medicinal Crops. She obtained her Ph.D. degree from Tamil Nadu Agricultural University and completed her PDF at TNAU in 2015. She has authored a total of 15 research papers, one book and 8 book chapters, published in both national and international journals. She has received both the coveted Professor M.S. Swaminathan Best Young Scientist Award and the Best Young Scientist Award.

Dr. B. Gopu is presently working as an Assistant Professor and Head, Department of Fruit Science, SRM College of Agricultural Sciences, SRM Institute of Science and Technology, Chengalpattu, Tamil Nadu. He obtained his B.Sc. (Hort.), M.Sc. (Hort.) and Ph.D. in Fruit Science degrees (2009, 2011 and 2015) from Tamil Nadu Agricultural University (TNAU), Coimbatore. His significant contribution in the field of Horticulture has bagged him many awards, and proudly he is the recipient of the Jain Research Fellowship for both of his M.Sc. (Horticulture) and Ph.D. (Horticulture) program. In addition, he has received Kanwar Virender Singh Memorial All India Best Publication Award in Fruit Science (2018) and Young Scientist Award (2016). He has also qualified ICAR–NET. He has published 18 research papers on various National and International Journals and has authored one book and nine book chapters. His work has appeared in reputed magazines as popular articles.

Dr. Deepa Thangjam is currently working as an Assistant Professor, Department of Rural Development and Planning, School of Agricultural Sciences (SAS), Nagaland University. She has served as a Research Assistant and SRF in ICSSR funded project and under IDP-NAHEP project at College of Agriculture, Iroisemba, CAU, Imphal. She did her B.Sc. Agriculture from Assam Agricultural University, Jorhat, M.Sc. and Ph.D. from CAU, Imphal. She has qualified both UGC-NET and ICAR-NET. She has attended many national seminars, training programmes and workshops. She has published 10 research papers, 5 book chapters in many national journals.

Dr. Sashilila Ao is currently working as a Guest Faculty, Department of Agricultural Extension, School of Agricultural Sciences (SAS), Nagaland University. She completed her M.Sc. and Ph.D. from SAS, Nagaland University. She had qualified ICAR-NET.

Dr. Mary N Odyuo is current working as an Associate Professor in the Department of Agriculture Extension and Education, SAS, Nagaland University. She is a gold medalist in academics. She is actively involved in teaching and research since last 11 years. She has more than 40 research papers in reputed journals and a number of book chapters to her credit. She has been awarded Young Scientist Award by Extension Society of Agra and also best presenter award for research articles in many national and international occasions.

Mr. Vivek Kamble is currently a Master's student, Department of Soil Science and Agricultural Chemistry, Lovely Professional University (LPU), Punjab. He did his B.Sc. from Padmabhushan Vasant Dada Patil College of agriculture, Ambi, Talegaon Dabhade, Pune (Maharashtra). He had attended 3 international conferences and published 5 articles in magazines.

Dr. Meraj Ahmed, Assistant Professor, Department of Soil Science & Agricultural Chemistry, Lovely Professional University (LPU), Punjab. He has also worked as Guest Lecturer in Sam Higginbottom University of Agriculture and Technology (SHIATS) (2012-2014). He has graduated from Allahabad Agriculture Institute Deemed University (AAIDU) 2008, completed his Master's and Doctorate degree in Soil Science from SHIATS. He has 20 publications in reputed journals, 100+ magazine articles, 3 edited books, Young Scientist award, Best Paper Award, Scientist of the year 2022, Powerful Orator (2023) awards and 20 minutes talk on "Krishi Darshan" program of "Doordarshan" of PRASAR BHARTI (India's Public Service Broadcaster).

Dr. Neeraj Singh Parihar, Assistant Professor, Department of Soil Science & Agricultural Chemistry, Lovely Professional University (LPU), Punjab. He has served as a Field Investigator under the IFPRI project, Skuast-J and Assistant Professor at Shoolini University. He did his B.Tech (Agril. Engg) from SKUAST-K and his M.Tech & Ph.D. from SKUAST-J. He is the recipient of different awards such as Best Ph.D. thesis Award, Young Scientist Award, etc. He has attended many national seminars, training programmes, and workshops. He published 20+ research papers, including book chapters and articles in national and international journals.

Dr. Akshita Barthwal is serving as a Testing and Operations Manager, R&D Bayer Crop Science Ltd. Central U.P. She did her Ph.D. and M.Sc. from Sam Higginbottom University of Agriculture, Technology and Sciences (SHUATS), Prayagraj (U.P) and B.Sc. (Agriculture) from H.N.B. Garhwal University, Srinagar in 2017. She had 20 publications in reputed journals, 9 Magazine Articles, 8 book Chapters and Best Paper Award.

Dr. Abhijit Das is working as an Assistant Professor in the Department of Agricultural Economics and Extension, School of Agriculture, Lovely Professional University Punjab. He has been involved in teaching and research in Agricultural Economics. He has published more than 30 research papers, patents, popular articles and book chapters in various journals of repute, books and magazines.

Mr. Sahin Aktar Munshi is pursuing Ph.D. in the Division of Agricultural Economics at Indian Agricultural Research Institute (IARI), New Delhi. He has published more than five research papers, popular articles and book chapters in various journals of repute, books and magazines.

Mr. Bikram Barman is pursuing Ph.D. in the Division of Agricultural Extension at Indian Agricultural Research Institute (IARI), New Delhi. He has published more five research papers, popular articles and book chapters in various journals of repute, books and magazines.

Mrs. Anushka Mishra is a Ph.D. Research Scholar at Department of Soil Science and Agricultural Chemistry, Lovely Professional University (LPU), Punjab. She did her M.Sc. from Bundelkhand University, Jhansi, UP. She has attended two international conferences and has received Best Paper Presentation Award. She has published research papers and articles in national and international journals.

Dr. Raghu Nandan Singh Khatana graduated from Maharana Pratap University of Agriculture and Technology, Udaipur 2017, completed his Master's and Doctorate degrees in Soil Science and Agricultural Chemistry from Sam Higginbottom University of Agriculture, Technology and Sciences (SHUATS) in 2019 and 2022. He has also served as Teaching Assistant and worked as a Senior Research Fellow in NICRA project in KVK, Jhunjunu (SKRAU, Bikaner). He is presently working as an Assistant Professor in the Department of Soil Science, Suresh Gyan Vihar University, Jaipur, Rajasthan. He has published various books, articles, research papers, book chapters in

International and National Journals. Also awarded with best Ph.D. Thesis Award, best M.Sc. Thesis Award, Young Agriculturist Award and Best Oral Presentation Award nominated by different national societies of India at University level.

Dr. Meenakshi Anand, Gold medalist, M.Sc, M.ed, NET, SLET, did her Ph.D. in Human Development, Department of Home Science from University of Jammu, J&K, India. She has published several research articles including book chapters in reputed national as well as international journals and books, respectively. Presently she is working as an Assistant Professor in Lakshmibai College, University of Delhi, India.

